AF803941

Studies in Applied Philosophy, Epistemology and Rational Ethics

Volume 19

Series editor

Lorenzo Magnani, University of Pavia, Pavia, Italy
e-mail: lmagnani@unipv.it

Editorial Board

Atocha Aliseda
Universidad Nacional Autónoma de México (UNAM), Coyoacan, Mexico

Giuseppe Longo
Centre Cavaillès, CNRS—Ecole Normale Supérieure, Paris, France

Chris Sinha
Lund University, Lund, Sweden

Paul Thagard
Waterloo University, Waterloo, ON, Canada

John Woods
University of British Columbia, Vancouver, BC, Canada

About this Series

Studies in Applied Philosophy, Epistemology and Rational Ethics (SAPERE) publishes new developments and advances in all the fields of philosophy, epistemology, and ethics, bringing them together with a cluster of scientific disciplines and technological outcomes: from computer science to life sciences, from economics, law,and education to engineering, logic, and mathematics, from medicine to physics,human sciences, and politics. It aims at covering all the challenging philosophical and ethical themes of contemporary society, making them appropriately applicable to contemporary theoretical, methodological, and practical problems, impasses,controversies, and conflicts. The series includes monographs, lecture notes, selected contributions from specialized conferences and workshops as well as selected Ph.D. theses.

Advisory Board

A. Abe, Chiba, Japan
H. Andersen, Aarhus, Denmark
O. Bueno, Coral Gables, USA
S. Chandrasekharan, Mumbai, India
M. Dascal, Tel Aviv, Israel
G.D. Crnkovic, Västerås, Sweden
M. Ghins, Lovain-la-Neuve, Belgium
M. Guarini, Windsor, Canada
R. Gudwin, Campinas, Brazil
A. Heeffer, Ghent, Belgium
M. Hildebrandt, Rotterdam, The Netherlands
K.E. Himma, Seattle, USA
M. Hoffmann, Atlanta, USA
P. Li, Guangzhou, P.R. China
G. Minnameier, Frankfurt, Germany
M. Morrison, Toronto, Canada
Y. Ohsawa, Tokyo, Japan
S. Paavola, Helsinki, Finland
W. Park, Daejeon, South Korea
A. Pereira, São Paulo, Brazil
L.M. Pereira, Caparica, Portugal
A.-V. Pietarinen, Helsinki, Finland
D. Portides, Nicosia, Cyprus
D. Provijn, Ghent, Belgium
J. Queiroz, Juiz de Fora, Brazil
A. Raftopoulos, Nicosia, Cyprus
C. Sakama, Wakayama, Japan
C. Schmidt, Le Mans, France
G. Schurz, Dusseldorf, Germany
N. Schwartz, Buenos Aires, Argentina
C. Shelley, Waterloo, Canada
F. Stjernfelt, Aarhus, Denmark
M. Suarez, Madrid, Spain
J. van den Hoven, Delft, The Netherlands
P.-P. Verbeek, Enschede, The Netherlands
R. Viale, Milan, Italy
M. Vorms, Paris, France

More information about this series at http://www.springer.com/series/10087

Tommaso Bertolotti

Patterns of Rationality

Recurring Inferences in Science, Social Cognition and Religious Thinking

Springer

Tommaso Bertolotti
Department of Humanities,
Philosophy Section
University of Pavia
Pavia
Italy

ISSN 2192-6255 ISSN 2192-6263 (electronic)
Studies in Applied Philosophy, Epistemology and Rational Ethics
ISBN 978-3-319-17785-4 ISBN 978-3-319-17786-1 (eBook)
DOI 10.1007/978-3-319-17786-1

Library of Congress Control Number: 2015937367

Springer Cham Heidelberg New York Dordrecht London
© Springer International Publishing Switzerland 2015
This work is subject to copyright. All rights are reserved by the Publisher, whether the whole or part of the material is concerned, specifically the rights of translation, reprinting, reuse of illustrations, recitation, broadcasting, reproduction on microfilms or in any other physical way, and transmission or information storage and retrieval, electronic adaptation, computer software, or by similar or dissimilar methodology now known or hereafter developed.
The use of general descriptive names, registered names, trademarks, service marks, etc. in this publication does not imply, even in the absence of a specific statement, that such names are exempt from the relevant protective laws and regulations and therefore free for general use.
The publisher, the authors and the editors are safe to assume that the advice and information in this book are believed to be true and accurate at the date of publication. Neither the publisher nor the authors or the editors give a warranty, express or implied, with respect to the material contained herein or for any errors or omissions that may have been made.

Printed on acid-free paper

Springer International Publishing AG Switzerland is part of Springer Science+Business Media (www.springer.com)

To Anne-Claire

Acknowledgments

This book is the product of five years of philosophical research that I revised, refined, and recollected in tranquillity. During these five years, which also included my Ph.D. program at the University of Pavia, I could benefit from the personal and intellectual interaction with a number of most precious people, to the extent that I can say how, had I not met them, my work and my research would now be completely different (and much poorer).

The first acknowledgment goes to Prof. Lorenzo Magnani, for guiding, constraining, and supporting my research with ever-constructive and loyal liberality. I am most grateful for being able to think of him as a master and a friend: he has been a priceless enabler at the intellectual, professional, and personal level.

I want to express my gratitude to the University of Pavia and to the Italian Ministry of Education, University and Research (MIUR) for providing the funds for my research: the former as a Ph.D. student at the Ph.D. Program in Philosophy, and the latter as a postdoctoral fellow at the same university, supported by the 2012 National Research Grant (PRIN) "Models and Inferences in Science: Logical, Epistemological, and Cognitive Aspects." I am also extremely grateful to the faculty and the students at my Department, because the relationships I established with many of them fostered priceless occasions for my intellectual growth.

While writing this book I revised and enriched many of the papers I authored and coauthored in the past years: while working on my writings, I stressed and extended my personal contributions to these papers, but I must nevertheless express my gratitude toward the people on whose collaboration I could count, first as coauthors, and then as providers of insightful comments for the preparation of this book.[1]

[1]This project was conceived as a whole, but as it developed various parts have become articles, which have now been excerpted, revised, and integrated into the current text. I am grateful to Springer for permission to include portions of previously published articles. As part of this book includes the revision and expansion of previously published material, this is the list of original publications: For Part I: T. Bertolotti (2012). From mindless modeling to scientific models: the case of emerging models. In L. Magnani and P. Li (eds.), *Philosophy and Cognitive Science, Western and Eastern Studies*, pages 75–104. Springer, Heidelberg/Berlin; T. Bertolotti (2013).

I also express my recognition for the three commissioners of my Ph.D. defense, Profs. Claudia Casadio, Elisa Paganini, and Andrea Bottani, for the interest and congeniality they displayed in assessing my thesis—in which many of the ideas included in this book were first tested—and for the helpful comments they offered.

The development and consolidation of my thesis is heavily indebted toward the rich and rewarding conversations I had with my friends Emanuele Bardone, Silvia Baio, Mario Matsukura, Marianna Girlando, Carolina Cinerari, Simone Pace. A special "thank you" must go to Selene Arfini, for all the unyielding philosophical arguments. Cody Kommers is a very promising young cognitive scientist and I am glad I enjoyed his company and conversation in Quebec City, at the Cognitive Science Society Conference in 2014. I am also grateful to him and his colleagues for letting me test some of the ideas of this book at a seminar at UCLA. I thank Patrick Damien Williams for his extensive comments on the third part of this book.

I believe there is little profit in separating the intellectual and personal development of a person, I am therefore both intellectually and personally grateful to a number of people who contributed and still contribute, each in their own ways and extents, to making my life so rich and stimulating: these are Jaqueline Beauchere, Caroline Rizza, Simone Regazzoni, Nicola Polloni, Alessandra Stocchetti, Giovanna Magnani, Federica Airoldi, Caterina Sforzini, Chiara Lo Passo, Virginie Brigitte Gallati, Filippo Cristini, María Ajora Bernal, Luca Ferrari, and, of course, my family.

I had the privilege to achieve this book in Paris, taking advantage of the beautiful facilities and libraries of the Ecole Normale Superieure, where I was hosted as a

(Footnote 1 continued)
Generative and demonstrative experiments, in L. Magnani (Ed.), *Model-Based Reasoning in Science and Technology. Theoretical and Cognitive Issues*, pages 479–498. Springer, Heidelberg/Berlin. For Part II: L. Magnani, T. Bertolotti (2013). Selecting chance curation strategies: Is chance curation related to the richness of a cognitive niche? *International Journal of Knowledge and Systems Science* vol 4(1) (pp. 50–61); T. Bertolotti, L. Magnani (2013). A philosophical and evolutionary approach to cyber-bullying: Social networks and the disruption of sub-moralities. *Ethics and Information Technology* vol 15(4) (pp. 285–299); T. Bertolotti, L. Magnani (2014). An epistemological analysis of gossip and gossip-based knowledge. *Synthese* vol 191(17) (pp. 4037–4067); T. Bertolotti, L. Magnani (2013). Terminator niches. In S. Richir (Ed.): *Virtual Reality International Conference–Laval Virtual, VRIC 2013, Laval, France, March 20–22, 2013*, ACM, New York. For Part III: L. Magnani, T. Bertolotti (2011). Cognitive bubbles and firewalls: epistemic immunizations in human reasoning. In L. Carlson, C. Hoelscher, & T.F. Shipley (Eds.), *Proceedings of the 33rd Annual Conference of the Cognitive Science Society* (pp. 3370–3375). Austin, TX: Cognitive Science Society; T. Bertolotti, L. Magnani (2010). The role of agency detection in the invention of supernatural beings: an abductive approach, in L. Magnani, W. Carnielli, C. Pizzi (eds), *Model-Based Reasoning in Science and Technology. Abduction, Logic, and Computational Discovery*, pages 195–213. Springer, Heidelberg/Berlin; T. Bertolotti, C. Cinerari (2013). The Gospel according to Google: The future of religious niches and technological spirituality, *The European Journal of Science and Theology* vol. 9(1) (pp. 41–53); L. Magnani, T. Bertolotti (2011). Co-writing of sections 6.4 Sacred and Sacrifice: the Violent Boundaries of Magic; 6.6.2 Moral Viscosity and the Ultimate Paradox of Forgiveness, in L. Magnani (2011), *Understanding Violence. Morality, Religion, and Violence Intertwined: A Philosophical Stance*. Springer, Heidelberg/Berlin.

visiting postdoc fellow by Prof. Giuseppe Longo: I thank him for making me feel welcome to the encounters of his research group, and for the thought-provoking discussions we had. All of the insights afforded by my Parisian stay proved crucial for the final outcome of this book.

I am well aware that the completion of a book requires much more than intellectual constance and fertility: this book would have been much harder to achieve had I not been able to rely on the most competent and precious collaboration of Leontina Di Cecco at Springer.

Last but not least, I want to thank Anne-Claire Podlipski, whose unexpected appearance in my life made me realize that the appraisal of the product of five years of research is not the sign of an end but of a new beginning, and that if philosophers cannot make their ideas clear enough to be understood, appreciated, and criticized by laypeople, then they are not true philosophers. I am also grateful to her for giving me the chance to put the very final notes on this book in Los Angeles, CA, filled with the most beautiful inspiration.

Pavia, Italy
February 2015

Tommaso Bertolotti

Contents

Chapter 1
Introduction: For an Epistemology of the Human Being

Homo sum, humani nihil a me alienum puto
(I am a human being, I consider nothing that is human alien to me)
Publius Terentius Afer, *The Self-Tormentor*, II Century BCE

1.1 About this Book

During the research that lead to this book, I focused on three topics that I found pivotal for shaping human reasoning: the mechanisms underlying the development of scientific modeling, the distributions of knowledge in the environment amounting to cognitive niche construction, and the "epistemic immunizations" that, compared with higher regimes of rationality such as the ones displayed by science and logic, produce what is defined as irrationality. The structure of the book consists in three parts, each devoted to one of the main topics, for a total of twelve specific chapters, plus an Introduction—which you are reading—and some Concluding Remarks.

The first part concerns the analysis of *scientific rationality* carried out through a study of scientific *modeling and experimentation*, developing a perspective rooted in the so called eco-cognitive (ecological/cognitive) tradition. The intent is to frame scientific endeavor within a *naturalistic and pragmatic* outlook: the first objective is to spell out a conceptual line connecting the use of rudimentary mental models in natural cognition to the use of models characterizing modern and contemporary science; the second goal is to demonstrate how scientific experimentation can be seen as aiming at two different possible targets, one is the increase of scientific knowledge, and the other one the dissemination of precedent results in a given social context.

The second part deals with the *distribution of knowledge* onto the environment operated by human agents (a phenomenon that can be described as "cognitive niche construction"), and also with how this distribution, seen as a collective effort, affects and is affected by *social cognition*. Albeit the interdisciplinary theory of cognitive niche construction robustly accounts for human advancements (including the possibility of science itself), I will tackle the pivotal need to understand the *social*

© Springer International Publishing Switzerland 2015

T. Bertolotti, *Patterns of Rationality*, Studies in Applied Philosophy, Epistemology and Rational Ethics 19, DOI 10.1007/978-3-319-17786-1_1

mechanisms, for instance the simple fact of supporting group living, allowing the coordinated effort of constructing and managing such niches: I will illustrate how this knowledge is also necessary to understand the *crises* impacting cognitive niches when new *technological* structures significantly change basic social dynamics within the niche itself, especially when these structures mediate established forms of communication and decision-making.

The third part, connecting with the first two ones in a kind of dialectic opposition, focuses on what is commonly labeled as "irrational" by *contrasting* the standard ecological and social inferences analyzed so far (that is those usually employed to cope with the environment and other human beings) with another inferential regime characterized as an epistemic, or cognitive, "immunizations." The epistemological and cognitive interest will lie in the analysis of *religious and magical thinking*, in order to uncover other regimes of rationality that allow agents relying on them to cope *in a different way* with the demands of *ecological-cognitive problem solving*, both in past and contemporary/technological cognitive niches.

The global aim of this book is to achieve a *coherent and unitary perspective on human rationality* out of the three thematic chores. The outlook is not diachronic, as the interest is not to draw the emergence of a progress in reasoning. Instead, I will provide arguments, and examples, showing that it can be fruitful to engage in a research that, sampling out certain features of scientific rationality such as the necessity to cope with *constraints of the surrounding environment* (both social and ecological), highlights the *recurrence* of similar *inferential patterns* in forms of cognition rationally characterized as less demanding (i.e. as social, and religious cognition), thus legitimizing original conceptual connections and theoretical speculations.

1.2 About the Subdivision

As explained by the title, *Science and Strategic Cognition*, the first part frames epistemology in an ecological and cognitive perspective. The two first chapters focus on the nature and use of models, while the third on the operationalization of models through experimentation: this is coherent with Hacking's philosophical milestone *Representing and Intervening* (Hacking 1983). Since the debate on models has been crucial to philosophy of science for the past four decades, I deemed it was apt to begin my investigation with scientific models, first of all by investigating their origin. The next chapter approaches the theme of models as mental models by a survey of animal cognition studies linked to camouflage, to sustain the claim that biological camouflage can be seen as the operationalization—also in extremely rudimentary cognitive systems—of mental models representing the other's cognitive system. In the same chapter, by analyzing the inferential operations (supported by the aforementioned modeling activity) underpinning camouflage-breaking strategies, I will try to explain how the same tacit use of models representing the other's cognitive abilities is at play in human communication, when enacting and uncovering linguistic deception. The third chapter explicitly focuses on scientific modeling, drawing from some conclusions made explicit in the previous chapter. The interlocutors of my argumentation will be the contemporary philosophical actors of the debate about scientific models, especially advocates of *fictionalism* (Frigg 2010; Suárez 2010), that is the view according to which scientific models are best understood by the tools analyzing literary fictions, and can therefore be described as displaying a *fictional* nature. I will reject this view by appealing to the biological origins of modeling—described in the previous chapter—and arguing that the construction of scientific models can be conceptualized as an *emergence* of the transparent, and careful, use of models (mental, or otherwise distributed) made by scientists, which is quantitatively but not qualitatively different than the *biological* reliance on models displayed by any cognitive agent: assessing the instrumental use of some fictional elements does not entail that the whole system should be labeled as fiction. Otherwise, to assess the self-aware fictional nature of science would compel us into defining as fictional also the proto-models displayed by animal cognition: the next step would be to advocate a randomized fictional behavior in animals and humans.[1] In Chap. 4, I will turn my attention at scientific practice and consider one of the pillars of scientific rationality, the experiment. The idea behind the analysis is to merge Hacking's analysis of scientific *intervention* (Hacking 1983) with the eco-cognitive perspective (Magnani 2009) and some insights from the anthropological and ethnographic approach to science (Latour 1987; Knorr Cetina 1999): the result of this research is the theoretical subdivision of experimentation in two broad areas, namely *generative* and *demonstrative* experiments, each with its own epistemological and cognitive characterization.

The second part of this book, entitled *Cognitive Niches and Social Cognition: Using Knowledge as a Tool*, will analyze cognitive niche construction and, in particular, the social/collaborative grounding of such activities. After an introduction on the theory of cognitive niche construction (5), I will shift my attention on how a collectivity can select techniques for enriching and maintaining cognitive niches (6), stressing how the selection of the best curation method is a non-monotonic activity and indeed depends on the original condition of the cognitive niche itself. The two subsequent chapters will be devoted to the analysis of gossip (epistemological in Chap. 7 and more on the perspective of philosophy of technology in Chap. 8). Considering the relevance of language for cognitive niche construction (Clark 2005), it was interesting to study a widespread phenomenon such as gossip from a perspective different from anthropology, psychology and evolutionary studies (Dunbar 2004). In my first research I studied gossip as a social-knowledge management tool, offering an epistemological modeling that made sense of the socio-biological and evolutionary bias towards gossip as an effective knowledge management system (Wilson et al. 2002) on the one hand, and the low esteem it had traditionally received within the philosophical tradition. The second analysis partly rests on the philosophical investigation of violence and conflict offered in Magnani (2011), and aims at setting

[1] Echoing Giere's commitment (2009), philosophy of science should convey an appreciation of science that must not be taken for granted (especially in the actual socio-economical ochlocracy): on top of being theoretically weak, a fictionalist account might offer Science's uncovered flank to her foes, from hard-creationists, to neo-liberal fund-shrinkers and reborn Lullists.

up a proper theoretical framework to understand how a mechanism as widespread as gossip—which can be seen as a powerful niche construction tool—alters when immersed in a radically different cognitive niche, for instance in computer-mediated social communication. The last Chap. 9 introduces the notion of *terminator niche*, a techno-cognitive niche in which the modifications introduces to foster improvements backfire, and become the direct cause of diminished welfare, impoverishment and potential collapse of the niche itself.

The third and last part *The Eco-Cognitive Epistemology of Counterfactual Beliefs* deals with the limits or, so to say, with the *twilight zone* of my epistemological approach to human cognition, starting from the models of epistemic immunization proposed by logician John Woods in the form of the *epistemic bubble* (Woods 2005). At this point of my research it was possible to frame religion with the conceptual tools developed so far, such as the modeling of external agencies and the construction of cognitive niches. As I said in the beginning of this introductory chapter, I was not interested at drawing a history of cognitive "progress," which emerges through a juxtaposition of different "stages" of thought (Barnes 2000): then it would have made sense to begin with religion and the origins of culture and climb our way all the way up to science. Thus, in Chap. 10 I analyzed religion as a model of a class of inferences, traditionally perceived as irrational (or having to do with counterfactual beliefs), but which can be very interestingly studied through our epistemological "flooding" of traditionally nonscientific domains. Such outlook can be regarded as analogous to the psycho-anthropo-cognitive effort to frame religion "as a natural phenomenon" (Boyer 2001; Atran 2002; Dennett 2006): similarly, my aim was to investigate religion as a "philosophical phenomenon" and yet without adhering too much to what is traditionally understood as *philosophy of religion*. My interest was in fact to apply an epistemological toolbox in order to research and model the religious inferential style and how religion could be defined an ecological-cognitive activity: indeed, the part's second chapter deals with an abductive modeling of how "supernatural" beliefs are generated starting from signs available in real environments, the following chapters applies to the religious phenomenon some insights developed in the previous part about cognitive niches, and focuses on the development of religious inferences in high-tech cognitive niches. The following Chap. 12 focuses precisely on the epistemic bubble in order to show how its conceptualization can be further expanded to a broader "cognitive bubble:" by this I mean to show how the kind of mechanism allowing our *thinking of knowing* what we *do not actually know* is crucial to many dimensions of human cognition, and not just to the sentential ones (as suggested by Woods' approach). Relying on the knowledge about "irrationality" gathered so far in the book, the final Chap. 13 focuses on religious pragmatics and deals with two dense philosophical topics: forgiveness (in a perspective that merges anthropological and cognitive studies) and (self-)sacrifice, analyzed through the lenses of epistemic violence, Girardian studies and some icons of popular culture.

1.3 About the Theoretical Framework

1.3.1 Eco-Cognitive Epistemology

As stated at the beginning of this Introduction, this book and the research it comprehends belong to what can be defined as *eco-cognitive epistemology*, a mode of epistemology introduced, pursued and advocated by Lorenzo Magnani (2009). The two keywords defining this approach to epistemology are *ecological* and *cognitive*.

- **Ecological**: We try to avoid as much as possible the abstraction of agents out of their environmental constraints, and we are suspicious about the use of idealized agents. All processes that are viable for an epistemological analysis are, in our view, goal-oriented and nested in a particular environment, upon which the epistemic agent relies in order to achieve her goal—and which may hinder at the same time the agent's efforts. Any epistemological enquiry must, according to this view, face this issues as a starting point and not as a future or occasional care.
- **Cognitive**: We put forward an ideal of epistemology that must not be cognition-blind. As we are interested in the epistemic processes enacted by real agents, firstly our hypotheses must be in accordance with the evidence offered by the cognitive sciences, and secondly our epistemological attitude welcomes results and insights from cognitive science as a primary source for reflection.

The aim of eco-cognitive epistemology is to analyze thought and reasoning in its different forms, so it must assume the characteristics of philosophy of science, social epistemology, applied epistemology, etc., depending on the issue at stake. Eco-cognitive epistemology has the following enormous advantage, which is coherent with the unitary (albeit multifaceted) nature displayed by human cognition: one single theoretical framework, without a multiplication of jargons, can cover a range of issues varying from scientific modeling and laboratory science all the way to religious inference and language-mediated social cognition (gossip). This will also be crucial in order to understand the emergence of similarly configured *patterns of rationality* across apparently much different domains.

Eco-cognitive epistemology is not *veritistic*, and virtually not *prescriptive*.[2] The two things come usually together, since an epistemological outlook aimed at describing what inferential processes guarantee reaching the Truth would make little sense if it did not prescribe to follow those inferences. Eco-cognitive epistemology has a chiefly descriptive aim, and it is empowering to the extent that, for every different fields it surveys, it may spell out the concerned inferential schemas and thus highlight where there is greater possibility of errors occurring. As for the *veritistic* aspect, it is lacking because eco-cognitive epistemology traces back to Peirce's pragmaticism, and sees as one of its main explanatory elements the notion of *abduction*, which is traditionally received as a non-truth-preserving inference.

[2] I will go back to this issue in the Concluding Remarks of this book.

1.3.2 Abductive Reasoning

Abductive processes will be referred to all along the book, pointing out their epistemological and cognitive relevance. The single occurrences will each specify different aspects of abductive reasoning, but it might be good to introduce from now the unacquainted reader with the notion at stake: prior to entering the subject, though, it is mandatory to say that this will be a very small introduction to the matter, and the reader who wishes to get a fuller idea about abduction should refer to Lorenzo Magnani's monograph *Abductive Cognition: The Epistemological and Eco-Cognitive Dimensions of Hypothetical Reasoning* (Magnani 2009), and to Dov Gabbay and John Woods's *The Reach of Abduction* (Gabbay and Woods 2005).

Abduction, as understood within the Peircean framework I am referring to, can be accounted for as the process of inferring certain facts and/or laws and hypotheses that render some sentences plausible, that explain (and also sometimes discover) a certain (eventually new) phenomenon or observation: it is also the process of reasoning in which hypotheses are formed and evaluated.

The notion of abduction was introduced by Aristoteles, as *epagoghe*, to indicate a retroductive reasoning that is different from induction and deduction.[3] In the Middle Ages, studies on syllogism relegated abduction to its role of fallacy (namely the fallacy of affirming the consequent), as it is still classified today by classical logic, being a non-truth preserving inference.

- **Premise**: If a then b
- **Premise**: And b
- **Abduction**: Then a

It does not take to be any logician to understand why this inference is fallacious. Had it been a *modus ponens*, had the second premise been a, then we would have known for sure that the right conclusion was to be b: assuming both premises were true, the conclusion would have been true as well, as in the following case.

- **Premise**: If a then b
- **Premise**: And a
- **Deduction**: Then b

Whereas in the case of abduction, there is no such certainty. Of course, there is some degree of plausibility: if we assume that the first premise is true, and that the second premise is true as well, there is some logical possibility that the conclusion obtains, but no certainty in spite of both premises being true. Consider this small example:

[3]On the relationship between abductive and deductive reasoning in Aristotle refer to Magnani (2015).

- **Premise**: If *the cat climbs on the table* then *it makes the crystal centerpiece fall and break.*
- **Premise**: And *the crystal centerpiece fell and broke.*
- **Abduction**: Then *the cat climbed on the table.* (*Abduction*)

Before taking our poor cat to the shelter (or worse) for breaking grandma's precious centerpiece, we should appeal to our best logical conscience. The inference we performed to find the culprit was an abduction: it is of course plausibile to some extent that the cat did that, but our two premises, as true as they can be, do not entail the truthfulness of the conclusion. It might have been one of the neighbor's kids who came in while the door was unlocked, or a gush of window that made the window open and bash against the table, or something else might have fallen before (a bottle?) and pushed the centerpiece off the table.

Had the first premise been like "If *the crystal centerpiece falls and breaks* then *the cat climbed on the table*," and the same second premise, then indeed our inference would not have been a (no matter how informed) *guess* in the attempt to provide an explanation for an event that struck us as remarkable, but a decent—yet less useful—deduction.

- **Premise**: If *the crystal centerpiece falls and breaks* then *the cat climbed on the table.*[4]
- **Premise**: And *the crystal centerpiece fell and broke.*
- **Deduction**: Then *the cat climbed on the table.*

Should this little example, explaining why abduction is traditionally considered a fallacy and a non-truth-preserving inference, make us set it aside as something irrelevant and undesired? Not at all! As Charles Sanders Peirce (1839–1914) started to show in the second half of XIX century, abduction is actually the most powerful and vital mode of inferencing any cognizing organism may rely upon. Notwithstanding all the formal deficiencies individuated by classical logic, abduction is crucial for reaching, from what we know, what we don't know yet. This is how C. S. Peirce essentially schematized abduction (Peirce 1931–1958, 5.189):

1. The surprising fact C is observed.
2. But if A were true, C would be a matter of course.
3. Hence there is reason to suspect that A is true.

In this sense, abduction can indeed be considered as the most powerful inference aimed at *making sense*. And this *making sense* of one's surroundings through abductive reasoning is what will make abduction studies a fundamental building block of this book. When I introduced abduction I said that it is the process of inferring certain facts and/or laws and hypotheses that render some sentences plausible, that explain

[4]This example is just for the sake of explaining the difference between deduction and abduction using a similarly structured narrative. Accept this premise to be true for the sake of understanding, even if as I just showed there are a number of reasons why this premise should be questioned, that is other reasons causing the fall of the centerpiece.

(and also sometimes discover) a certain (eventually new) phenomenon or observation: it is also the process of reasoning in which hypotheses are formed and evaluated. This may of course take place at a sentential, propositional and explicit level, and this is the case when scientific hypotheses are at stake (as I will examine in Chap. 3), or when some explanation must be produced about someone's erratic behavior (as it happens in gossip, which I will analyze in Chap. 7), but a great part of the abductive inferences carried out my human beings (and other cognizing organisms) are neither sentential nor explicit. Pierce himself, for instance, considered perception as something clearly abductive (and hence inferential, mediate). Consider the following passage, which Peirce originally intended to include in the last of his seven Harvard Lectures (May 1903): "A mass of facts is before us. We go through them. We examine them. We find them a confused snarl, an impenetrable jungle. We are unable to hold them in our minds. [...] But suddenly, while we are poring over our digest of the facts and are endeavoring to set them into order, it occurs to us that if we were to assume something to be true that we do not know to be true, these facts would arrange themselves luminously. That is *abduction* [...]".[5] This passage seems to classify abduction as emerging in "perceiving" facts and experiences, and not only in the conclusions of an "inference" (Hoffmann 1999, pp. 279–280), intended in the classical sense, as expressed by symbols carrying propositional content. Indeed, Peirce was quite attached to the idea that no strict demarcation could be drawn between a perceptual judgement and an abductive inference, to the extent that even perception partakes of this *sense-making* activity that is necessary to all living organisms.

Most of the references to abduction in this book will be about the intuitive, cognitive role of abduction as sense-making, even if I will occasionally deal with and refer to typologies of abduction that are rather propositional and language based as in the case of the production of scientific models (which I will nevertheless show as a direct descendant of intuitive cognitive modeling in Chap. 3).

Different typologies of abduction can be isolate and analyzed. Whereas an exhaustive analysis can be found in Magnani (2009), and some specific remarks are to be found along this book, it is proper to provide the readers with an essential subdivision that might help them further understand some of the claims I will be putting forward.[6]

- **Selective abduction** is the process in which a hypothesis is abductively selected from a pre-stored encyclopedia of "abducibles." An example of selective abduction is diagnostic reasoning in medicine: it starts from patient data that is abstracted into clinical features to be explained. Then, selective abduction generates plausible explanatory diagnostic hypotheses.

[5] cf. "Pragmatism as the logic of abduction", in Peirce (1992–1998), pp. 227–241, the quotation is from footnote 12, pp. 531–532.

[6] The following definitions are adapted from the *Lexicon of Abductive Cognition* (Magnani 2009).

- **Creative abduction** is the process in which a completely new hypothesis is created. An example of creative abduction is scientific discovery: the discovery of a new disease and the manifestations it causes in the field of medical knowledge or the Kepler's discovery of the elliptic orbit of the planets.
- **Manipulative abduction** is a process in which a hypothesis is formed and evaluated resorting to a basically extra-theoretical and extra-sentential behavior that aims at creating communicable accounts of new experiences to integrate them into previously existing systems of experimental and linguistic (theoretical) practices. Manipulative abduction represents a kind of redistribution of the epistemic and cognitive effort to manage objects and information that cannot be immediately represented or found internally. An example of manipulative abduction is the case of the human use of the construction of external diagrams in geometrical reasoning, useful to make observations and "experiments" to transform one cognitive state into another for example to discover new properties and theorems.
- **Model-based reasoning** can be an important form of abduction. In such reasoning, information about one thing is supplied from something else. E.g., a person reasoning about how a cat got stuck in a tree might draw upon a time when a different cat got stuck in the same tree, or in a similar tree elsewhere.[7]
- **Multimodal abduction** depicts hybrid aspects of abductive reasoning. Abductive inference can be visual as well as verbal, and consequently we have to acknowledge the sentential, model-based, and manipulative nature of abduction. Both evidence and hypotheses can be represented using various sensory modalities. Some basic aspects of this constitutive hybrid nature of multimodal abduction involve words, sights, images, smells, etc. but also kinesthetic experiences and other feelings such as pain, and thus all sensory modalities.

One last thing should be said before letting the reader commence the real matter: this book is not moved by a furious desire for definitions. It is about discovering the different and recurring patterns enacted by human rationality in different settings, in order to move towards an epistemology of the human being. In order to achieve this, definitions need to be flexible to serve the purpose of the book, and not constrain it. Some definitions and concepts may overlap, and the reader should not be alarmed by this: one cognitive phenomenon may be better framed through one concept in a given perspective, and through another concept in a different one without this leading to the perception of a contradiction. For instance, a given inferential process could be at a certain point analyzed as a creative abduction, while at a later point it will be better to study it as a form of model-based reasoning: those two things do not contradict each other. Human cognition is about inferential richness, otherwise we would live in a merely deductive world, with nothing to make sense of, and not much to take pleasure in.

[7]This simple yet amazingly effective definition is proposed by Cameron Shelley in his abductive analysis of biomorphism (Shelley 2015).

References

Atran S (2002) In Gods we trust: the evolutionary landscape of religion. Oxford University Press, Cambridge

Barnes MH (2000) Stages of thought: the co-evolution of religious thought and science. Oxford University Press, Oxford

Boyer P (2001) Religion explained. Vintage U.K. Random House, London

Clark A (2005) Word, niche and super-niche: how language makes minds matter more. Theoria 54:255–268

Dennett D (2006) Breaking the spell. Viking, New York

Dunbar R (2004) Gossip in an evolutionary perspective. Rev Gen Psychol 8(2):100–110

Frigg R (2010) Fiction in science. In: Woods J (ed) Fictions and models: new essays. Philosophia Verlag, Munich, pp 247–287

Gabbay DM, Woods J (2005) The reach of abduction. A practical logic of cognitive systems, vol 2. North-Holland, Amsterdam

Giere R (2009) Why scientific models should not be regarded as works of fiction. In: Suárez M (ed) Fictions in science. Philosophical essays on modeling and idealization. Routledge, London, pp 248–258

Hacking I (1983) Representing and intervening. Introductory topics in the philosophy of natural science. Cambridge University Press, Cambridge

Hoffmann MHG (1999) Problems with Peirce's concept of abduction. Found Sci 4(3):271–305

Knorr Cetina K (1999) Epistemic cultures. How sciences make knowledge Harvard University Press, Cambridge

Latour J (1987) Science in action: how to follow scientists and engineers through society. Harvard University Press, Cambridge

Magnani L (2009) Abductive cognition: the epistemological and eco-cognitive dimensions of hypothetical reasoning. Springer, Berlin

Magnani L (2011) Understanding violence. morality, religion, and violence intertwined: a philosophical stance. Springer, Berlin

Magnani L (2015) The eco-cognitive model of abduction. Naturalizing the logic of abduction (Forthcoming)

Peirce CS (1931–1958) Collected papers of Charles Sanders Peirce. Harvard University Press, Cambridge, vols. 1–6, Hartshorne C, Weiss P (eds) vols. 7–8, Burks AW (ed)

Peirce CS (1992–1998) The essential Peirce. Selected philosophical writings. Indiana University Press, Bloomington and Indianapolis, vol. 1 (1867–1893), ed. by N. Houser & C. Kloesel; vol. 2 (1893–1913) ed. by the Peirce Edition Project

Shelley CP (2015) Biomorphism and models in design. In: Magnani L, Li P, Park W (eds) Philosophy and cognitive science, western and eastern studies II. Springer, Berlin (Forthcoming)

Suárez M (2010) Fictions, inference, and realism. In: Woods J (ed) Fictions and models: new essays. Philosophia Verlag, Munich, pp 225–245

Wilson D, Wilczynski C, Wells A, Weiser L (2002) Gossip and other aspects of language as group-level adaptations. In: Hayes C, Huber L (eds) The evolution of cognition. The MIT Press, Cambridge, pp 347–365

Woods J (2005) Epistemic bubbles. In: Artemov S, Barringer H, Garcez A, Lamb L, Woods J (eds) We will show them: essays in honour of Dov Gabbay, vol II. College Publications, London, pp 731–774

Part I
Science and Strategic Cognition

The first part concerns the analysis of scientific rationality carried out through a study of scientific modeling and experimentation, developing a perspective rooted in the so-called eco-cognitive (ecological/cognitive) tradition. The intent is to frame scientific endeavor within a naturalistic and pragmatic outlook: the first intent is to spell out a conceptual line connecting the use of rudimentary mental models in natural cognition to the use of models characterizing modern and contemporary science; the second goal is to demonstrate how scientific experimentation can be seen as aiming at two different possible targets, one, the increase of scientific knowledge, and the other, the dissemination of precedent results in a given social context.

Chapter 2
On Biological and Verbal Camouflage: The Strategic Use of Models in Non-Scientific Thinking

Abstract The chapter approaches the topic of models *as mental models* by a survey of animal cognition studies linked to camouflage, to sustain the claim that biological camouflage can be seen as the operationalization—also in extremely rudimentary cognitive systems—of mental models representing the other's cognitive system. In this same chapter, by analyzing the inferential operations (supported by the aforementioned modeling activity) underpinning camouflage-breaking strategies, I will try to explain how the same tacit use of models representing the other's cognitive abilities is at play in human communication, when enacting and uncovering linguistic deception.

2.1 Introduction

The second half of past century witnessed a flourishing of ethological and biological studies concerning the issue of camouflage as related to animal cognition.[1] Still, it is important to notice how biological and cognitive studies have been focusing on the neurological and physiological correlates of camouflage dynamics but less on the inferential grounding.[2]

Concentrating on the inferential ground underpinning camouflage mechanisms can be of extreme interest because of the richness of the afforded theoretical implications, which include analogical suggestions that might be developed in an epistemological framework, in order to uncover new but similar *pattern* in other fields: this will allow me to rely on considerations of biological concern as a tentative theoretical tool for further investigations transcending biology and ethology to land in a more philosophical framework.

---

[1] A recent special issue of the *Philosophical Transactions of the Royal Society B: Biological Sciences*, introduced by Stevens and Merilaita (2009), does provide both the state of the art and new insights in the field.

[2] If all inference is, in fact, a form of sign activity—as received from the Peircean tradition—and we use the word sign to include feelings, images, conceptions, and other representations, then we must include unconscious thought among the model-based ways of moral thinking.

© Springer International Publishing Switzerland 2015
T. Bertolotti, *Patterns of Rationality*, Studies in Applied Philosophy, Epistemology and Rational Ethics 19, DOI 10.1007/978-3-319-17786-1_2

Indeed, while the notion of camouflage I will start from originates from biological studies and describes a range of strategies used by organisms to dissimulate their presence in the environment, it has been frequently borrowed by other semantic fields as it is possible to camouflage one's position, intentions, opinion, etc.: an interesting conceptual continuum between the multiple denotations of camouflage seems to emerge from the multiple homologies. The etymology of the word itself suggests that the origin is french French, from *camoufler* 'to disguise' (originally thieves' slang), from Italian *camuffare* 'disguise, deceive,' perhaps by association with French *camouflet* 'whiff of smoke in the face.' The goal I set before the reader and myself in this chapter is to vindicate the strategic nature of camouflage, both in its ethological and human connotation, through a full appreciation of its theoretical foundations.

Following this insight, the beginning of this chapter aims at sketching out the main forms of camouflage as understood within their biological framework, insisting on the inferential dynamics underdetermined and allowing camouflage, making use of the concept of abduction as received from the Peircean heritage. Then, I will explore some of the most relevant occurrences of camouflage in dialectical and rhetorical perspectives. Finally, I will draw the sums of the comparison between linguistic and biological camouflage, showing how strategies aimed at debunking verbal camouflage correspond to their respective countermeasures in biologically-intended camouflage.

2.2 Understanding Camouflage as Inferential Warfare

2.2.1 Situating Camouflage in the Environment

It might be good to start by considering how animated beings do not merely "contemplate" their environment in an uninterested fashion, but as *survival machines* they cannot separate their perceiving from continuous activities of problem-solving—which could be ultimately described as cognition. An organism's surroundings are not uniformly relevant for the organism itself.

An important part of a living agent's ecology is composed by *other agents*: other agents are constituted of matter as any other part of the environment—rocks, plants, dirt, water, etc.—but on top of the efficient causation displayed by non-living elements they exhibit the possibility of *semiotic causation* (Hoffmeyer 2008), as they are capable of producing effects guided by inferences they operate on other elements of *their* own environment.[3]

[3]It might be unfair to acknowledge plants only as passive elements being part of an environment, only passible of efficient causation: it has been suggested that even plants can be described as displaying a kind of embodied cognition (Calvo and Keijzer 2009) and are therefore concerned by semiotic causation as well. The perceptual and inferential horizon at play is of course radically incommunicable with respect to ours and to that of non-human animals we able to refer to.

This is true for human agents too, as they look at a natural landscape, for instance, their attention is automatically driven to search for and investigate little movements, glitters, shadows that could signify the presence of life forms.[4]

The ecological problem concerning external agency that I am beginning to delineate is quite complex, and it will be at the core of my interest for spelling out the *patterns of rationality* setting the title of this book: to begin with, the notions of *external* environment and that of *other* agents are not absolute but rather immediately rise the "with respect to?" question. It is easy to understand that every agent can be part of any other agent's ecology: even human beings are, as individuals, constituents of each other's environment, we are part of each other's "surroundings."

To introduce this investigation, it might be interesting to rely on the semiotic concept of *semiosphere*, modeled upon that of biosphere: as contended by Hoffmeyer, "[...] this semiosphere truly is a sphere like the atmosphere, the hydrosphere or the biosphere, in that it penetrates these spheres for living organisms and consists in communication: sounds, odors, movements, colors, electric fields, waves of any kind, chemical signals, touch, etc." (Hoffmeyer 2008, p. 153). Every organism has only a *partial* access to this semiosphere, constrained by its situatedness and biological endowments. Such a description is complementary with the eco-psychological concept of affordance[5] (Gibson 1979), which provides an alternative account of the role of the environment and external—also artifactual—objects and devices, as the source of action possibilities (constraints for allowable actions).

As far as biological camouflage is concerned, interspecific dynamics will matter the most and therefore, wherever possible, ignore the relevance of intraspecific dynamics affecting the definition of environment.[6] This leads to the formulation of the following working hypotheses, considering as "other agents" organisms that do belong to different species and are therefore potential predators or preys, and not rivals:

1. Every organism normally attempts to detect the presence of other agents and hide its own presence from other agents in the surroundings.

---

[4]The issue of the relationship between the cognizant and her surroundings will be tackled again in the following chapter, but especially in Part III, when dealing with the study of religion as a cognitive phenomenon (Chaps. 10, 11).

[5]Originally belonging to the conceptual toolbox of ecological psychology, an affordance is a resource or chance that the environment presents to the "specific" organism, such as the availability of water or of finding recovery and concealment. Of course the same part of the environment offers different affordances to different organisms. Part II will further rely on the notion of *affordance*, and hence provide a deeper understanding.

[6]When dealing with intraspecific predation, competition over sexual mates or available resources (such as food or nesting room), it seems apter to consider the external environment as related to every single organism; conversely, if we want to frame interspecific dynamics such as predator-prey ones, symbiotic relationships and so on, it might be simpler to consider the notion of environment as related to a species or at least to a localized population.

2. *Both* predators and preys simultaneously behave according to (1), as organisms tend to avoid recognition by both their predators and their preys.[7]

2.2.2 Truth and Survival in Agency Detection and Recognition: The Importance of Animal Abduction

Sensorial perception is what organisms must rely on in order to recognize the presence of other organisms in their proximities. What senses pick up is not an immediate picture of external agency, but a more or less rich complex of signs: these signs mostly partake of the senses of sight, smell and hearing (taste separated from smell, and touch not aimed at picking up vibrations in the ground, seem to provide cues that are more useful to proximally investigate the nature of an organism rather than to infer its presence). The resulting situation is somehow like this: an animal agent must manage to detect the presence of other agents in order to maximize its own chance of survival, and such detection can only be inferred by operating upon meaningful signs. Once the other agent is individuated, the following step consists in the operationalization of the correct "affordances" concerning the detected organism: i.e., the *detector* should not know whether to attack, flee, ignore, etc. the *detected.*

In order to get a better understanding of such cognitive phenomena, we can rely on the powerful vision proposed by philosopher of biology Ruth G. Millikan. She suggests that internal representations of animals might mostly consists of PPR (*"push-me pull-you" representations*), meaning that they are *both aimed at representing a state of affairs and at producing another*, thus suggesting a "chance" for behavior (as

[7]One could start by arguing that these hypotheses could be questioned by referring to the famous *handicap principle* (Zahavi and Zahavi 1993) and *honest costly signaling* theories: part of the handicap principle theory concerns cases in which sign suppression is abandoned favoring a loud semiotic activity by which the predator or the prey signals to its counterpart that the latter has been spotted and will not manage at catching the former off-guard. The (maybe over-)notorious example is that of the gazelle's "stotting" (i.e. jumping several times up and down): biologists following the handicap principle theory maintain that the stotting behavior is *de facto* a waste of energy that could be employed to run away immediately, but instead this waste of energy (that is, the handicap) is afforded by energetic specimens that therefore convey the message "I am so full of energy that I can even waste it like this, I am not going to get tired that easily!": the aim is to achieve a win-win balance so that both the predator and the prey avoid an energy-consuming chase or struggle whose outcome is not foreseeable. Nevertheless, it should be considered that honest signaling is enacted only *after* recognition is accomplished: *stealth* attack and defense remain the highest-success strategies for both predators and preys. If this was not the case, it would be legitimate to expect from all living creatures to be flashily colored in orange and pink and extremely loud, all the time, while even the long-time favorite gazelle displays colors useful to blend in the savanna grass. Furthermore, tt should be considered that the factuality of the handicap principle has become a hotly debated topic over the past few decades. Since its introduction, it witnessed alternating periods of popularity and periods of decline: such alternation was caused on the one hand by a growth in popularity in humanities and economics separate and unmatched by its biological counterpart, which was on the other hand strongly opposed by influential biologists such as Maynard-Smith, who claimed—among several others—the impossibility to find actual evidence of the handicap principle in nature (Grose 2011).

received by the Gibsonian/affordance tradition). Therefore, the content of a PPR mental representation will never be of a mere contemplative nature as far as animals are concerned (it is not the place to argue whether that can ever be the case with human beings), but it will always propose and enact a behavioral pattern connected with the very same representation.

> An animal's action has to be initiated from the animal's own location. So in order to act, the animal has to take account of how the things to be acted on are related to itself, not just how they are related to one another. In the simplest cases, the relevant relation may consist merely in the affording situation's occurring in roughly the same location and at the same time as the animal's perception and consequent action. More typically, it will include a more specific relation to an affording object, such as a spatial relation, or a size relative to the animal's size, or a weight relative to the animal's weight or strength, and so forth (Millikan 2004, p. 19).

In the perspective I have adopted, an abductive model is the fittest one to describe and investigate the formation of those internal representations which animals produce inferring them on the basis of those signs they are able to recover from the environment. It seems therefore legitimate to speak of "animal abduction" (Magnani 2007a).

Abduction, as understood within the Peircean framework, can be accounted for as the process of inferring certain facts and/or laws and hypotheses that render some sentences plausible, that explain (and also sometimes discover) some (eventually new) phenomenon or observation: it is the process of reasoning in which hypotheses are formed and evaluated. Abductive reasoning is active in many scientific disciplines but also in everyday rationality: it is essential in scientific discovery, medical and non medical diagnosis, generation of causal explanations, generations of explanations for the behaviors of others, minds interplay, when for example we attribute intentions to others, empathy, analogy, emotions, as an appraisal of a given situation endowed with an explanatory or instrumental power, etc.

In fact, abduction must not be regarded as a merely sentential inferential process: indeed, many studies explored the existence of "model-based" abductive processes, concerning the exploitation of internalized (or to the manipulation of external) models of diagrams, pictures and so on. Recent studies on abduction opened a much wider field of investigation concerning these multi-modal inferences: survival, for any animate organism, is a matter of coping with the environment and the relationship with the environment is mediated by a series of cues the organism must make sense of in order to generate, even if tacitly, some knowledge it did not possess before.[8]

Traditionally, studies have concentrated on the human dimension of reasoning, nevertheless Peirce himself had stressed several times how the concept of abduction was to be held relevant for a biologically wide description of cognition.

The *making sense of signs* we are dealing with is indeed an abductive activity that human beings share with any organism endowed with a nervous system or, on an even bigger perspective, any organism capable of reacting actively to modifications of its environment. From this perspective humans, and the most part of non-human

[8]Refer Sect. 1.3.2 for a better introduction to the issue of abduction and further references.

animals possess what can be defined as "semiotic brains" (Magnani 2007b), which make up a series of signs and which are engaged in making or manifesting or reacting to a series of signs: through this semiotic activity they are occasionally engaged in being "cognitive agents" (as in the clear case of human beings) or at least in thinking intelligently.[9] As far as the biological and pre-linguistic levels are concerned, it can be argued that such "mental" representations do not matter for their *truth-reliability* but rather for their *fitness-reliability*[10] (Sage 2004). While our human language-dominated world informs the fact that we are used to consider the notion of truth, naïvely, as correspondence,[11] from a biological perspective (which is often engaged by human beings as well) the *favored* inference is the *most successful* inference, the one leading to survival.

Consider this: according to our common standards of epistemic decency, entertaining a true belief is always better than not entertaining it, especially if the belief concerns an agent's immediate surroundings. We would say that, for an organism, to be entertain a correspondentist true belief about the presence of a predator is the best way of surviving it. Nevertheless, it is not impossible to imagine a kind of opposite situation, where not detecting a predator allows the prey to go by unharmed, in case the predator is for instance sleeping: should the prey notice the predator and "freak out", it would make its presence clear and potentially be killed. Thus, not noticing the presence of a predator, not entertaining any form of PPR representation concerning it—and thus not reacting—might be the best way to avoid being noticed in turn and killed: this is clearly a limit example, still it is credible and it shows how in some cases, a potential proto-belief is clearly false, and yet successful.[12] This is similar to how

> […] cautious cognitive faculty that "over detects" dangerous predators (frequently generating the false belief that a predator is nearby) may generate an abundance of false beliefs, though it may turn out to be adaptive because these false beliefs increase an organism's inclusive fitness (p. 97). […] The abundance of adaptive false beliefs gives us reason to doubt that true beliefs are more likely to increase an organism's inclusive fitness than are false beliefs (p. 102) (Sage 2004).

A fundamental feature of abduction is crucial for this discourse. As shown in the Introduction of this book, abduction is not a truth-preserving inference: setting off from a number of true premises, the resulting abductive inference will not be

[9]Semiotic brains, their role in defining human cognition and supporting many kinds of inference will be a pivotal topic in Part III, especially in Chap. 10.

[10]Especially when comparing animal fitness and cultural evolution, the concept should be understood in a "loosely Darwinian" connotation. In this book, when I refer to fitness I intend a very informal notion, hinting towards both a rigorous definition of *fitness* and to the one of *welfare*, the latter being less geared towards reproduction and inheritance and more towards the well-being of an organism.

[11]A belief can be true inasmuch it corresponds to a state of affairs in reality, and we can communicate this belief, build further inferences on it and so on, and we expect the positive or negative outcome of those processes to depend on the truthfulness of the original beliefs.

[12]This argument is akin to Gigerenzer's famous treatment of *more is less* heuristics (Gigerenzer and Brighton 2009).

necessarily true, but merely *plausible* (it is not the case with deduction, which is a locally correct, truth-preserving reasoning). In this case, abductive mechanisms allow to set off from a number of (phenomenologically) true signs, and come up with an explanation that may not be true (poor epistemic-reliability) and yet enable the survival of the organism (high fitness reliability).

To sum up so far, I am arguing, coherently with Millikan's authoritative observations, that animals' internal representations concerning agency are always strictly related to the agent performing the inference, and cannot be considered from an absolute perspective: we can say that what an animal operationalizes is the other agent's *affordances*, that is what organism *a* can do with organism *b*. As it will be pointed out several times across this book, perception is necessarily agent-dependent, it necessarily concerns the individuation of a proper course of action, and it is not immediate but always mediated, as it ultimately consists in a kind of abductive sense-making representations.[13]

The inference intending the presence of an agent is in fact not a deductive one: if signs *a*, *b* and *c* *necessarily* signified the presence of agent *A*, then natural dynamics as we know them would be extremely different, especially as far as predation is concerned. Conversely, errors and misperceptions can always happen in abductive inferences: this allowed the emergence of a series of gene-phenotypical characteristics, such as the ones we will analyze in the next section, to exploit the more or less narrow *semiotic gap* between the presence of an agent in an environment and the actual possibility to infer its presence from a certain sign configuration.

Such abductive representations seem to be the product of situated abductive inferences and they are in fact peculiar inasmuch "they tell in one undifferentiated breath both what the case is and what to do about it" and they "represent the relation of the representing animal itself to whatever else they also represent" (Millikan 2004, p. 20). This kind of inferential process, residing in the *coupling* of the detector and the detected, is not based upon a random appraisal of an animal's semiotic cloud, but specific sign configurations match certain affordances, which ultimately trace back to the desired property. Jacob and Jannerod's description seems particularly illuminating:

> Property *G* matters to the survival of the animal (e.g. a sexually active male competitor or an insect to capture). The animal's sensory mechanism, however, responds to instantiations of property *F*, not property *G*. Often enough in the animal's ecology, instantiations of *F* coincide with instantiations of *G*. So detecting an *F* is a good cue if what enhances the animal's fitness is to produce a behavioral response in the presence of a *G* (Jacob and Jeannerod 2003, p. 8).

The hypothesis about the presence of an agent who detains the property *G* is *abduced* on the basis on one or more *perceptible* properties *F* that usually signify the relevant properties. If an organism is hunted as a prey or avoided as a predator because of a property *G*, it must try to reduce the occurrences of the properties signaling their characteristic, and this varies widely from organism to organism. It is

---

[13] I will analyze the abductive inferences informing the whole complex of perception in the first part of Chap. 10, dedicated to religious cognition.

inconvenient to hypothesize for a very basic animal mind to have a complex internal representation to deal with a notion such as *agency*, but we can easily postulate that its PPR representations may intend properties that depend on the fact that the apperceived complex of signs *A* is an agent: such as, "*A* can eat me", "I can eat *A*", "*A* can thread on me", "I do not have to worry about *A*". The actuality of these traits must be abducted by the subject from the series of signs it is able to perceive: they can be visual, kinesthetic, auditory, tactile or olfactive-chemical. What I described as *A*'s affordances are not always straightforward as they rely on perception which is potentially erroneous by itself. Affordances can abductively activate the right behavioral response, but they can fail as well. Everybody has had more or less direct experience with the misperception of an affordance. In those cases, an external sign configuration leads to formulating a poor abduction which usually leads to regrettable consequences: the most painful outcomes are entailed by dealing with false negatives rather than false positives, e.g. flirting with a resisting object of our loving desire, bashing one's head against a low ceiling, eating a poisonous berry or poking a sleeping crocodile assuming it was dead.

2.2.3 Pragmatic-Semiotic Models of Camouflage

At this point, it is legitimate to advance the claim that every agent has a *twofold inferential relevance*, an active and a passive one: on the one hand, it disperses signs *out* in its environment, on the other hand it receives and processes signs *from* other organisms, and former process must be minimized while the latter maximized either to counteract predation or to avoid being spotted by a potential prey. As we will see in the next section, something similar could be said also as far as "machiavellian" communicative-pragmatic interplays are at stake: a strategic agents uses her words to deceive others and, at the same time, avoid being deceived herself. So, organisms try to produce signs that fall out of their preys' and predators' agency-detection mechanisms, but they can nevertheless be extremely loud with respect to other systems: for instance birds in the rainforest can be colorful and noisy insofar as those signs are not a valuable clue for predators to infer their presence and position and conclude "There is a prey."

If we assume that organisms are endowed with abductive cognitive systems aimed at the detection and identification of other agents in their surroundings, we can suppose that these systems operate within a determinate threshold selecting semi-encapsulated stimuli which activate the inferential processing. Elements of the *semiosphere* that fall within these abductive thresholds are likely to produce in the organism an internal representation involving some kind of awareness about a particular nearby agent, and subsequently determine its behavior.[14]

[14]This view can be surely related to the discourse on modularity (Fodor 1983; Barrett and Kurzban 2006; Carruthers 2007), and it would probably be coherent with a postulation of an "agency detection module," but I would rather not tackle that (formerly?) hotly debated issue if not, again rather marginally, in Part III, Chap. 10.

In order to maximize their chances of not being discovered by agency recognition systems, certain organisms were favored by natural selection into modulating their semiotic footstep and let out signs that can be few and deceiving (falling under the inferential threshold of other agents, so that they do not trigger any positive agency-detection response) or meant to overwhelm and saturate the agent's abductive threshold.

This is the point where we cannot avoid talking about camouflage anymore. According to Stevens and Merilaita (2009), camouflage comprehends "all strategies involved in concealment, including prevention of detection and recognition" (p. 424), and they maintain that camouflage should be analyzed with respect to its *function* and *mechanism*, thereby stressing the relevance of local semiotic and pragmatic interactions: "in defining different forms of camouflage, we use the term 'function' to describe broadly what the adaptation may do (e.g. breaking up form, distracting attention), and the term 'mechanism' to refer to specific perceptual processes (e.g. exploiting edge detection mechanisms, lateral inhibition)" (Stevens and Merilaita 2009, p. 424). As a consequence, most dynamics broadly labeled as camouflage are seemingly aimed at preventing detection, avoiding recognition or averting the opponent from operationalizing a PPR representation (in other words, to prevent the other agent from correctly exploiting the affordances of that particular representation). Let us analyze from within the framework we developed so far some of the most widespread camouflage techniques in the animal kingdom.

Crypsis usually individuates those processes in which the initial attempt is to prevent detection. When we intuitively think of camouflage, we usually think of *crypsis*. In crypsis, the semiotic-abductive informational exchange is altered so that an organism attempts to "go stealth" by minimizing the extent to which the signs of its agency contrast against the background environment. To mention a few examples, stonefishes (*Synanceiidae*) shaped their appearance so to be inapparent from the sea bottom, while the famous peppered moth (*Biston betularia*) makes itself virtually invisible to its avian predators by blending, in plain sight, with likens covering birch trees (Majerus et al. 2000). Cryptical tactics are usually confined to one perceptual dimension, as in the case of the Australian frog (*Limnodynastes tasmaniensis*) and its main predators, i.e. snakes (Osorio and Srinivasan 1991): visual crypsis can protect the frog as long as it is not detected and recognized from its chemical-olfactory trace. Sharks are another fitting example of crypsis: their darker back blends with the background when seen from above, while their clear bottoms are hard to tell from below, as they merge with the clearer shade of water light from the Sun. If we mean to describe crypsis in a semiotic, abductive and pragmatic framework, we could say it works by downplaying signs so that they not activate other organisms' agency detectors: those signs do not nudge the cognitive system into reacting and abducing their origin, and therefore do not trigger the production of a PPR representation that could prove lethal for the camouflaged organism, or alert the prey if crypsis is enacted by a predator.

Masquerade is a semiotically different kind of camouflage, inasmuch as organisms do not attempt to merge with the background: conversely, they provide into the semiosphere signs that make them easily detectable, but "their bearers are misidentified as either inedible objects by their predators, or as innocuous objects by their prey:" to make a few examples "plants from the genus *Lithops* look remarkably like stones; stick insects are easily mistaken for the twigs of the branches on which they sit[15]; the *Ornithoscatoides decipiens* spider closely resembles bird-droppings; the leafy sea dragon *Phyllopteryx eques* is often misidentified as seaweed; the Amazon fish *Monocirrhus polycanthus* is visually almost indistinguishable from leaves, and birds from the family *Nyctibiidae* bear an uncanny likeness to tree stumps" (Skelhorn et al. 2010, p. 1).

Other forms of camouflage exist, such as the *kinesthetic camouflage*, which relies on the alteration of a given subpart of the organism's semiotic shadow: their aim is not to prevent an organism from being detected nor to be recognized, but to prevent an effective prediction of their spacial bearings (Srinivasan and Davey 1995). "Motion camouflage is a strategy whereby an aggressor moves towards a target while appearing stationary to the target except for the inevitable change in perceived size of the aggressor as it approaches" (Glendinning 2004, p. 477).

To sum up, if *crypsis* produces signs that are not configured as cues for possible abductions, *masquerade* tactics offer indeed a profusion of signs likely to be picked up by other agents that are not to be processed as relevant for agency recognition but are instead actively acknowledged as inert objects belonging to the environment. What is at stake is not the possibility of performing abductions upon a configuration (or non-configuration) of signs in the semiosphere, but the quality of such abductive inference, and the reliability of the consequent PPR representation. Similarly, *kinesthetic camouflage* does not aim at impairing an agent's abducibility as far as its detection or recognition are concerned, but rather compromises the quality of the PPR representation fostered by the agent's semiotic shadow.

It is important to note, both for our present discourse and for the one I will carry out towards the end of this book in Part III about the origins of supernatural belief,[16] that a certain *counter-factuality* could be ascribed to the kinds of PPR representation (or the lack thereof, i.e. when a predator or prey is not spotted) triggered by camouflage, insofar as they either depict organisms differently from their real nature or they fail to depict them at all, when they are present. Some particular semiotic configurations are selected to overwhelm an organism's abductive thresholds, so that it is likely to entertain particular kinds of counterfactual internal representations, concerning agents that disappear (after ink-blindness in dark environments, immobility in motion-detecting systems, camouflaged organisms) or strange, fearsome agents which inhere to but do not comply with already known agents. These representations can either appear as menacing predators or as awe-inspiring preys who cannot be overwhelmed. A radically unknown agent can in fact mesmerize or discomfort agents in its proximity.

[15]The stick insect, *Phasmatodea*, enriches its structural camouflage by faking a typically *atmospherically-induced* way of moving, shaking and trembling like a small branch moved by gushes of wind (Bedford 1978).

[16]Chapter 10.

2.3 Argumentation, Truth and Survival: Human Beings and Linguistic Camouflage

After having delineated a decent inferential model of the main forms of camouflage, I will now focus on those situations to which the concept of camouflage can be extended metaphorically, that is involving no physical perceptual deception, and try to demonstrate how such extension is particularly legitimate inasmuch as exactly the same kind of dynamics, *mutatis mutandis*, can be individuated in argumentative as in biological camouflage.

By achieving "ecological dominance" (Flinn et al. 2005), human beings obtained a progressive increase in safety from natural predators, making the necessity to detect life-threatening animals a more and more obsolete part of their evolutionary endowment. Similarly, it can be reasonably suggested that the diffusion of farming in most populations reduced human beings' dependance on their predatory skills (and thus on their skills for coping with camouflaging preys). Conversely, increased levels of sociability and civilization established humans as the highest threat for their conspecifics' survival and welfare, fostering clashes not necessarily involving physical violence but equally dangerous and likely to affect one's future development, for instance concerning accessibility to feeding resources and sexual mates (Boehm 2012). The dimension permitting this unprecedented level of sociability is of course language. Over the past few decades, scholars have intensively explored the artifactual and externalized dimension of language, and how it provides a scaffolding dimension for human activities (Clark 2005, 2006). As I will abundantly examine in the following part, several studies stressed how the very origin of language can be traced down to its social relevance, as a policing tool aimed at enforcing moral (and proto-moral) norms and coping with free-riders (Dessalles 2000; Dunbar 2004; Ross 2007; Sperber and Mercier 2010).

Today it is widely acknowledged that language—in its various manifestations including arguing—can subtly support violent aggression and oppression (Magnani 2011): the development of language would not make aggression, conflict and fighting disappear but simply moved them to another plane. For instance, Hample and colleagues—discussing the particular case of playful argument—suggested that arguing is closely related to *verbal combat* and *verbal force*. More precisely, they claim that verbal combat is the *base appearance* of arguing (Hample et al. 2010). It follows that the more "civilized" and sophisticated approaches to arguing—i.e. those assuming it as a means for finding a mutually accepted solution, for persuading, or for exchanging valuable information—are basically new avatars of its ancestral form.

As illustrated by Magnani in his recent book "Understanding Violence" (Magnani 2011), the idea that arguing is verbal combat is related to the expression "military intelligence" coined by the French mathematician René Thom. It refers to all those situations in which arguing is deployed to maintain the structure of societies (Thom 1988): that is, language can efficiently transmit *vital* pieces of information about the fundamental biological oppositions (life–death, good–bad). It is from this perspective that we can clearly see how human language—even at the level of more complicated syntactical expressions—always carries information about moral qualities of persons, things, groups, and events. In this sense, arguing as verbal combat may be considered an indispensable maneuver that serves the purpose of managing the various coalitions and groups that are *facing in the battlefield*. Information warfare is the most visible example, yet not the only one. Indeed, the battlefield that is tacitly projected through verbal combat does not involve fangs, claws and spears anymore. Conversely, such a battlefield is populated by narratives created by the different coalitions that are confronting with each other.[17] People exercise their argumentative skills in both rhetoric and dialectic settings, and they aim at prevailing one over the other or, at least, protecting and preserving their own integrity. Therefore, one of the main functions of arguing and reasoning is eminently social, meaning that it deals with the management of the coalition. That is, persons argue not only because they exclusively aim at finding meaning to understand a thing in its relationship with other things, and thus successfully gain control over the environment: it is clear that person may also start arguing for influencing and manipulating other people's thoughts and actions (Malle et al. 2001). Managing social interaction seems to be an evolutionarily primitive function of language comparing with more sophisticated ones, as suggested, for instance, by Sperber and Mercier (2010): they support the view according to which reasoning is a social competence, that is, an ability to "convince others and to evaluate arguments others use in order to convince us".

When the exchange of reasoning aims at persuading or, more generally, managing social interaction, even fallacies become a fundamental part of the toolbox we have at disposal.[18] They are particularly good in one specific respect: following Malle, I suggest that arguments that make us gain control over the environment have to be true, whereas those which help us manage social interaction only have to be *compelling* (Malle et al. 2001). Fallacies are poor at the former task while being much better at the latter. The main reason why fallacies are particularly good at persuading is that they usually enlarge the information base by introducing some (apparent) irrelevancies. For instance, that is what characterizes a specific class of fallacies labeled by traditionally-minded fallacy theorists as *ignoratio elenchi*. Introducing irrelevant information serves two main purposes. First of all, it contributes to manipulating people's attention by producing a shift in issue: relevance is no longer a logical criterium for evaluating an argument, but the result of a social dynamics in which people continuously strive to prevail one over the other (Dessalles 2000). Secondly, loading a discussion with apparent irrelevancies has a cognitive meaning: enlarging the information base makes it easy for a person to have an opinion even

[17] This issue will be crucial in the next part, especially in Chaps. 7 and 8.

[18] The use of fallacies will be a crucial topic when analyzing gossip in Chap. 7, refer especially to Sect. 7.5.2.

in presence of patent ignorance about the topic in discussion (Bardone and Magnani 2010; Bardone 2011).[19]

We can start by acknowledging that whereas in biological camouflage the aim is to provide an observer with a series of signs that lead into making the wrong inference, communicative camouflage differs inasmuch as it involves the production of semantical and performative acts likely to mislead one's interlocutors, by shielding from intellection their actual beliefs, intentions, etc. In a situation of lack of information and knowledge (constitutive "ignorance") abductive reasoning is usually the best cognitive tool human beings can adopt to relatively quickly reach explanatory, non-explanatory, and instrumental hypotheses/conjectures, exactly as it happens when the cognitive target is to guess the presence of other agents in the environment.[20]

Setting off from this my contention is that, in argumentative dynamics, *masquerade* seems the to be the most easily individuated camouflage analogue. Crypsis is not totally absent but concerns rather subtle, subliminal argumentative devices: Shakespeare put indeed in the mouth of a young lover the words "She speaks, yet she says nothing", but while a peppered moth can indeed *go stealth* by blending with the bark, it is hard to conceive how an utterance or a writing can be actually produced, be efficacious at a pragmatic level, and still not be noticed. *Subliminal* communication strategies aim at influencing the receiver's decision-making processes (so to persuade her into acting in a certain way, buying certain products, approving of a certain policy and so on...) by targeting her with an intense fire of signals that stand out from background noise but are "low" enough so that the receiver's consciousness does not engage and evaluate them (Krosnick et al. 1992; Pessiglione et al. 2008). Nevertheless, voluntary subliminal signaling has often been associated with practices akin to hypnosis, or affecting corporate interests, national propaganda and so on, and are less frequently deployed in ordinary communication.[21]

Nevertheless, in ordinary argumentative settings the figure of speech named "*praeteritio*" could indeed be identified as a particular kind of *crypsis*. Such figure—already described in the *Rhetorica ad Herennium*—involves a speaker stating her intention not to mention something, but by this she *precisely* refers to what she said she would not say (Snoeck Henkemans 2009): for a brutally simple example think of something along the lines of "I do not want to remind you of about the indecent behavior you held at the party." *Praeteritio* is thus a way to push and "smuggle" contentions towards the other speaker without her being able to openly rebate. If a semantic camouflage is harder to achieve with *praeteritio* (that is, it does not affect

[19] I will discuss a similar point in the next Part, Chap. 6: specifically, I will analyze how the knowledge-richness of a given environment affects the attitude one should adopt towards different regimes of rationality, for instance fallacies and what is commonly, but also academically (sic!) known as *bullshit* (Frankfurt 2005).

[20] Gabbay and Woods (2005), Magnani (2013), Woods (2013) recently dealt with the connotation of abductive reasoning as "ignorance-preserving," to which I will resort several times along this book.

[21] Here, I am not addressing the whole range of subliminal *conversation* where signals are subliminally—involuntarily—produced and not only received, as far as pheromones, body language or even lapsuses for instance are concerned.

the *meaning* of the words), it is easier to push through the performative one effect, as the speaker affirms she is *not* going to perform the speech act that she immediately performs, leaving it in a kind of suspended state. A skilled use of *praeteritio*, such as "I am not saying that the Pope is ignorant, I am saying the Pope *knows nothing about Islam*" can strategically maneuver the discourse frame by (deceitfully) insinuating a series of assumptions which become "invisible" (inasmuch as undeclared) constraints for the other speaker. Resorting to Thom's aforementioned notion of "military intelligence", *praeteritio* allows an arguer to stealthily introduce in the battlefield a number of strategic or offensive pawns on which either she can rely to reinforce her position against the opponent or that can directly fire against the opponent's position: by use of this figure of speech those pawns are indeed deceitfully disguised in order to deflect recognition and most of all proper engagement.

Moving on to the broader family of rhetorical devices defined as "fallacies", it seems proper to say that truth is frequently distorted by means of *argumentative masquerade*. As a matter of fact, most of informal fallacies are usually labelled as *red herrings*, stressing their fundamental utility as attention traps, aimed at diverting the interlocutor's attention from the matter at stake onto something else: many of these can be labeled "gossiping fallacies," considering their crucial role in many socially-oriented linguistic exchanges (Bardone and Magnani 2010). For the sake of our argumentation, it is curious to reflect on the origins of the label: a "red herring" was tied to the tail of captive foxes that were then released in a field to the scope of training fox hounds in following smell tracks in open country. Common sense seems to be powerfully aware of how the discourse on biological and argumentative camouflage is perfectly interchangeable, and such awareness is witnessed by the continuous etymological borrowing between the two fields.

I maintain that various kinds of appeals to emotion reenact the dynamics of masquerade, and in these cases it is much easier to spot the analogies. Consider the *ad baculum* fallacy, also know as "appeal to force," by which an arguer is invited to accept or reject a claim chiefly because of some kind of threat, or the appeal to consequences (*argumentum ad consequentiam*) by which the likelihood that a claim is true is linked to its consequences or the consequences of its acceptance: the biological analogue is the *masquerade* technique involving the display of false eyespots or the unexpected (and thus frightening) modification of physical features, offering false affordances to the observer who subsequently activates a behavioral response which is likely to negatively impact its welfare or survival possibility.[22] Similarly, the *ad verecundiam* and other "appeals to authority" provide a series of semantical signs aimed at tampering with the quality of the abduction the interlocutor should make concerning what is relevant in the discourse. In all of these cases, the argument can be more or less skillfully "wrapped" so that the receiver processes it and abductively evaluates it by making use of appraisal systems that are not relevant with respect to what is at stake. Analogously to masquerade, the argument (like the camouflaging

[22]Blowfishes, ink-shedding cephalopods such as cuttlefishes, squids and octopi, frill-necked lizards are all masers of *ad baculum* and *ad ignorantiam* equivalents in the animal kingdom.

organism) is in plain sight, the arguer does not try to dissimulate it, but presents is as *something else* from what she actually knows is the case!

The fallacy known as "straw man" is also very fitting to explore the analogy between argumentative devices and biological masquerade systems: the *construction* of a straw man which grotesquely (or inaccurately) represents the opponent's position is strongly connected to the devising of a configuration of external signs aimed at puzzling and deceiving other agents. What results is that the interlocutor of the fallacious reasoner is trapped in a biased perspective, which she might transpose into other argumentative settings thus becoming intrinsically biased herself. The interesting point in the comparison between "straw man" fallacies and masquerade technique is that whereas masquerade can be usually reduced down to a strategic engagement between a predator and a prey, the straw man has such a strongly rhetorical-argumentative connotation that it structures environmental engagements among several actors, involving the arguers and a third party assessing the discussion. The rhetorical device allows *arguer A* to camouflage not her own semantic production but that of *arguer B*. The altered semantic shadow (i.e. the straw man, the caricatural production of the arguer) is directed towards the third party: they can use it, combined with their knowledge base, to abductively infer a *character B** which does not coincide with that of *arguer B*. It can be said that *arguer A*—resorting to the straw man—works like an *argumentative prism*, breaking down her interlocutor's argument and then providing a decomposed and deformed version of it to some other party. Biologically, it can be compared to parasitic techniques such as the Old World cuckoo's, which lays the egg in the nest of another bird so that the fledgling cuckoo is nurtured by its "foster parents" and outcompetes its step-sisters and brothers, curbing both the welfare and the fitness of the nest it parasites upon.

Of course, as already highlighted the conception of *truth* intended by biological and argumentative camouflage is, at least *prima facie*, not the same: whereas a "good" behavioral response in a biological framework means the animal was able to survive (and reproduce), debunking a fallacious argument has instead to do with the assessment of whether what is being uttered *corresponds* to a state of things in the world. Two different conceptions of fitness[23] can be at stake here as well: a pragmatic conception of truth, related to survival, echoes an idea of fitness embodying one's possibility to have descendants, whereas in argumentative settings the failure to appreciate truth in the argumentation of one's interlocutor usually does not impair her possibility to give birth to a progeny, but rather her overall welfare.

At this point, it seems legitimate to resort to the same definitions I had use to describe the dynamics of biological camouflage: *some argumentative devices do make use of strategies aimed at by-passing the threshold of abductive appraisal of a claim; on the other hand some other devices are structured so to overwhelm the appraisal threshold and have the claim processed by an irrelevant system* (i.e. emotional rather than rational and so on).

---

[23]Here again *fitness* should not be intended in a strict Darwinian connotation: but rather be considered as a local trait, relating to the survival and the welfare of the single individual, without long-term evolutionary implications.

2.4 Biological and Argumentative Camouflage are Debunked by Similar Methods

The next step I mean to take is to demonstrate how the strength of the analogy between the natural and the argumentative framework can be further explored by individuating and analyzing some of the possible counteracting strategies against biological camouflage, and showing how they can be enacted in the argumentative framework as well.

Our hunting ancestors knew that preys (and potentially dangerous animals as well) were hiding in the bushes, even if their presence was not evident at a first glance. The target was to acquire and operationalize an un-biased perspective on the environment, that is, of the potential preys to be hunted and predators which might threaten the hunters. That is to say, as far as camouflaging biological organisms' are concerned, hunting could be considered as the first systematic attempt to go beyond Nature's appearance—an outlook on things that would be assumed by philosophy after almost 200,000 years, and that would inform modern science more than twenty centuries after the birth of philosophy!

It is a fact that hunting strategies became more and more sophisticated, and they were reflected by an always stricter dependance on the artifactual apparatus (Bingham 2000). Such development structured hunting as part of the cognitive niche, an externalization of knowledge onto the environment able to modify some of the natural selection pressure present in their local selective environments, as well as in the selective environments of other organisms.[24] It could be speculatively suggested that hunting structures a cognitive niche also according to the definition of cognitive niche as a "set of affordances" (Gibson 1979; Magnani 2009): camouflage turns down many of an organism's ecological and practical affordances in order to avoid detection, recognition and the subsequent pragmatical effects. Hunting applies onto the environment a series of affordances counteracting the disruption operated by camouflaging animals.[25]

Most human hunting practices could be considered within the inferential-abductive warfare dynamics that were put forward in the first part of this chapter. Recapitulating, I showed that *camouflage and other techniques meant to avoid detection and recognition operate either by overwhelming or by not meeting the agency recognition mechanism's abductive threshold.* We also just briefly reviewed the fact that "man the hunter", thanks to the development of consciousness and intentionality, could actively provide its agency detection systems with cues that would have

[24]Most of the following Part will be devoted to cognitive niche construction and maintenance, this definition is therefore just a stub for the sake of present discourse.

[25]Cognitive science, evolutionary psychology and paleoanthropology have thoroughly studied the pivotal role played by hunting in the development of many contemporary human endowments. Epistemologist Giuseppe Longo puts forward a fascinating hypothesis linking the development of an abstract concepts such as Euclides' *line without thickness* to the rapid eye movements from one point to the other (saccade) by which a hunter precedes the trajectory of the prey in order to capture it (Longo 2005).

otherwise been overlooked. Considered within this perspective, the following excerpt by C S. Peirce might be particularly illuminating: "[...] A mass of facts is before us. We go through them. We examine them. We find them a confused snarl, an impenetrable jungle. We are unable to hold them in our minds. [...] But suddenly, while we are poring over our digest of the facts and are endeavoring to set them into order, it occurs to us that if we were to *assume something to be true that we do not know to be true, these facts would arrange themselves luminously*. That is abduction [...]".[26]

Peirce's description of abuction, with all of its natural strength and the sense of *being alive* it conveys, fundamentally informs how abductive warfare can be counteracted so that the detecting agent can individuate the camouflaged one: is it possible to spot a similarity between what happens in biological context and in an argumentative one? Relying on Peirce's intuition, I will make it clear in the following subsections.

In order to tackle the matter at stake, it is necessary to introduce the reader to an incredibly useful notion, that is Egon Brunswik's "lens model." The *lens model* (Brunswik 1952; Hammond and Steward 2001) is based on the idea that the relationship between the organism and the environment is mediated by the use of the so-called *proximal stimuli*, from which the organism can infer the *distal* state of the environment, which brought it about. *Ecological validity* is the term introduced by Brunswik to refer to the situation in which a given proximal stimulus acts as a valuable indicator of a certain distal state or event; ecological validity is a normative measure about how diagnostic—reliable—certain proximal stimuli are with respect to a given distal event (Vicente 2003; Kirlik and Storkerson 2010). By referring to Brunswik's *lens model*, it could be argued that the debunking of camouflage involves a manipulation of proximal stimuli by the observer, aiming at the reconstruction of the distal event to which proximal stimuli lead.

Thus, employing the lens model lexicon, we can say that agents must assess the traces available in their environment in order to maximize their *ecological validity*. We can correlate the notion of ecological validity to the possibility and the quality of abductive inferences concerning the presence of hidden organisms (as far as biological camouflage is concerned) or the truthfulness and trustworthiness of one interlocutor's claim (in case of argumentative camouflage). Since camouflage aims at tampering with abductive mechanisms allowing for the detection of external organisms, or the appraisal of claims made by one's interlocutor, it can be proposed that strategies to counteract camouflage rely on the improvement of the starting set of cues from which the abductive inference sets off. This manipulation can happen both internally and externally, and can be briefly modeled as follows: in order to disrupt a camouflage pattern, a cognitive agent introduces a belief to be used as a lever to see if her perceptual impression holds or breaks by assuming—in Peirce's words—as true something she does not know to be true and witness the possible reconfiguration of the perceptual-argumentative experience.

[26]Italics not in the original. cf. Peirce's "Pragmatism as the logic of abduction", in Peirce (1992–1998), pp. 227–241, the quotation is from footnote 12, pp. 531–532.

2.4.1 Countermeasures for Animal Camouflage

Let us consider biological camouflage first, and apply Peirce's description of abduction mentioned above: if *crypsis* works indeed by downplaying signs so that they do not alert agency recognition mechanisms, the abductive counteraction may consist in the assumption of a certain cue to be *more relevant* than it seems to be in the configuration which actually impacts the receiving perceptual system. An example could be to consider a particular spot as if it was an *eye*: assuming such a detail as true might afford the reconfiguration of the visual perception so that the rest of cues become *then* meaningful signs of the presence of an agent (this corresponds to Peirce's "but if *A* was true, then *C* would be matter of fact"). Furthermore, if this additional hypothesis proves to be meaningful, a cascade of positive inferences may follow: i.e. once the eye of the organism is located, the observer can detect the face, which leads to the appearance of the body, the legs etc.

Masquerade dynamics, meant to impair the detector's abductive ability to recognize and identify an agent (rather than preventing it from being individuated *tout court*), can be disrupted in a similar way: by assuming something to be different from its perceptual appearance and letting this assumption redefine the way in which the semiotic configuration is perceived by our senses.

Also by intervening about what is the origin of perceptual judgements, it is sometimes possible to break down a camouflage attempt by triggering the manifestation of the distal event: kinesthetic perceptions and physical manipulation are possible methods. Kinesthetic perceptions relate to the enriched perceptual imagery offered to an observer who moves in the environment so to change her point of view and match the different aspects she perceives at different stages.[27] The other kind of environmental manipulation (sometimes complimentary to other types of investigations) involves active manipulation—probing—of the environment. Such practices can consist in approaching the potential camouflaging organism in order to trigger a reaction of some kind, or, in a more artifactual dimension, the use of fire and smoke to reveal the presence of animals hiding in trees, caves etc. Such practices clearly demonstrate how the debunking of camouflage can also rely on

[27]Phenomenology's toolbox includes the pivotal concept of *adumbration* (Husserl 1960, §17), referring to how only one partial aspect of an objects is manifest to the observer at one time, and this single aspect *foreshadows* the rest of the thing by an interplay of hinting and hiding. Such a concept is interesting for the philosophical investigation of camouflage inasmuch as to spell out a camouflage instance could also mean to look for the most relevant possible adumbration, and this is achieved also by the kinesthetic control of perception. The kinesthetic control of perception is related to the problem of generating the objective notion of three-dimensional space, that is, to the phenomenological constitution of a thing as a single body unified through the multiplicity of its appearances (Husserl 1960, §44). The "meaning identity" of a thing is of course related to the continuous flow of adumbrations: given the fact that the incompleteness of adumbrations implies their synthetic consideration in a temporal way, the synthesis in this case—kinetic—involves eyes, body, and objects. This kinesthetic synthesis of adumbrations increases the inferential knowledge-base on which agents can perform abductions concerning possible camouflaged agents, by noticing for instance parts of the—initially—irrelevant background that are mismatching, moving or unexpected in other ways.

abductive manipulations consisting in "discovering through doing",[28] which are energy-consuming but effective when the inferential knowledge base concerning potential camouflaging organisms is poor and difficult to improve.

2.4.2 Countermeasures for Verbal Camouflage

So far we explored the modeling of heuristics meant to contrast biological camouflage. What about linguistic camouflage? My contention is that coping with occurrences of camouflage in argumentative settings can rely on the same *lever model* that seemed the most plausible for dealing with biological camouflage. Since linguistic camouflage downplays the possibility and quality of abductive appraisals concerning particular claims (or clusters of claims), the aim of counteracting strategies is to assess and improve the ecological validity of single pieces of information and so to *verify or reshape the distal event* (for instance the interlocutor's true intentions etc.).

Received information is appraised by matching it with either *context-dependent* or *context-independent* beliefs that were already acquired by the agent, in order to improve the possibility of managing successful abductive inferences. Context-independent beliefs are thought of as true in every situation (apart from playful occurrences, jokes etc.), and concern the way one expects a discussion to be carried out (think for instance of Grice's conversational implicatures (Grice 1975)), and inform the kinds of (context-dependent) checks one agent should perform upon the received information in order to spell out arguments attempting at camouflaging or biasing truth. The main context-independent beliefs used to scan arguments usually resemble to "Arguments and claims should be coherent" and "Information provided should be relevant with respect to the context." It is possible to use context-independent beliefs as levers to test whether the structure holds or not, or to perform some checks upon the interlocutor's single claims in order to understand whether her utterances correspond to how things are.[29]

As far as relevance is concerned, checks are usually aimed at scanning for fallacies. Fallacies—as already suggested—present the claim and sometimes make it effective by coating it with irrelevant information. If the option if pragmatically viable,[30] the receiver of the fallacious claim analyzes it expanding the inferential knowledge base resorting to the belief that "Information should be relevant with respect to the context." If the information provided only apparently complies with this belief, the camouflage attempt is debunked and rebutted. The gradient of irrelevant information

[28]On cognitive processes relating to manipulative abduction and "discovery through doing" (see Magnani 2001, 2009). I will also take advantage of this notion in the next part, specifically in Chap. 4.

[29]I will analyze the role of repeated experimentation as a manipulative way to conquer true beliefs in the final chapter of this Part, when dealing with a new framework to understand experimentation (Chap. 4).

[30]As suggested by Herbert Simon, human beings have to cope with limited mental processing capabilities taking place in limited temporal settings (Simon 1955).

affecting the exchange is not necessarily an indicator of malicious intentions, but might depend on a lack of competence on either one or both poles of the dialogue. In such cases, the abductive appraisal of the interlocutor's claim becomes harder because of the systematic and often purposeless use of fallacies.

Consider that fallacies, even if they do not rely on context-dependent information, can lead us to solve the task we are supposed to face. Paradoxically, and yet enforcing this analysis of camouflage (both biological and argumentative) as abductive warfare, it can be claimed that *the very strategies aimed at debunking camouflaged fallacies are fallacious themselves*, inasmuch as an abductive appraisal involves, according to Peirce, to assume as true and relevant something we do not know to be true: abduction, it is worth reminding, is a formally fallacious argument and therefore non truth-preserving. True premises will not necessarily lead the reasoner to a true conclusion. In the sentential perspective of classical logic, abduction is classified as the fallacy of affirming the consequent. In abductive reasoning, this kind of appraisal is linked to evaluating various inferred explanatory, non-explanatory, and instrumental hypotheses/reasons, and, of course, it varies depending on the concrete cognitive and/or epistemological situation.

If an agent relies on relevant information as if it was an abductive lever, it is usually quite easy to debunk and rebut a claim that was camouflaged by means of fallacies. From a theoretical point of view, fallacies are therefore a rather weak system of argumentative camouflage: nevertheless, they remain an extremely widespread tool available to the vast majority of arguers. Indeed, time and computational resources to perform the disrupting strategies I just described might not always be available, while there's always time to commit a quick and effective fallacy. This factor makes fallacies a cheap-and-fast camouflage method whose potential benefits (for the fallacious arguer) are far higher than its costs.[31]

Checks on the relevance of an interlocutor's claim still correspond to a manipulation of *proximal* stimuli (this time linguistic) to produce a higher quality inferential base to make the distal phenomenon appear clearly, or at least better abducible. Exploring the relevance of a piece of information helps assessing its ecological validity and whether it leads to a positive or negative appraisal of the claim it relates to. A check on proximal stimuli may affect the perception of the distal event also when stimuli are compared one to the other (and not taken as isolated pieces of information) in order to assess their coherence. A correct appraisal of the coherence of an argument deeply affects the quality of an inferential knowledge base and the quality of the abductions one can operate upon it. As a matter of fact, the belief "Arguments and claims should be coherent" is another important context-independent rule that interlocutors are expected to observe. It is therefore possible to try and lure an interlocutor into the acceptance of a claim only apparently coherent.

[31]I will analyze the environmental deployment of fallacious (and sometimes careless) knowledge into one's epistemic environment in Part II, Chap. 6. It is already possible to argue, though, that phenomena such as *bullshitting* might be analogous to engaging in meaningless linguistic camouflage efforts.

As far as coherence is concerned, a strategy to break argumentative camouflage may consist in the analysis of whether the various claims put forward by one's arguer display a truly explanatory coherence or merely an *emotional* one.[32] *Emotionally* coherent arguments might in fact be put forward instead of *explanatorily* coherent ones, so to push the receiver into accepting their claims more easily. Explanatory coherence and emotional coherence neither mutually imply nor mutually exclude each other. An explanatorily coherent argument might or might not be emotionally coherent, and vice versa. It must be noted that systematic failures in appreciating the difference between explanations that are explanatorily coherent vs. emotionally coherent might drag the whole system down to a state of *terminator niche*, where modifications meant to improve the total welfare of the system are indeed the prime causes for its ruin (see Part II, Chap. 9).

By assessing the argument's explanatory coherence, the receiver does in fact improve the possibility to make a correct abductive appraisal of the distal event (e.g. the speaker's true beliefs, intentions, etc.), and avoid the possibility of being entangled in a web of emotionally self-assessing beliefs.[33]

To provide an argument with many appeals to emotional coherence is a widespread technique which informs yet another kind of "masquerade," inviting one's interlocutor to make the wrong inference based on an ambiguous semantic configuration. To enact countermeasures, for instance by checking whether a piece of information that the receiver already holds can be used to successfully deepen the explanation brought forward (or if a new belief can be obtained to this scope) is not always a viable strategy, especially if the topic at stake is emotionally dear to the interlocutor himself (Thagard 2007). As a matter of fact, recent results from neuroscience seem to corroborate the thesis considering the importance of emotional

---

[32]According to Paul Thagard, from an epistemological outlook we should prefer beliefs displaying a greater explanatory coherence, that is connecting in a deep and consilient way. A computational algorithm can be easily programmed to select explanations with the highest explanatory coherence, nevertheless human cognitive behavior can be better approximated making use of an algorithm which also takes into consideration emotional coherence: "according to the theory of emotional coherence, inferences about what to do and believe are affected not only by hypotheses and evidence, but also by the emotional values that are attached to representations whose coherence is assessed" (Thagard 2005, p. 62).

[33]The analysis of this issue could be widened resorting to the concept of *epistemic bubble* (Woods 2005), that I will elaborate in the Part III (Chap. 12). To introduce the matter very briefly, let us remember that there is no causal correlation between explanatory and emotional coherence: therefore, there is no guarantee that an emotionally coherent argument will be explanatorily consistent as well, and vice versa an explanatorily coherent argument might not spark any particular emotional preference. It is also true that particular mechanisms affect the attainment of truth: "truth is a fugitive property. That is, one can never attain it without thinking that one has done so; but thinking that one has attained it is not attaining it" (Woods 2005, p. 746). Therefore an emotionally coherent claim, camouflaged as an explanatory coherent one, can easily trigger the receiver's acceptance and thus entrap her in an epistemic bubble. Once in an epistemic bubble, the agent is unable to commend her real knowledge from her ignorance concerning the subject at stake (that is, she cannot tell her *knowing P* from her *thinking that she knows P*, inasmuch as her emotional-abductive appraisal is satisfied by the argument, and she is consequently unable to effectively revise her beliefs concerning that topic.

coherence: a whole area of cerebral activation seems dedicated to whether one has to take an argument "personally" or not (Schwartz and Gladding 2011). Such systems can be easily highjacked by argumentative camouflage so to make an interlocutor react in a particular way to certain arguments. This kind of device is frequently used in political rhetorics, advertisement, etc.: Slade (2002) maintains that advertisement works inasmuch as they are not directed at a merely concupiscent cognition but rather present themselves as "rational" and thus compel us into behaving in ways that are, at least prima facie in our own judgement, rational.

A provocative comment should conclude this excursus from biology to argumentation, lest we idealize human beings for achieving a sublimation of conflict and predation by shifting all clashes onto the inferential-epistemic level: most of the countermeasures I analyzed so far involve the weapons of abductive-inferential warfare, that is a selective inferential manipulation of proximal stimuli, but it should not be forgotten that such checks might not concern only pieces of knowledge but knowledge-carriers as well, human beings' possessions, their psyches and their bodies. Just as biological camouflage can be challenged with brute force attacks such as random physical probings or making use of smoke, fire, etc., *similarly* human beings have always displayed a peculiar taste for performing brute-force checks on knowledge carriers (i.e. human beings—human bodies) by means of threat, aggression and torture (Magnani 2011). Notwithstanding the debates concerning their moral legitimacy and their actual usefulness (dating back to the 18th century and still ongoing), these practices seem to prove, with a certain bitter irony, the fundamental continuum between biological and argumentative camouflage.

2.5 Conclusion

The aim of this chapter was to explore and compare camouflage strategies belonging to different frameworks, but whose common goal consists in a struggle to make things appear as different from how they are "in truth." Truth, naïvely conceived as what is the case in a local environment, can be camouflaged by different means, but all of those means could be modeled considering the interplay between the camouflaging object and the target of camouflage, that is the potential detector.

The concept of abduction, as received from the Peircean tradition, proved to be the best explanatory tool to analyze the inferential ground that underdetermines camouflage: it even explains the continuity of inferential interplays, from those characterizing natural predation to the ones typical of human forms such as arguments, claims and dialogues. Focusing our epistemological outlook on data provided by the tradition of natural studies, I sketched out models of "military intelligence" at play in biological camouflage: organisms aim at tampering with each others' instinctual agency recognition mechanisms. As a result, they are prevented from correctly detecting and identifying the presence of other organisms in the surroundings. At this point I could move the enquiry one step further, and consider some occurrences of strategic communicative behavior among human beings: was it possible to characterize them as camouflage as well? Many human activities such as fallacious argumentation seem in fact to tamper with the detector-detecting inferential mechanisms just as much as natural camouflage does: following this idea, it was actually possible to apply the same abductive models to both natural and argumentative camouflage, thus proving the legitimacy of extending the concept of camouflage to frameworks other than biology in a non-metaphorical way. The final point consisted in showing how countermeasures against both kinds of camouflage could be described using the same models: strategies aiming at debunking camouflage involve internal or external manipulations that are used as levers to appraise the ultimate validity of the received perceptual imagery, or of the claim put forward by one's interlocutor. The fertility of this approach does not reside in its being yet another consilience theory, showing homologies between different settings, but rather in the possibility of augmenting our understanding of both natural and human strategic dynamics, using one conceptual framework to foster new reflections about the other.

The next chapter will start from the same "naturalistic" framework, as I will shift my focus from a specific kind of mental representation (the one about agency, analyzed at the beginning of this chapter), to something broader that is the nature of mental modeling. That is to say, I will frame the contemporary debate about scientific models in the naturalistic one about the *emergence* of models as *mental models* at play in biological processes.[34]

References

Bardone E (2011) Seeking chances: from biased rationality to distributed cognition. Springer, Berlin

Bardone E, Magnani L (2010) The appeal of gossiping fallacies and its eco-logical roots. Pragmat Cogn 18(2):365–396

Barrett HC, Kurzban R (2006) Modularity in cognition: framing the debate. Psychol Rev 113(3):628–647

Bedford GO (1978) Biology and ecology of the phasmatodea. Annu Rev Entomol 23(1):125–149

Bingham PM (2000) Human evolution and human history: a complete theory. Evol Anthropol 9:248–257

Boehm C (2012) Moral origins: the evolution of virtue, altruism, and shame. Basic Books, New York

Brunswik E (1952) The conceptual framework of psychology. University of Chicago Press, Chicago

Calvo P, Keijzer F (2009) Cognition in plants. In: Baluška F (ed) Plant-environment interactions: from sensory plant biology to active plant behavior. Springer, Berlin, pp 247–266

Carruthers P (2007) Simple heuristics meet massive modularity. In: Carruthers P, Laurence S, Stich S (eds) The innate mind: volume 2: culture and cognition. Oxford University Press, Oxford

Clark A (2005) Word, niche and super-niche: how language makes minds matter more. Theoria 54:255–268

Clark A (2006) Language, embodiment, and the cognitive niche. Trends Cogn Sci 10(8):370–374

[34]Transposed in a different philosophical outlook, the debate about the strategic use of camouflage is strictly connected with the construction of cognitive niches (Part II), and the selection of ecological chances. As camouflage is about either hiding real chances/affordances, or simulating non existent ones, some interesting connections can be drawn between linguistic camouflage and what is commonly referred to as "bullshit" (cf. Sect. 6.2.1).

Dessalles JL (2000) Language and hominid politics. In: C Knight JH, Suddert-Kennedy M (ed). The evolutionary emergence of language, social function and the origin of linguistic form. Cambridge University Press, Cambridge, pp 62–79

Dunbar R (2004) Gossip in an evolutionary perspective. Rev Gen Psychol 8(2):100–110

Flinn MV, Geary DC, Ward CV (2005) Ecological dominance, social competition, and coalitionary arms races: why humans evolved extraordinary intelligence. Evol Hum Behav 26:10–46

Fodor J (1983) The modularity of the mind. The MIT Press, Cambridge

Frankfurt H (2005) On bullshit. Princeton University Press, Princeton

Gabbay DM, Woods J (2005) The reach of abduction. A practical logic of cognitive systems, vol 2. North-Holland, Amsterdam

Gibson JJ (1979) The ecological approach to visual perception. Houghton Mifflin, Boston

Gigerenzer G, Brighton H (2009) Homo heuristicus: why biased minds make better inferences. Top Cogn Sci 1:107–143

Glendinning P (2004) The mathematics of motion camouflage. Proc R Soc Lond Ser B: Biol Sci 271(1538):477–481

Grice H (1975) Logic and conversation. In: Sternberg R, Kaufman K (eds) Syntax and semantics 3: speech acts. Academic Press, New York

Grose J (2011) Modelling and the fall and rise of the handicap principle. Biol Philos 26(5):677–696

Hammond KR, Steward TR (eds) (2001) The essential brunswik beginnings, explications, applications. Oxford University Press, Oxford

Hample D, Han B, Payne D (2010) The aggressiveness of playful arguments. Argumentation 24(4):245–254

Hoffmeyer J (2008) Semiotic scaffolding of living systems. In: Barbieri M (ed) Introduction to biosemiotics. Springer, Berlin, pp 149–166

Husserl E (1960) Cartesian meditations [1931]. Martinus Nijhoff Publishers, The Hague

Jacob P, Jeannerod M (2003) Ways of seeing: the scope and limits of visual cognition. Oxford University Press, Oxford

Kirlik A, Storkerson P (2010) Naturalizing peirce's semiotics: ecological psychology's solution to the problem of creative abduction. In: Magnani L, Carnielli W, Pizzi C (eds) Model-based reasoning in science and technology. abduction, logic, and computational discovery. Springer, Berlin, pp 25–39

Krosnick J, Betz A, Jussim L, Lynn A (1992) Subliminal conditioning of attitudes. Personal Soc Psychol Bull 18(2):152–162

Longo G (2005) The cognitive foundations of mathematics: human gestures in proofs and mathematical incompleteness of formalisms. In: Grialou P, Longo G, Okada M (eds) Images and reasoning. Keio University, Tokyo, pp 105–134

Magnani L (2001) Abduction, reason, and science. Processes of discovery and explanation. Kluwer Academic/Plenum Publishers, New York

Magnani L (2007a) Animal abduction. From mindless organisms to artifactual mediators. In: Magnani L, Li P (eds) Model-based reasoning in science, technology, and medicine. Springer, Berlin, pp 3–37

Magnani L (2007b) Semiotic brains and artificial minds. How brains make up material cognitive systems. In: Gudwin R, Queiroz J (eds) Semiotics and intelligent systems development. Idea group Inc., Hershey, pp 1–41

Magnani L (2009) Abductive cognition: the epistemological and eco-cognitive dimensions of hypothetical reasoning. Springer, Berlin

Magnani L (2011) Understanding Violence. Morality, religion, and violence intertwined: a philosophical stance. Springer, Berlin

Magnani L (2013) Is abduction ignorance-preserving? Conventions, models, and fictions in science. Log J IGPL 21(6):882–914

Majerus M, Brunton C, Stalker J (2000) A bird's eye view of the peppered moth. J Evol Biol 13(2):155–159

Malle BF, Moses LJ, Baldwin DA (2001) Introduction: the significance of intentionality. In: Malle BF, Moses LJ, Baldwin DA (eds) Intentions and intentionality foundations of social cognition. The MIT Press, Cambridge, pp 1–24
Millikan R (2004) On reading signs: some differences between us and the others. In: Kimbrough Oller D, Griebel U (eds) Evolution of communication systems: a comparative approach. The MIT Press, Cambridge, pp 15–29
Osorio D, Srinivasan M (1991) Camouflage by edge enhancement in animal coloration patterns and its implications for visual mechanisms. Proc. Biol. Sci. 244(1310):81–85
Peirce CS (1992–1998) The Essential Peirce. Selected Philosophical Writings. Indiana University Press, Bloomington, vol. 1 (1867–1893), ed. by N. Houser and C. Kloesel; vol. 2 (1893–1913) ed. by the Peirce Edition Project
Pessiglione M, Petrovic P, Daunizeau J, Palminteri S, Dolan R, Frith C (2008) Subliminal instrumental conditioning demonstrated in the human brain. Neuron 59(4):561–567
Ross D (2007) H. sapiens as ecologically special: what does language contribute? Lang Sci 29(5):710–731
Sage J (2004) Truth-reliability and the evolution of human cognitive faculties. Philos Stud 117:95–106
Schwartz J, Gladding R (2011) You are not your brain. Avery, New York
Simon H (1955) A behavioral model of rational choice. Am Econ Rev 69(1):99–118
Skelhorn J, Rowland H, Ruxton G (2010) The evolution and ecology of masquerade. Biol J Linn Soc 99(1):1–8
Slade C (2002) Reasons to buy: the logic of advertisements. Argumentation 16:157–158
Snoeck Henkemans F (2009) Maneuvering strategically with praeteritio. Argumentation 23:339–350
Sperber D, Mercier H (2010) Reasoning as a social competence. In: Landemore H, Elster J (eds) Collective wisdom. Cambridge University Press, Cambridge
Srinivasan M, Davey M (1995) Strategies for active camouflage of motion. Proc. Biol. Sci. 259(1354):19–25
Stevens M, Merilaita S (2009) Animal camouflage: current issues and new perspectives. Philos Trans R Soc B: Biol Sci 364:423–427
Thagard P (2005) The emotional coherence of religion. J Cogn Cult 5(1–2):58–74
Thagard P (2007) Coherence, truth and the development of scientific knowledge. Philos Sci 74: 28–74
Thom R (1988) Esquisse d'une sémiophysique. InterEditions, Paris, translated by Meyer V, Semio physics: a sketch. Addison Wesley, Redwood City, 1990
Vicente KJ (2003) Beyond the lens model and direct perception: toward a broader ecological psychology. Ecol Psychol 15(3):241–267
Woods J (2005) Epistemic bubbles. In: Artemov S, Barringer H, Garcez A, Lamb L, Woods J (eds) We will show them: essays in honour of Dov Gabbay (volume II). College Publications, London, pp 731–774
Woods J (2013) Errors of reasoning. naturalizing the logic of inference. College Publications, London, (forthcoming)
Zahavi A, Zahavi A (1993) The handicap principle: a missing piece of Darwin's puzzle. Oxford University Press, New York

Chapter 3
Proto-Models, Mental Models and Scientific Models

Abstract This chapter focuses explicitly on scientific modeling, drawing from some conclusions made explicit in the previous chapter. The interlocutors of my argumentation will be the contemporary philosophical actors of the debate about scientific models, especially advocates of *fictionalism* (Frigg, Fiction in science, 247–287, 2010b; Suárez, Fictions, inference, and realism, 225–245, 2010), that is the view according to which scientific models are best understood by the same tools used for analyzing literary fictions, and can therefore be described as displaying a *fictional* nature. I will reject this view by appealing to the biological origins of modeling—described in the previous chapter—and arguing that the construction of scientific models can be conceptualized as an *emergence* of the transparent, and careful, use of models (still mental—that is neurally objectified—or otherwise distributed representations) made by scientists, which is quantitatively but not qualitatively different than the *biological* reliance of models displayed by any cognitive system: assessing the instrumental use of some fictional elements does not entail that the whole system should be labeled as "fiction". Otherwise, to assess the self-aware fictional nature of science would compel us into defining as fictional also the proto-models displayed by animal cognition: the next step would be to advocate a randomized fictional behavior in animals and humans.

3.1 Introduction

To provide an initial definition, we can agree that a model is something we use in order to gain some benefit in order to understand or explain something else, which we shall call the target. A model lets us understand the target, and behave consequently, in a way that would not be possible without the model itself: different models usually optimize the understanding of different aspects of the target.

This definition of model should make it easy to appreciate how many situations we face everyday are tackled by making use of models: to deal with other people we make models of their minds and their intentions, to operate or repair a piece machinery we make models the way it functions, in the remote case of trying to escape from wildlife

© Springer International Publishing Switzerland 2015
T. Bertolotti, *Patterns of Rationality*, Studies in Applied Philosophy, Epistemology and Rational Ethics 19, DOI 10.1007/978-3-319-17786-1_3

we make models of their hunting strategies and perceptual systems,[1] to explore novel environments we make models of their spatial configurations, to mention only a few.

We make use of models in wide array of circumstances, but what all models actually share is a dimension of non-abstractness: we create them, or make use of models that were already constructed by other people, and models usually display a distributed nature, since they are either built on external, material supports (i.e. by means of artifacts, paper sheets, sound waves, body gestures) or, in case of mental models, encoded in brain wirings by synapses and chemicals (a mental map, for instance, is the mental simulation of the action of drawing a map—a powerful model construction activity—whose embodiment in the brain was made possible by the enhancement of human cognitive capabilities).

This is just as far as overt modeling is concerned: recent studies showed how the range of cognitive activities that can be classified as *model based cognition* is wider than expected, as many cases rely on forms of modeling that are not explicit to the agent's consciousness. Such use of tacit modeling is shared by animals as well, and is not a trait specific to human beings.

Conversely, a human-specific use of models seems to be displayed by scientific practice. In science, as I will show, models not only provide simplified descriptions of known phenomena, but often serve as an inferential tool to explore and constitute the target itself. As we will see, current epistemology is engaged in fierce debates about *what* models are and *why* we can trust their empirical successfulness: I will suggest that an insightful approach to these questions can be derived from the analysis of what scientific models share with all other forms of modeling (of which they can be considered a peculiar subset), what their peculiarities are and, on a final note, whether these peculiarities can be acknowledged legitimately or should be rather thought of as an attitude of scientific endeavor towards models and the peculiar use it makes of them.

3.2 Models Without Modelers?

"A model allows us to infer something about the thing modeled" is one of the most straightforward definitions of model available today (Holland 1995, p. 33).[2] The nature of scientific models has been one of the most debated topics in philosophy

[1] Movies offer many examples of this kind of modeling: fugitives sometimes cover their body in mud to prevent predators from targeting them because of their smell, other times their freeze to exploit the predator's—supposed—blindness to immobile objects: in all of these cases, in order to decide a course of action one has to construct a model of the predator he wants to avoid, considering relevant factors and factors that can be manipulated (for instance, a model suggesting that the predator can sense the prey's brainwaves is not put forward, as at the moment it would not allow to undertake any viable course of action).

[2] The simplicity of this definition must not sidetrack us: as I will contend in Sects. 3.3.2 and 3.3.3, together with Morrison (2009) and Magnani (2012): the model is the *conditio sine qua non* for poietically establishing a new scientific understandability which coincides with establishing the "borders" and the essence of the target-phenomenon itself.

of science over the past few decades. A number of interesting solutions have been put forward by different philosophers, but there seems to be a common issue: as if in a kind of name-fury, the common approach consists in branding the object of investigation with a new name, and consequently generate a new class of problems that do not relate specifically to scientific models but to their new "avatar".[3]

An alternative suggestion might be to adopt a kind of bottom-up, naturalistic approach considering scientific models first of all simply as "models" (almost as if this was a primitive concept), and once their "behavior" is assessed to a satisfactory extent, analyze what makes them "scientific".

The first bias that should be dispelled is the one characterizing the notion of model as associated with scientific modeling and thus intentional representation: models are often considered as the *intentional* output of high-level cognitive capacities, and their development requires the display of linguistic, mathematical and graphical abilities, plus a theoretical penchant towards explicit analogical reasoning and mental simulation, and a necessary ability to externalize and disembody knowledge in the production of artifacts that serve as external representations. It must be acknowledged, though, that studies in distributed cognition already showed that such intentionality only describes part of the endeavor in scientific modeling: Magnani (2009) argued that the process of manipulative abductive modeling embodies a dimension of thinking through doing that is shared by certain mammals and birds too, as I will show in this very chapter of my book.

Most models are considered to have a descriptive function of how (a particular) target system works, and serve the purpose of making successful prevision on future events based on causal relationships, whether they are held to be necessarily true accounts or merely effective "fictions".

As suggested by Cartwright, scientific models can be understood as "prepared descriptions". A model speaks to us in a different way than a non-actively modeled perception does: if compared to unprepared descriptions (that is, perception), our models seem to be wrong (Toon 2010), yet the descriptions put forward by the model *prepare* for the application of mathematical structures, for instance. The actual spring I am playing with right now might be a fuller and more accurate phenomenon than what presented by the model, but the model can illuminate us about some traits that will be shared by *all* springs much better than the contemplation of the single, present spring.

From a naturalistic perspective, the notion of model is intimately bound with the adoption of a future course of action, and linked with the notion of emergence as

[3]To say that scientific models are fictions leads us into examining the core problems of fiction (Woods 2010; Woods and Rosales 2010b; Contessa 2010; Frigg 2010a, b, c; Godfrey-Smith 2009; Woods and Rosales 2010a; Suárez 2009, 2010), to label them as representations opens the ancient issues of representation and mimesis (Chakravartty 2010): similar problems arise if we just apply to models classical definitions such as abstract entities (Giere 1988, 2009, 2010) and idealizations (Portides 2007; Weisberg 2007; Mizrahi 2009), to the more recent ones, as surrogates (Contessa 2007), credible worlds (Sugden 2009; Kuorikoski and Lehtinen 2009), missing systems (Mäki 2009; Thomson-Jones 2010), make-believe (Toon 2010), parables (Cartwright 2009b), epistemic actions (Magnani 2004a, b) or revealing capacities (Cartwright 2009a).

intended by Holland (1997): it signifies those biological and interactional phenomena characterized by *constrained complexity*, in which an extremely articulated and complex resulting state is triggered by a limited number of simple components and rules. The model is valued to make predictions, and—in the case of emerging models—it sometimes serves as a blueprint of the target system. The different regulation of a parameter in a model usually generates novel hypotheses, in an operation of eco-cognitive attuning between the external reality which is the target of the model and internal representation where the model is constructed, exploiting the common coding (in the sense of the expression received from Chandrasekharan (2009)) connecting the execution, perception and imagination of, for instance, motor impulses.

With this respect, and also to better asses certain claims about animal cognition I put forward in the previous chapter, some questions could be asked in order to contribute to the outgoing debate around the nature of models themselves: to what extent is it possible to produce models without the display of a conscious, intentional intelligence? Are human beings affected by this connotation of models? I will therefore set off by further elaborating what I developed in the previous chapter and show how a widespread biological feature such as the display of camouflage technique might suggest the emergence, even in organism poorly endowed at cognitive level, of actual *models* of how their predators' perceptual systems work.

As I will argue further on, making this claim explicit suggests that the spontaneous construction of models can actually *emerge* at a mindless stage (as in perception Magnani 2007a), and how some characteristics of such-conceived models do *a fortiori* apply to scientific modeling as well.

3.2.1 Embodied Models of Agency Recognition: An Eco-Cognitive Necessity

One of the greatest cognitive problems shared by all organism that are able to react (more or less plasticly) to external stimuli is external agency. A perceptual representation of one's external environment must as a matter of fact highlight the presence of external agents who might be predators, preys or competitors for available resources, in order to maximize one's chance of survival. Such *situated* and ecologically informed character of organisms' cognition is reverberated by the concept of affordance[4] (Gibson 1979), which provides a useful account of the role of the environment and external—also artifactual—objects and devices, as the source of action possibilities (constraints for allowable actions). Of course, different organisms apply to the kinds of local signs they can perceive a wide array of different modelings: while simpler organisms possess extremely simple models that encode an off-line

[4]Originally belonging to the conceptual toolbox of ecological psychology, an affordance is a resource or chance that the environment presents to the "specific" organism, such as the availability of water or of finding recovery and concealment. Of course the same part of the environment offers different affordances to different organisms.

representation of environmental affordances (Laurent 2003), organisms with more plastic cognitive capacities and properly semiotic brains can produce symbolic models (Magnani 2007a) or propositional-sentential models.[5]

The claim, that I will support in the following subsections, is that the construction of models emerges from low-level cognitive capacities, and can thus be thematized as an issue that is necessarily ecological in its nature, that is to say, it concerns the cognitive relationship of an organism and some aspects of its external reality. I will rely on contemporary studies concerning ethology and animal cognition already presented in the previous chapter to show that processes of agency recognition are basic forms of model-based cognition, and that a further proof of the emergence of these simple models is that they can be found as operationalized in (and therefore extracted from) camouflage mechanisms.

As I will soon prove, I think that an abductive framework is the fittest to describe and investigate the formation of those models which animals produce, emerging from signs they are able to recover from the environment. It seems therefore legitimate to speak of "animal abduction" also as far as the emergence of models is concerned.

3.2.2 Emerging Animal Models as Abductive Representations

As we saw in the previous chapter, and in better detail in the Introduction, abduction, as understood within the Peircean framework, can be accounted for as the process of inferring certain facts and/or laws and hypotheses that render some sentences plausible, that explain (and also sometimes discover) some (eventually new) phenomenon or observation: it is the cognitive process in which hypotheses are formed and evaluated. I already lengthily described in Sect. 2.2.2 the case for an abductive understanding of animal cognition, and I will therefore rely on that chapter in order to avoid unnecessary repetitions.

The core point was, and still is, not to regard abduction as a merely sentential inferential process: indeed, recent studies on abduction opened a much wider field of investigation concerning these multi-modal inferences. Survival is an eco-cognitive task, requiring organisms to engage in a relationship with the environment that is often a conflicting one—as I am claiming in this chapter—and the relationship with the environment is mediated by a series of cues the organism must make sense of in order to generate, even if tacitly, some knowledge it did not possess before (Magnani 2009).

Summing up very quickly, animals rely on their senses alone in order to recognize the presence of other organisms in their surrounding. What senses pick up is not an immediate picture of external agency but a more or less rich complex of signs: an animal must manage to detect the presence of other agents in order to maximize its own chance of survival, and such detection can only be inferred by operating upon

[5]Our social cognition is for instance aided by models of behavior called moral templates (Magnani 2007b).

meaningful signs. Signs are not associated randomly, but according to certain *models that emerge in the animal system* in connection with the stimuli it receives: they can be pre-wired or learnt. The following step consists in the operationalization of the correct affordances concerning the detected organism.

It is useful to remember Millikan's suggestion already adopted in the previous chapter: he contends that internal model-based representations in animal minds might mostly consists of PPR (*"Push-me pull-you" representations*), meaning they are both aimed at representing a state of affairs and at producing another, often suggesting a chance for behavior as received by the Gibsonian/affordance tradition (Millikan 2004). The indicative content of a PPR mental representation about external agent will therefore never be of the kind *Oh, look at that organism PERIOD* but rather *Look at that organism: should I attack/avoid/hurt/kill/eat it/mate with it?*: a striking connection between this kind of model and the "common coding" suggested by Chandrasekharan (2009), in his contention that:

> [...] the execution, perception, and imagination of movements share a common representation (coding) in the brain. This coding leads to any one of these three (say perception of an external movement), automatically triggering the other two (imagination and execution of movement). One effect of this mechanism is that it allows any perceived external movement to be instantaneously replicated in body coordinates, generating a dynamic movement trace that can be used to generate an action response. The trace can also be used later for cognitive operations involving movement (action simulations). In this view, movement crosses the internal/external boundary as movement, and thus movement could be seen as a "lingua franca" that is shared across internal and external models, if both have movement components, as *they tend to do in science and engineering* (Chandrasekaran 2009, p. 1061, added italics).

The way that animal modeling responds to the "common coding" criterion is clearly embryonic if compared to the use of models displayed by science and engineering: as for animal modeling, I would not go as far as claiming that "the trace can also be used later for cognitive operations involving movement (action simulations):" this can be true for animals displaying more plastic cognitive abilities and learning mechanisms, individually or socially. A point in case, nevertheless, is the centrality of movement in both scientific and biological, emergent modeling. Movement is, as a matter of fact, at the core of the idea of manipulation, and therefore of experiment[6]: manipulations impress movements on the external reality so that the resulting changes can work as "props" for the construction of models, since thinking through doing is often "thinking through moving,"[7] but movement is also the primary

[6]The role of experiments in bringing models to life will be the topic of the next chapter.

[7]Magnani (2009) stresses the centrality of manipulative abduction and the problem of thinking through doing in the scientific enterprise. The role of manipulation and thinking through doing is crucial also in the expression of the most advanced kinds of models displayed by animals: corvids, for instance, do not only exhibit exceptional ability in the creative use of tools, making smart use of non-natural items (e.g. aluminum strips Weir and Kacelnik 2006), but can also operationalize complex mental representations such as Archimedes' principle: Bird and Emery (2009) show how rooks can drop stones in a container filled with water so to raise the water level and attain a floating prey. They are also aware that larger stones cause a higher raise in water level than small ones. This seems to be a more sophisticated model-based activity, because if rooks are able to operationalize

building block in emerging models, as the first difference to be discriminated is the difference between biological and non-biological movement.

As already contended, a trait that is typically displayed by biological emerging models is their *tenacity*. The success of artifacts such as fishing baits and hunting traps depends on the fact that most animals either display limited capacities for learning and revising their inner models, or are not able to share their advancements with their conspecifics.[8] Emerging models display such tenacity, and change over long period of times, because their very persistence indicates evolutionary success (that is, if they did not cause the systematic death of the organisms who entertained certain models).

It is interesting to note how abduction, as it is not a truth-preserving inference, perfectly depicts such inferential scenarios: considering as premises beliefs held as true by the subjects, abduction generates emerging models which may not be true and yet be endowed with a powerful *fitness-reliability* for the organism's welfare.[9] The same happens in human reasoning: in peculiar settings we may produce models of a target without having gathered necessary evidence, as in the case of hasty generalizations. Such hastily generalized models (concerning for instance generalizations about women who are not able to drive through traffic, or concerning big felines that are afraid of water, or generalized models of bombs that can be defused by cutting the yellow cable—no wait, was it the blue one?) can be valuable for their contribution to the agent's fitness inasmuch they can help her make a decision that saves her life, but do not benefit the *epistemic welfare* of the agent herself.

3.2.3 Emerging Models: Useful Instruments or Fictions?

At this point of my analysis of emerging models, a proper demarcation between what I just described and scientific modeling should be introduced. If we transpose

(Footnote 7 continued)
a model corresponding to Archimedes' principle, then we should concede that they *possess* that model: they cannot relate to the model in theoretical-sentential way typical of humans beings, but it is nevertheless encoded in neural systems and accessed in instances of thinking through *through doing*.

[8] Conversely, some species (typically rodents and birds which share their habitats with human beings) are said to be endowed with a kind of "culture", inasmuch as they show a clear predisposition towards constructing models that are actively tuned with ecological necessities and sharing them with conspecifics by means of social learning, observation etc. (Heyes 1993).

[9] I have already pointed out in the previous chapter that, as far as the biological and pre-linguistic levels are concerned, it can be argued that those emerging models do not matter for their *truth-reliability* but rather for their *fitness-reliability*, understanding fitness as welfare (Sage 2004). From a biological outlook (which is often engaged by human beings as well) the favored inference is the most successful inference, the one leading to survival. For instance, not noticing the presence of a predator, not entertaining any form of PPR representation concerning it—and thus not reacting—might be the best way to avoid being noticed in turn and killed: in this case, the potential prey's proto-belief is clearly false, and yet successful. This aspect was dealt with in greater detail in Sect. 2.2.2.

into the domain of scientific models the importance of *fitness-reliability*, one could think that it is indeed possible—to paraphrase a verse of Shakespeare's that is indeed much loved in the debate on the nature of models—to catch a "carp of truth" with a "bait of falsehood". One the one hand, as contended by Giere, the instrumental use of some models that are known to be fictional does not entail that the whole system should be labeled as fiction. Some models explicitly aim at simplifying calculations by offering a different systems to refer to:

> Applying the method of image charges, one replaces the original model with a model in which the infinite metal plate is replaced by a "fictional" negative charge placed symmetrically on the other side of where the surface had been in the original model. The solution to the problem using the new model, in full accord with electrostatic theory, is exactly the same as if one had solved the mathematically more difficult problem using the initially suggested model. What is meant by calling the negative charge in the second model "fictional"? As a component of a model, the image charge in the second model is no more and no less fictional than the positive point charge and infinite metal surface in the original model. It is telling that textbooks do not refer to the latter as fictional although they are clearly physically impossible entities. My analysis of the situation is that the original model is understood to be an idealized representation of a concrete system. The concrete system would only have counterparts to the original positive charge and conducting surface. Relative to this suggested concrete system, the negative charge in the second model is called "fictional" because it would have no counterpart in the assumed concrete system. On this understanding of the situation, there is no basis for calling either model as a whole a work of fiction (Giere 2009).

If we compare the solution of a problem (maybe even the same problem, say a physics test) resorting to a hasty generalization ("My teacher always follows the same pattern in distributing the right answer in quizzes: the right answer must be the third!") or by using the system described by Giere, we can easily say which one of the two is closer to the idea of fiction. The first model just provides a (highly fallacious) *fitness-reliable* way to plug an ignorance leak with no commitment towards the relevant target system, while the second model projects a scientific and *truth-reliable* structure of understandability on the target system in order to make the problem more easily solvable.

If we go back to the origin of the demarcation, it could be said that I drew an analogy between animals' fitness-reliable models and hasty generalizations, but it would be actually unfair and wrong to depreciate the dignity of emerging animal models by comparing them to hasty generalizations: coherently with Millikan's observations, I argue that animals' emerging models concerning external agency strictly depend on the *extreme situatedness* of their cognitive capacities. Most emerging models cannot therefore be separated from the *here and now* relationship between the organism who entertains the model and the target of the model (that is, its predators or preys and the way they relate to the cognizing agent). With this respect, it could be suggested that models, even from a nearly mindless perspective, can be seen in turn be seen as a "model" of the target system's affordances. As maintained by Laurent (2003) and coherently with what I have suggested so far, an emerging model can be conceived as simulation of the target's affordances, thus operationalized to allow the prediction of future events connecting to those very affordances. For instance, a model could represent a property of the target such as "being able to detect only moving agents", by the related affordance "escape from the predator by remaining immobile", and direct the behavior of the cognizing agent. Analogously with Chandrasekharan's use of the concept of "common coding", such abductive representations appear in fact to be the product of situated abductive inferences, peculiar inasmuch "they tell in *one undifferentiated breath* both what the case is and what to do about it" and they "represent the relation of the representing animal itself to whatever else they also represent" (Millikan 2004, p. 20, added italics). This kind of inferential process, residing in the coupling of the detector and the detected, is not based upon a random appraisal of an animal's semiotic cloud, but specific sign configurations match certain affordances, which ultimately trace back to the desired property. Jacob and Jannerod's description seems particularly illuminating:

> Property *G* matters to the survival of the animal (e.g. a sexually active male competitor or an insect to capture). The animal's sensory mechanism, however, responds to instantiations of property *F*, not property *G*. Often enough in the animal's ecology, instantiations of *F* coincide with instantiations of *G*. So detecting an *F* is a good cue if what enhances the animal's fitness is to produce a behavioral response in the presence of a *G* (Jacob and Jeannerod 2003, p. 8).

The hypothesis about the presence of an agent who detains the property *G* is *abduced* on the basis of one or more *perceptible* properties *F* that usually signify the relevant properties according to the emerging model. If an organism is hunted as a prey or avoided as a predator because of a property *G*, it must try to reduce the occurrences of the properties signaling their characteristic, and this varies widely from organism to organism: the emerging model prescribes that something is taken to be symptomatic of something else on the base of some regularity, and this *cognition* can be instinctual or plastically shaped by learning.

Before seeing some camouflage mechanisms described in the previous chapter in the light of *emerging strategic modeling*, we should take advantage of the last considerations about emerging models to reflect about the impossible ontological abstractness of models: as I showed, emerging models are necessarily situated and have often a clear operative and heuristic role (they guide the cognizant's behavior) while we cannot say much about their representational scope (it would be rather speculative to say something like "The sepia represents anything moving as a possible prey": this is about *our* model of the sepia's cognitive capacities, and not about the actual sepia's model). Our simple models, on the other hand, do have an abstracting quality: a simple map, or a model of a person's way of reacting to some news, can be accessed and constructed independently from the proximity and the actuality of the target, but they are nevertheless *connected* with their target, produced by a *peculiar* cognition and externalized by a modification of some support (be it neural, physical, interactional etc.): in brief, from emerging models we can already learn that models are *necessarily* someone's models of something.

3.2.4 Camouflage as the Strategic Use of Models in Nature

At this point of the argumentation, it seems proper to revise some observations made in Sect. 2.2.3 in the light of the operationalization of "emerging models": the discussion on natural models that I began to lay out in Sect. 3.2.2 seems to legitimize the claim that every agent has a twofold inferential relevance, active and passive: on the one hand, it disperses signs *out* in its environment, on the other hand it receives and processes signs *from* other organisms: these signs are received and used by model-based agency recognition cognitive processes. The first mechanism must be minimized (spreading out signs) while the second (recovering signs) maximized either to counteract predation either to avoid being spotted by a potential prey. If by now we exclude that model-building is an activity engaged by human beings alone, we can concur in saying that every organisms (more or less explicitly) attempts at producing valuable models of other organisms' behaviors.

If we assume that an organism is endowed with—mostly model based—abductive cognitive systems aimed at the detection and identification of other agents in its surroundings, we can suppose that these systems operate within a determinate threshold selecting stimuli which activate their inferential processing. Signs that fall within these abductive thresholds are likely to produce in the organism an internal representation involving some kind of awareness about a particular nearby agent, which emerges as a model of that agency.

As we saw, in order to maximize their chances of not being discovered by agency recognition systems, certain organisms modulate their semiotic footstep so that the signs of their presence either fall under the inferential threshold of other agents and not trigger any positive agency-detection response, or they are meant to overwhelm and saturate the other agent's abductive threshold. This is about a strategic use of models: the peculiar display of signs is meant to hinder the application of an unsuitable model, or block the development of any model at all. Let us briefly consider the two most widespread types of biological camouflage.

Crypsis individuates those processes in which the initial attempt is to prevent detection. As explained in Sect. 2.2.3, crypsis is very effective inasmuch as it is intuitively hard to develop an emerging model upon something which is simply *not there*: no prop is provided at all for the construction of the model. *Masquerade* camouflage provides into the environment signs that make them easily detectable, but "their bearers are misidentified as either inedible objects by their predators, or as innocuous objects by their prey" (Skelhorn et al. 2010, p. 1): they provide, that is, patent props to operationalize recognition models that are actually misleading.

The strategic model-based warfare incorporated by camouflage dynamics concludes when the model employed or constructed by one of the organisms fails in the representation of the target, thus instancing an occurrence of what could be seen as "negation as death": negation of the model because of the death of the modeler.[10]

---

[10]Magnani (2012) examines the relationship between fictionality and falsehood in terms of the logical concept of "negation as failure" (Clark 1978). Negation as failure is a *weak* form of falsification, compared for instance to a counterexample that would negate the model in its matter: the failure

To put it another way, it can be said that a successfully camouflaged agent managed rearranged its imprint of signs according to the model of agency-recognition of its prey/predator: its predators and preys, in turn, were not able to spot it (or failed to recognize it) because their model of agency-recognition was not in tune with the target. But if we consider this kind of models in their relationship between the representation and the target system, can it be said that they display a fictional nature, even when they contribute to producing clearly counterfactual beliefs? I would not say so: contemporary philosophical tradition—in particular phenomenology—has stressed how, despite the fact that perception can trick us and become misperception, it can never be said to be *wrong*. If perception cannot be wrong, then it cannot be fictional either, inasmuch as the fictionality of a model can be—if ever—assessed necessarily *a posteriori*.

3.3 The Naturalness of Scientific Models

In the previous chapter I showed how the presence of emerging models in basic animal cognitive faculties can be inferred from the analysis of eco-cognitive mechanisms such as camouflage interactions. I now intend to go to the heart of the matter and generalize to human modeling part of the scenario described so far, highlighting the continuum between emerging modeling and scientific models. In Sect. 3.3.1 I will contend that the relevance of models in many cognitive tasks makes unviable their reduction to fictions: this applies, *a fortiori*, to scientific models. Section 3.3.2 briefly analyzes how both emerging and scientific modeling is the preparing step for mathematical abstraction.[11] Finally, in Sect. 3.3.3 I will deal with the peculiar nature of scientific models analyzing their origins in the scientific revolution, contending that their specificity relies not in the models *per se* but in the attitude towards them in scientific practice.

3.3.1 All Human Knowledge Is a (Sometimes) Virtuous Distortion (and a Model Too)

In the third part of this book, dealing with the role of our cognitive endowment in the generation of "religious" beliefs (Chap. 10), I will provide an extended excursus on the building blocks of inferential model-based reasoning, that is on how human beings process raw information in their environment according to model-based cognitive

---

(Footnote 10 continued)
of the model means that it just stops working, even if it was not *proven* wrong: emerging models admit another kind of (paradoxically) *weak* negation, that is "negation as death" of the cognizing agent.

[11] As meant by Morrison (2009): a projection of meaning which reconfigures the phenomena, as opposed to models by idealization of the target.

heuristics that result in the complex cognitive artifact that is our perception: this should make us consider how, even such models that could be called "emerging models" in animals—to stress their instinctual and immediate nature—are shared by human beings as well. As suggested by Woods and Rosales (2010b), there is a continuum connecting basic mental models and the most complex scientific models so that the analysis of the latter benefits the former and vice versa: they all share the dimension of being a virtuous distortion. Even perceptual knowledge for human beings (though at a very basic level, before being influenced by higher structures such as symbols, languages, theories etc.) is a model based activity, whose data processing rate "is in the neighborhood of 11 million bits per second. For any of those seconds, something fewer than 40 bits make their way into consciousness. Consciousness therefore is highly entropic, a thermodynamically costly state for a human system to be in. At any given time, there is an extraordinary quantity of information processed by the human system, which consciousness cannot gain access to", and the result is that, fundamentally, knowledge is not only a product of "information-processing" but also an "information suppressor" [p. 17].

It is a model-based activity as far as the picture accessed by consciousness is both an abstraction (some true stimuli are suppressed) and an idealization (some fake stimuli are introduced, as in optic illusions): this phenomenon is referred to in cognitive science as semi-encapsulation of information (Raftopoulos 2001a, b; Albertazzi et al. 2011).

In this case, if we compare perceptual models with scientific models, the former are clearly not the descriptions of missing systems, even if "competent practitioners" could assure us that albeit perception being an apparently "accurate description of an actual, concrete system (or kind of system) from the domain of inquiry, [...] there are no actual, concrete systems in the world around us fitting the description it contains" (Thomson-Jones 2010, p. 283). Perception could match this description, but such a categorization falls short of any value if the system is the only available one, produced in order to make sense of the world, and operationalized *as if* it coincided with the target system. Other animals embody different perceptual models than ours. For instance, mosquitos are endowed with a heat-and-CO_2-seeking sensorial apparatus, while bats are known for their ability to perceive ultrasounds: clearly their models of external reality differ from ours, but this should not lead us into defining every perceptual model as a description of a missing system (albeit the definition would fit, it would be of little profit).

Clearly, with respect to scientific models, the kind of models we just described is affected by two fundamental properties: dominant opacity (the agent cannot be aware of them, as in Musgrave's famous analogy of Kantian categories as pink glasses one cannot remove and hence is not aware of) and a consequent lack of plasticity. The models of perception enacted by our cognitive systems are *not negotiable*: we cannot affect our perception naturally, but by means of technological artifacts it is possible to integrate out embodied perceptual models, for instance by use of heat-sensing (infrared) goggles.

Notwithstanding the existence of differences between individuals, perception considered as an unintentional model is usually *de facto* shared by all organisms belonging to the same species, but lacks a characteristic that is typical of scientific models: the latter are in fact basically *intentionally* shared (even if characterized by embodied cognitive acts transparent to consciousness), and their plasticity derives from the fact of being arguable, revisable and withdrawable, just as the inferences that underpin them.

Woods and Rosales seem to have pinned the issue in a very clear way:

> Here, then, is the basic picture. Knowledge is the fruit of information-processing. But it is also an information suppressor. There is a basic reason for this. Consider the particular case of conscious representational knowledge. If what is suppressed by our cognitive processes were admitted to consciousness and placed in the relevant representational state, it would overload awareness and crash the representation [...] This supports the abstraction thesis: *A cognitive state is an abstraction from an information state.* There is another way of saying the same thing: *A cognitive state is a model of an informational environment* (Woods and Rosales 2010b, p. 17).

Our knowledge-gathering sensorial experience emerged naturally as an informat ion-processing activity preventing *overloads* that would just *crash* the representation. And, since we are not prone to give up our assent to our *now appearing heavily mediated* percepts, we should—out of epistemic honesty—refrain from refusing scientific models the basic, unquestioned, acceptance we accept perception with, because the underlaying structure is the same (perchance better controlled and executed in scientific models).

> What I now propose is that scientific models mimic the structures and processes of cognition quite generally, and that a fully worked up model-based scientific theory would capture with some precision the constructive impossibility of knowing a thing through an awareness of most of what's true of it. With perception again as our guide, knowing of the world what you do when you see the robin in the tree is, in comparative terms, knowing hardly anything that's true of it. Such knowledge a conscious awareness of the disclosures of your five senses beckons paradoxically. *It supplies you with ample knowledge on practically no information.* [...] In these respects, the abstractions of model-based science mimic the abstractive character of knowledge in general. *If, as I suppose, the abstractive character of perceptual states doesn't deny them cognitive value, why would the abstractive character of model-based reasoning deny it cognitive value?* After all, shouldn't it be that what's sauce for the goose is sauce for the gander? (Woods and Rosales 2010b, pp. 18–19, added italics)

Woods and Rosales' argument is philosophically powerful: they do not utterly deny that there might be some kind of a fictional nature in model-based reasoning, but to assume the fictional dimension as its *characterizing* trait would imply to drastically and methodically question the trust we place in something as basic as perception and every specimen of knowledge. This is a clearly feasible philosophical endeavor which has often been practiced in the history of philosophy, but its value in questioning a process that science has *proved* to be effective for the past four centuries is a whole different kettle of fish.

Of course, their (and my) claim is not that there are *no* fictional models whatsoever. First of all, as explained by Giere, we already have a splendid and non-ambiguous word to define fictional models, that are conceived and shared as such: *fiction*. A novel, classified in libraries and bookstores under the label "fiction", is a *work of modeling* that clearly creates a universe of meaning, goes even as far as to *constitute* a whole new series of phenomena (think of Tolkien's *Lord of the Rings* saga, for instance), but the author's intention was never to describe an actual part of external reality, as science does instead. Clearly, as contended by both Giere and Magnani, the purpose of consciously fictional modeling is different from that of scientific—and perceptual—modeling. In the latter case, the aim is to construct models of reality that *directly mediate* the course of our understanding and of our action about that *particular* reality: as brilliantly resumed by Hacking, the key feature of scientific modeling is that, by a peculiar kind of *representation*, it allows *intervention* (Hacking 1983). Conversely, in fictional modeling there might be indeed an attempt to *refer* to an existing external reality, for instance adopting a moral framework, but this is not the main purpose of the fictional model and, at its best, it is not straightforward but *inferentially rich* (if not openly *ambiguous*).

This matter is worth a small digression: if we take as an example Watson and Crick's double-helix model of the DNA, we know it was constructed in order to explain a certain phenomenon—the base-pairing in DNA molecules.[12] It is an abstraction (as understood by Morrison 2009), because the model is not achieved by analyzing the external phenomenon and subtracting local constraints to achieve a functional description of the general case (e.g. the ideal pendulum, ideal spring etc.), but the model is what instantiates the fact that the phenomenon can be understood and subject to theory *as such*: to say it another way, it is the model that traces the theoretical borderlines of its target, which is in turn defined *through* the model, and subsequently proved to be true (as successful) by its engagement in scientific practice.

In this sense, a scientific model creates a phenomenon by *abductively* configuring it out of external reality (but, as we saw, is not this what perception does?) and then suggests how to look for confirmations of the model itself by investigating the consistency of phenomenon, now isolated and defined from reality.[13]

Let us compare it to something which is a model *and* a work of fiction at the same time: *Animal Farm* by George Orwell. The book is intended as a dystopian satirical novel, and focuses on a model society ruled by anthropomorphic animals. If we analyze the structure of the model in a strict way, it is a fictional model of a fictive target, that is the farm ruled by anthropomorphic animals. To echo Giere's analysis,

[12]The invention of the particular model is just a stage within the *epistemic warfare* introduced by Magnani (2012), and taken in further consideration in the following Part of this book. It is not invented *ex nihilo* and found to be matching with the desired phenomenon, but it is produced by a continuous series of attempts, successes, failures that cause the revision of precedent models and lucky events that serve as a "prop" to invent the new model (a snake apparently gave Watson & Crick the intuition about the shape of the DNA molecule!).

[13]A fundamental problem of epistemology was nonchalantly touched here, heavily connected to the issue of the so-called theory-laden*ness* of scientific facts (contended by Hanson, Popper, Lakatos and Kuhn among many others), and partially explains why scientific truth must be considered as provisional: if a model entails at least partially then phenomena it refers to, then it is somewhat self-supporting. Not in a vicious, question begging fashion, but because the model configures the range of phenomena that are *readable* to serve as falsification or corroboration of the model itself. Models are found to be outdated when new facts emerge to show it was wrong, usually from new observations (because of technological improvements) or from "blind-spots" of the model itself.

if we limited ourselves to diss the novel basing on the fact that there is no such known thing as a farm ruled by animals, all we would display is poor understanding of the word "fiction".

Furthermore, one could argue that the model has an allegorical meaning, and that the target system it refers to is indeed human society: three observations should be made to this respect. First, the model would indeed refer to some *moral* characteristics of human societies and not to some exact traits, and it would display a descriptive role rather than the predictive/heuristic role that models play in science; second, even the process of inferring some facts from the model onto external realty is, as we said, *inferentially rich*—that is, according to one's personal sensibility a more or less grim picture of human society may emerge (or even an absolutely positive or physiological one!). Last, but not least, the correct pathway to be taken to appreciate the model is not unique: usually depending on personal predispositions and culture, one can appreciate the book as a fictional model of a fictive system (like children usually do), or as a fictional model allegorical relating to an existing target.[14]

The mechanisms I just briefly described do not apply to scientific modeling: what the model refers to is unambiguous and most of the time there are no concurrent ways of exploiting the same model to say different things about one target system, and it is hard to swing a model as it is between different targets and sometimes accept its applicability and sometimes not. This is why, considering the premises I have been following so far, Cartwright's claim about models being like parables is at best thought-provoking:

> [...] in many cases the correct lessons to be drawn may be more abstract than those described immediately in the concrete situation of the model. But seldom can we really cast the models as fables because the moral is not written in. They are rather like parables, where the prescription for drawing the right lesson must come from elsewhere. Theory can help here, as can a wealth of other cases to look to, and having a good set of well-understood more abstract concepts to hand will play a big role. So the good news that one can move from falsehood in a model to truth by climbing up the ladder of abstraction is considerably dampened by the fact that the model generally does not tell us which ladder of abstraction to use and how far to climb (Cartwright 2010).

As Cartwright suggests in the paper, parables indeed differ from fables inasmuch as fables usually end with an explicitly stated moral, that is more or less entailed by the development of the fable. I would say that the similarity between parables and models stops here, or rather, could be expressed as *if models are fictions they are not fictional in the same way as a fable*. Conversely, the idea according to which there is a "right lesson"—and therefore a "wrong lesson"—to be drawn from a model is interesting but must be compared with a correct *dynamic* understanding of scientific endeavor.

As I have already suggested, scientific enterprise is rightly identified by Magnani as a state of *epistemic warfare* which sees scientists engaged in an aggressive battle *against* nature: the idea of warfare nicely captures the dynamic dimension of

[14]This is the key to the success of recent computer animation major pictures among public of different age: children appreciate them as 100 % fiction, while grow-ups can enjoy both the fictive nature, and the extent to which they mirror existing systems.

scientific endeavor, aimed at producing valuable knowledge about the various fields of investigation. This specificity of science can be traced back to the famous quotation by Francis Bacon about the "vexation" of Nature by the scientists themselves: Pesic (1990) argues in fact that Bacon's ideal was not of science violently aggressing an unarmed victim, but rather of a "heroic mutual struggle" (p. 81). Models are some of the most used and useful weapons of this struggle. As in any state of warfare, it can sometimes be the case to choose a preexistent weapon (i.e. model), while some circumstances might require the development of completely new weapons (i.e. models).

If we consider Cartwright's conclusion in this perspective, something seems to be slightly puzzling: "Theory can help here," she said "as can a wealth of other cases to look to, and having a good set of well-understood more abstract concepts to hand will play a big role." According to the line of thought I followed so far, this claim is odd. Applying a charity principle, I suggest that—if we consider the dynamic nature of science—Cartwright's claim is actually a self-evident truth. It goes without saying that theory helps in the selection and the construction of a good model (with the addiction of all other more or less accidental factors), and it can also happen that according to their level of expertise two scientists can make different sense of the same model (usually with the help of additional manipulation). What I contend is that the production of a model, which in turn as I stressed little earlier produces much of the target phenomenon itself, cannot be separated from the act of interpretation of the model. To draw the "right lesson" from a model is just another way of saying one developed a successful model, while drawing the "wrong lesson" means that one developed and applied an unfruitful model, which did not provide any reliable understanding of the target (nor configured the target as properly understandable). What I find a little bit puzzling is that such a distinction would make sense in a rather unrealistic static conception of science, where the modeler and who makes use of the model are not the same person—nor they belong to the same party like the laboratory group—as it happens in parables!

If we consider Jesus' preaching in a pragmatic-historical framework, it can be easily seen that Jesus did not admit a good lesson *and* a bad lesson to be drawn from his parables, and at least in one occasion he severely rebuked his disciples when they would not understand the meaning of a parable.[15] Today, as we cannot ask Jesus to explain parables to us, they are sometimes straightforward, sometimes inferentially rich, and some other times they are ambiguous *tout court*: different interpretations of famous parables such as the one of the *workers in the vine* or that of the *buried talents* played a role in opposing different Christian confessions over the centuries.

Coherently, we can say that the strength of a parable resides partly in their being inferentially rich (they have been able to tell something new to Christians spreading over five continents and twenty centuries), while this is not necessarily a quality in scientific models, whose desired qualities concern more the possibility of individuating

[15]As it happens in Mark 15:15–16, "Peter said, 'Explain the parable to us.' 'Are you still so dull?' Jesus asked them."

capacities, so that models can guarantee fruitful predictions.[16] The discovery of capacities is ultimately linked to the development of models according to Cartwright herself (Cartwright 2009a), and is echoed by Sugden: a "[...] satisfactory isolation, then, allows a real relationship of cause and effect to be demonstrated in an environment in which this relationship is stable. In more natural conditions, this relationship is only a latent capacity which may be switched on or off by other factors; but the capacity itself is stable across a range of possible circumstances. Thus, the model provides a 'theoretical grounding' for a general hypothesis about the world" (Sugden 2009, p. 20). I think I managed to explain why parables and other kinds of consciously fictional accounts of real or fictive targets do not help to isolate any capacity.

3.3.2 Both Emerging Models and Scientific Models Prepare for Mathematical Abstraction

A fundamental trait of contemporary scientific modeling, as stressed by Morrison (2009), is their being a support for mathematical abstraction: albeit neo-pythagorean intuitions possess an unmistakable philosophically romantic connotation, the mathematization of perception is necessarily mediated by a modeling structure and cannot be *naturally* given. As I will contend, the fact that even simple percepts often offer a significant mathematical meaning is a sign of how both emerging and scientific models are what supports the *creation* of meaning, for instance by mathematical abstraction.

The origin and status of numbers is indeed a fundamental problem of philosophy and philosophy of mathematics, but it will suffice for this analysis to agree with Holland (1995) in his claim that numbering is one of the most basic examples of emerging models: numbers emerge from a model of external reality that affords the isolation of quantities and the abstracting step that lets the cognizant grasp that quantities are the same even if the actual objects are different.

As proven by recent cognitive research, organisms' basic modeling capabilities (that already offer a what (Cartwright 1983) would call "prepared descriptions") afford more elaborate inferential processes, in spite of their being situated at a low cognitive level (Dehaene et al. 1999): as suggested by De Cruz (2006, p. 157), "the human capacity for mathematics is a category-specific domain of knowledge, hard-wired in the brain, which can be explained as the result of natural selection." Mathematical modeling could therefore be seen as a step in the evolution of human cognition, which had risen before the full development of conscience as we know it (considered as a necessary condition for scientific endeavor). Significant research was recently lead on a phenomenon called "subitization" (Davis and Holmes 2005): it relates to the numerical estimations that our cognitive systems can perform without

[16]This feature is indeed typical of science: as I will show in the next chapter, the aim of repeating experiments is precisely to assess that they are not inferentially rich but rather that, considering a different yet similar beginning state, the final one does not vary.

actual counting. Usually, human beings are able to recognize by subitization quantities that amount up to four units. In a loosely Pythagorean speculation, this kind of phenomenon could be understood as a tacit modeling connecting even and uneven quantities to agency detection:

> Crucially, most biologically important objects, such as predator or prey, are symmetrical and, in this respect, sensitivity to symmetry may have evolved because it is crucial for discriminating living organisms from inanimate objects. In fact, symmetry seems to act as an early warning system that directs the visual system to further scrutinise an object until full recognition has occurred. Mirror symmetry is thought to have special status in human perception, precisely because it is such an important cue as to the presence of natural organisms (Hodgson 2009, p. 94).

Subitization mechanisms were subsequently hybridized with other kinds of externalized modeling, either very complex or as simple as using a pencil to count a line of dots dividing it into groups of three or four units (Kirsh 1995).

As I already suggested and as I will further stress in the final section, we can agree with Cartwright in seeing models as an intentional, emerging or hybrid "prepared description" of the target which lets us perform inferential activities about new features of the target.[17] Morrison (2009) extends this insight by showing how this cognitive preparation of a mediating structure to understand the target is fundamentally creative in those cases calling for the mathematization of the phenomenon.

> In situations like this where we have mathematical abstractions that are *necessary* for arriving at a certain result there is no question of relaxing or correcting the assumptions in the way we de-idealize cases like frictionless planes and so on; the abstractions are what make the model work (p. 110).

Morrison sharply contrasts two kinds of models that are typical of scientific endeavor: the first is *idealization*, the more intuitive one, which occurs when a "model idealizes or leaves out a particular property but allows for the addition of correction factors that bring the model system closer (in representational terms) to the physical system being modeled or described" (p. 111). The ladder of idealization is very easy to individuate and to climb up and down, and I suggest that many models we rely on in non-scientific practice partake of this nature: easy representational schemes for instance, maps, etc. allow us to perform inferences on idealized systems, and being able to perform these inferences is automatically associated with the ability to opportunely transfer the results to the target system.

The second kind of modeling according to Morrison is *abstraction*,[18] which plays a conceptually pivotal role in scientific endeavor: it is the process of modeling, as I already said, by which the target phenomenon is essentially explained and constructed as such. This is especially the case for models that allow a massive mathematization of the target system[19]:

---

[17]It is the same activity of *making sense of signs* that I described in Sects. 2.2.2 and 3.2.2.

[18]Morrison connotes abstraction with a different meaning than (Woods and Rosales 2010b) do. Their distinction between abstraction and idealization is comprised by Morrison's definition of idealization.

[19]The constitution of the target through the model is not a matter of developing strategically useful

> [...] abstraction (typically mathematical in nature) *introduces* a specific type of representation that is not amenable to correction and is necessary for explanation/prediction of the target system. What is crucial about abstraction, characterized in this way, is that it highlights the fact that the process is not simply one of adding back and taking away as characterized in the literature; instead it shows how certain kinds of mathematical representations are essential for explaining/predicting concrete phenomena (p. 112).[20]

Morrison explains this by resorting to Maxwell's theory of electromagnetism, the development of which required a new model that supported the mathematization and the application of already known concepts, and this could not be worked out by idealization processes: "the foundation for electromagnetism emerged from the molecular vortex model and was in fact determined by it. But the important issue here is not that Maxwell was capable of deriving a set of field equations from a false model, but rather what it was about the model that underscored the applicability of the equations." Another example, the equations explaining the occurrence of phase transition in thermodynamics, had to be developed on similar models representing physically unrealizable situations, which are "required to explain a physically realizable one" (p. 130).

What these examples aim at showing is that to stress and investigate the fictional nature of scientific models equals to look at the finger pointing at the moon and not at the moon itself, and—as usual—to ignore the fundamentally dynamic nature of scientific practice: the construction of abstract models (which can even reverberate in the concrete exploitation of mediating artifacts, as showed by Faraday's experiments in the discovery of the first metallic colloid (Tweney 2006) plays a fundamental role in determining the target system itself. Resemblance cannot be used as a value guiding the development of the model (and holding the failure to comply with it as a reason to judge the model as fictional): this is the case because resemblance is instituted aprioristically inasmuch as the phenomenon is individuated by the model that describes it, in a mutual engagement fitting with the idea of epistemic warfare. Then it can also turn out that the model does not resemble the target at all, by this does not necessarily cause the failure of the abstracting model, inasmuch as it receives some valuable feedback from the target system (e.g. accurate prevision, consistence with other models etc.).[21]

(Footnote 19 continued)
fictional accounts: "Introducing a mathematical abstraction that is necessary for obtaining certain results involves a different type of activity than constructing a model you know to be false in order to see whether certain analogies or similarities can be established" (Morrison 2009, p. 111).

[20]Even though mathematization is the most straightforward example of creation of meaning subsequent abstracting modeling, other kinds of attributions of meaning exist: consider for instance Darwin's models of natural selection, which supported a significant new amount of meaning and individuated new features of the target even if not resorting to a massive use of advanced mathematics.

[21]Consider what already stressed in footnote 13. In the Part III, I will analyze some instances of "irrationality" within an eco-cognitive perspective, mainly referring to the inferences underpinning religious and magical thinking: as I will contend, supernatural explanations might be considered as a kind of modeling. The more primitive ones, involving hybrids and super-agents, can be considered as idealizations obtained by editing some traits of the biological agent (human—angel/giant, lizard—dragon, etc.), conversely properly religious, theological systems can be understood as abstractions

As I suggested at the beginning of this subsection, also on the basis of previous arguments, this *gnoseologically poietic* dimension of scientific modeling is indeed shared by emerging modeling which, sometimes, acquires a creative force in the assessment of external reality, especially when they set the ground for low-level abstractions such as basic mathematical ones (think of subitization). Of course, as I will show in the next and final subsection, this is not to say that emerging modeling and scientific models are exactly the same, but it might be suggested that a fundamental difference does not originate from the nature of models themselves but from the attitude by which models are conceived and used in scientific practice.

3.3.3 From Emerging Models to Scientific Models

The aim of this conclusive part of this chapter is to suggest that part of the impetus of the Scientific Revolution resided in the new attention that was given to modeling, conferring them a new function (hence a new status) that allowed models to better relate to (and individuate) the laws of nature that science would aim at discovering. The concept of *epistemic warfare* will be pivotal to understand this claim.

If we frame the question in the argument so far, it appears that what is at stake is clearly not the invention of models, but of scientific models as we know them. My claim is therefore twofold: Galileo was aware that his models were indeed a prepared, and preparing, description (fist claim) that supported the application of an advanced inferential systems and language, such as mathematics (second claim). It is interesting to read, with this project in mind, one of Galileo's most famous quotations:

> In Sarsi I seem to discern the firm belief that in philosophizing one must support oneself upon the opinion of some celebrated author, as if our minds ought to remain completely sterile and barren unless wedded to the reasoning of some other person. Possibly he thinks that philosophy is a book of fiction by some writer, like the Iliad or Orlando Furioso, productions in which the least important thing is whether what is written there is true. Well, Sarsi, that is not how matters stand. Philosophy is written in this grand book, the universe, which stands continually open to our [p. 238] gaze. But the book cannot be understood unless one first learns to comprehend the language and read the letters in which it is composed. It is written in the language of mathematics, and its characters are triangles, circles, and other geometric figures without which it is humanly impossible to understand a single word of it; without these, one wanders about in a dark labyrinth (Galilei 1957, pp. 237–238).

Before I carry on, it is important to bear in mind that Galileo was referring to natural philosophy (which would be known as science) as a whole, and not specifically to scientific models. Therefore, out of honesty, the quotation should not be used *sic et simpliciter* as an authority weapon against those who advocate the fictional nature of scientific models: I would be committing nothing but a false implicature and a straw

(Footnote 21 continued)
bearing no immediate similarity with their targets. Of course, those models are self-enforced by heritage, "faith" and authority, since it is not easy to display any non-doxastic feedback or corroboration.

man fallacy, since to contend that scientific models are fiction does not coincide with affirming that the whole scientific endeavor has a fictional nature.[22] Nevertheless, as I am about to contend, the non-fictional character of scientific models is necessarily implied by Galileo's conception of natural philosophy.

Until the scientific revolution, natural philosophy would mostly perpetuate received models, which had a chiefly descriptive function: this was the essence of the Aristotelian "science" (the theory of natural places, for instance, would rely on descriptive, idealizing models that would provide a simplified vision of external reality). The new conception of natural philosophy (i.e. science) could not be satisfied with models that were after all just arguable descriptions, favoring intrinsic qualitative and not quantitative analysis. The newborn science, in order to become intrinsically different than a "book of fiction"—where truthfulness is not a fundamental character for the appreciation of the work itself—had to rely on the construction of models that could grasp and produce an actual relationship between the model and the target system (external reality), even at the price of constructing this resemblance: scientists had to make the first move in the epistemic warfare against nature, and could not wait for nature to "amaze" them and direct their research, as prescribed by Aristoteles.

If we leave to science a chiefly descriptive function, then we witness the rise of two connected problems: on the one hand, models are prone to be nothing but descriptive accounts to be matched with a metaphysically rich external reality (it would be a mistake to call it phenomenon in this case), and hence always prone to be found fictional inasmuch as there is no clear criterion to define their truthfulness—the result is that truthfulness can be accounted on the bases of authority, especially when it is coupled with apparently self-evident truths, as in the case of Aristoteles affirming that lighter bodies fall slower. On the other hand, descriptive models as building blocks of a passive, descriptive science do indeed contribute to making science appear as a *book of fiction*, not because of its relationship with external reality, but because theories and models would be decided by likes and authority, as it was indeed the case when Galileo's heliocentric model was refuted inasmuch as it would endanger the Church's authority in the interpretation of the Scriptures.

Conversely, Galileo's conception of natural philosophy is that of an active quest, a true epistemic warfare: external reality begins to acquire its full dignity as a cluster of "phenomena," appearances where the self-evidence is not necessarily self-truthfulness, and a new conception of model is necessary. We need a kind of model that is conceptually poietic, that is to say, able to produce new phenomena by understanding and isolating them through a conceptually creative attribution of new meaning (connected to the discovery of new features).

In sum, the newly conceived model can be used to explain reality going beyond the simple received appearance, and this process is not the production an even grander fiction (if I cannot trust my senses, how can I trust a model, when it is even *further*

[22]Giere (2009) could be occasionally seen as slipping towards this fallacy, but it must be understood—as stated by Giere himself—as the will to preserve the dignity of science which has now to face extra-epistemic adversaries such as post-modern nihilism, aggressive creationism etc.: these actors are all too happy to commit the inverse fallacy and argue that if scientific models are fictional then science is fictional as a whole.

distant from reality?), because the fruitfulness of the abstracting quality of the model will be accounted for by the coherence of modeled (part of) reality itself, and the cluster of phenomena it entails. As I tried to demonstrate in this chapter, emerging modeling in biological organisms—and especially in human beings—does indeed possess an abstracting nature which goes beyond the "simple" idealization, but in the XVI century this attitude was for the first time brought to full awareness and used as such, a clear example of which is Kepler's discovery that the orbits of planets consisted in elliptical shapes (Gorman 1998): it was with an intentional act of poietic conceptualization that Kepler modeled the data he disposed so that they would fit in a novel geometrical pattern, and only (conceptually) subsequently this model pattern could support a successful mathematization, in the form of the ellipse. In this sense, therefore, I claim that for the first time Galileo acknowledged that models had to be used as prepared and preparing descriptions (Cartwright 1983), also to the aim of actively delineating the phenomenon out of external reality.

The next chapter will deal entirely with experimentation, but it can already be argued that within this conception it is possible to fully understand the experiment as the counterpart of the model as they make up two (theoretically) distinct stages of epistemic warfare. In the first stage, i.e. modeling, the scientists carry out their "attack" on nature; in the second stage, the experiment, scientific endeavor stages a "passive" disposition where, in the typically controlled environment, the natural phenomena is allowed to strike back and test the value of the model (that is, behaving as assumed by the model). Without the experiment, the poietic abstracting nature of the model would condemn science (and other modeling activities) to be nothing but a solipsistic delirium.

In this sense, the experiment acquires its fullest meaning: it is not a game, something to impress other people and to show one's skills, but a selective manipulation of a controlled environment that is artificially structured so to approximate the prepared description of reality embodied in the model. The experiment becomes therefore mutually bond with the model that had inspired it: the model affects the experiment and the experiment influences the model, together they manage to affect even the perception of reality.

Galileo's new attitude towards the model is emphasized by the development of models as bold as the *thought experiment*, which could be seen as the *bootstrap*[23] phase of epistemic warfare: the scientist models the phenomenon (and thus isolates it), and then sets off the next stage by enacting nature's response always within the model itself. Gendler (1998) shows how—*pace* Feyerabend—Galileo's abstract modeling of a target system into a thought experiment was not the mere reproduction

[23]My use of the concept of *bootstrap* is similar to Nersessian's as she contends that: "[...]the cognitive-historical method is the kind of bootstrapping procedure commonly used in science. The customary range of historical records, notebooks, diaries, correspondence, drafts, publications, and artifacts, such as instruments and physical models, serves as the source of empirical data on the scientific practices. The practices thought to be significant to the objectives of the analysis (in our case, creating concepts) are examined with respect to their cognitive bases. [...] The cognitive science research pertinent to analyzing the scientific practices comprises a wide range of investigations into how humans reason, represent, solve problems, and learn" (Nersessian 2010, pp. 6–7).

of a sophisticated but non-experimental argument: had this been the case, we would be back in another kind of metaphysical/theological modeling and have a merely doxastic reach (Faust 2008). Conversely, Galileo's model of the fall of two strapped bodies is structured to persuade "Aristotelians" as well, in a way that lets them persuade themselves.

> The thought experiment that Galileo presents leads the Aristotelian to a reconfiguration of his conceptual commitments of a kind that lets him see familiar phenomena in a novel way. What the Galilean does is provide the Aristotelian with conceptual space for a new notion of the kind of thing natural speed might be: an independently ascertainable constant rather than a function of something more primitive (that is, rather than as a function of weight). It is in this way, by allowing the Aristotelian to make sense of a previously incomprehensible concept,that the thought experiment has led him to a belief that is properly taken as new (Gendler 1998, p. 112).

The mental experiment can be rightly seen as bootstrapping the relationship between the model itself, the phenomenon it constructs and the reverberation of the experiment: it will of course require its enactment to surge to the status of a regular, physical experiment, but it plays nevertheless a fundamental role in the *épistemologie spontanée* embedded in Galileo's endeavor, that coincides with the spirit of epistemic warfare.

This new conception of the model is so powerful that it has to bend reality (effectively reduced to a phenomenon depending on the model) which ultimately recovers human beings' emerging way of making sense of their experience. Feyerabend (1993) provides an interesting hermeneutic of Galileo's lexicons, and captures how the mathematical model inverted the order of dignity between model and observation to the point of reducing appearances to mere fallacy against more counterintuitive truths.

> The senses alone, without the help of reason, cannot give us a true account of nature. What is needed for arriving at such a true account are "the...senses, *accompanied by reasoning*". Moreover, in the arguments dealing with the motion of the earth, it is this reasoning, it is the connotation of the observation terms and not the message of the senses or the appearance that causes trouble. "It is, therefore, better to put aside the appearance, on which we all agree, and to use the power of reason either to confirm its reality or to reveal its fallacy" (Feyerabend 1993, p. 57).

Feyerabend also stresses how, despite their eventual success, the scientist's initial claims are far from being evenly proved, but Galileo "uses *propaganda*. He uses psychological tricks in addition to whatever intellectual reasons he has to offer. These tricks are very successful: they lead him to victory. [...] They obscure the fact that the experience on which Galileo wants to base the Copernican view is nothing but the result of his own fertile imagination, that it has been invented" [p. 65].

I should be able here to vindicate my second claim: it must be remembered, in fact, that as I showed in Sect. 3.3.2, abstracting models are what support the subsequent mathematization of the phenomenon (within and by the model (Morrison 2009)). Therefore, Galileo's ambitious claim that the book of the universe "[...] cannot be understood unless one first learns to comprehend the language and read the letters in which it is composed. It is written in the language of mathematics, and its characters are triangles, circles, and other geometric figures without which it is humanly impossible to understand a single word of it," indeed points out to this *book* being constructed as a conceptual blend out of epistemic warfare.

It is necessarily a forced post-Kantian interpretation, but—on a careful reading—Galileo's inauguration of modern science does not seem to be echoing a Pythagorean conception of nature and its investigation, by which the laws emerge naturally and the scientist must just receive them. The newborn method was a conscious rationalization of how emerging modeling faculties could be turned into weapons to be used in a epistemic warfare between scientists and nature, in which the two opponents would *necessarily* taint each other but possibly in a virtuous way. The scientific intuition about the book of nature written in mathematical alphabet is that this is not a metaphysical given, but something acquired and projected by scientific endeavor.

On a provocative tone, it could almost be suggested that Galileo's was the first major successful attempt to lead philosophy to what could be called an eco-cognitive dimension (Magnani 2009), that is appreciating the non-dissolvable theoretical connection between cognizant agents and their ecology—the environment on which their cognitive faculties operate: if scientific models are indeed a self-aware and rationalized successor of emerging natural models, then natural philosophy was *naturalized* indeed (since it would recognize the continuous bond between the philosopher and the natural framework she investigates), and could finally give birth to modern science in the same conception we have now.

3.4 Conclusion

Summing up this chapter, I aimed at contributing to the ongoing epistemological debate on the nature of models by proposing an excursus from emerging/biological models to scientific modeling, in order to highlight the similarities between spontaneous forms of modeling and the careful elaboration of models witnessed in scientific settings. The analysis of basic forms of modeling tried to show how even the mindless processing of external reality does not provide passive descriptions but is rather a poietic, creative and constructive action which constitutes external reality as the organism perceives it, and in my argument I foreshadowed several times how this poietic character is indeed common to both emerging and scientific modeling. I significantly relied on the notion of "epistemic warfare" introduced by Magnani (2012): this concept is very useful to understand the difference between static and dynamic conceptions of science.[24] Paradoxically, to focus on the extent to which science can be seen as a warfare produces a double effect: on the one hand, it shows how the qualitative demarcation between the use of models in science and in the accomplishment of other cognitive tasks is often fuzzy, inasmuch as many prerogatives

[24] Lockhart (2008) contended that a static description of science is not recommendable even for didactical purpose as it completely spoils the nature, and thus the appeal, of scientific endeavor (Lockhart's contention specifically focuses on mathematics).

of scientific models (for instance their being *constitutive prepared descriptions that support further inferential activity*) are in fact widespread in model-based reasoning and appear to be shared by basic model-driven activities such as perception. On the other hand, the notion of epistemic warfare shows how science is indeed characterized by a peculiar and conscious attention towards models as inaugurated by Galileo and the founding fathers of modern science. Such awareness about the *power* of the model was immediately paired with the other pillar of science, that is the experiment (physical or mental), which served as a counterweight to the epistemically creative power of the model as seen within the epistemic warfare. The experiment, coupled with the model, contributed to correctly locate nature's response and provisionally assess the correctness of the model, so that the Baconian struggle is indeed a struggle between peers.

The next chapter, concluding this first epistemological part, will shift the focus towards the experiment. The scope of my analysis will be to engage the topic still within an ecological-cognitive attention, reflecting on the experiment as an epistemic device sometimes aimed at the epistemic, knowledge-related framework, some other time at the human framework, that is at the knowledge-carriers and operators supporting science.

References

Albertazzi L, van Tonder GJ, Vishwanath D (eds) (2011) Perception beyond inference: the information content of visual processes. The MIT Press, Cambridge
Bird CD, Emery NJ (2009) Rooks use stones to raise the water level to reach a floating worm. Curr Biol 19:1410–1414
Cartwright N (1983) How the laws of physics Lie. Oxford University Press, Oxford
Cartwright N (2009a) If no capacities then no credible worlds. But can models reveal capacities? Erkenntnis 70:45–58
Cartwright R (2009b) Models: parables v fables. Insights 1(8):2–10
Cartwright N (2010) Models: parables v fables. In: Frigg R, Hunter M (eds) Beyond mimesis and convention. Boston studies in the philosophy of science, vol 262. Springer, Netherlands, pp 19–31
Chakravartty A (2010) Informational versus functional theories of scientific representation. Synthese 172:197–213
Chandrasekharan S (2009) Building to discover: a common coding model. Cogn Sci 33:1059–1086
Clark KL (1978) Negation as failure. In: Gallaire H, Minker J (eds) Logic and data bases. Plenum, New York, pp 94–114
Contessa G (2007) Scientific representation, interpretation, and surrogative reasoning. Philos Sci 74:48–68
Contessa G (2010) Scientific models and fictional objects. Synthese 172:215–229
Davis G, Holmes A (2005) What is enumerated by subitization mechanisms? Percept Psychophys 67(7):1229–1241
De Cruz H (2006) Towards a Darwinian approach to mathematics. Found Sci 11:157–196
Dehaene S, Spelke E, Pinel P, Stanescu R, Tsivkin S (1999) Sources of mathematical thinking: behavioral and brain imaging evidence. Science 284(5416):970–974
Faust J (2008) Can religious arguments persuade? Int J Philos Relig 63:71–86
Feyerabend P (1993) Against method [1975], 3rd edn. Verso, London

Frigg R (2010a) Fiction and scientific representation. In: Frigg R, Hunter MC (eds) Beyond mimesis and nominalism: representation in art and science. Springer, Heidelberg, pp 97–138
Frigg R (2010b) Fiction in science. In: Woods J (ed) Fictions and models: new essays. Philosophia Verlag, Munich, pp 247–287
Frigg R (2010c) Models and fiction. Synthese 172:251–268
Galilei G (1957) The Assayer (1623). In: Discoveries and opinions of Galileo, pp 231–280 (trans: Drake S). Doubleday, New York
Gendler TS (1998) Galileo and the indispensability of scientific thought experiment. Br J Philos Sci 49(9):397–424
Gibson JJ (1979) The ecological approach to visual perception. Houghton Mifflin, Boston
Giere RN (1988) Explaining science: a cognitive approach. University of Chicago Press, Chicago
Giere R (2009) Why scientific models should not be regarded as works of fiction. In: Suárez M (ed) Fictions in science. Philosophical essays on modeling and idealization. Routledge, London, pp 248–258
Giere R (2010) An agent-based conception of models and scientific representation. Synthese 172:269–281
Godfrey-Smith P (2009) Models and fictions in science. Philos Stud 143:101–116
Gorman ME (1998) Transforming nature. Ethics, invention and discovery. Kluwer, Dordrecht
Hacking I (1983) Representing and intervening. Introductory topics in the philosophy of natural science. Cambridge University Press, Cambridge
Heyes CM (1993) Imitation, culture and cognition. Anim Behav 46:999–1010
Hodgson D (2009) Evolution of the visual cortex and the emergence of symmetry in the Acheulean techno-complex. Evolution 8:93–97
Holland J (1995) Hidden order. Addison-Wesley, Reading
Holland J (1997) Emergence: from chaos to order. Oxford University Press, Oxford
Jacob P, Jeannerod M (2003) Ways of seeing: the scope and limits of visual cognition. Oxford University Press, Oxford
Kirsh D (1995) Complementary strategies: why we use our hands when we think. In: Moore JD, Lehman JF (eds) Proceedings of the seventeenth annual conference of the cognitive science society. Lawrence Erlbaum, Mahwah, pp 212–217
Kuorikoski J, Lehtinen A (2009) Incredible worlds, credible results. Erkenntnis 70:119–131
Laurent E (2003) Mental representations as simulated affordances: not intrinsic, not so much functional, but intentionally-driven. Intellectica 1–2(36–37):385–387
Lockhart P (2008) Lockhart's Lament. http://www.maa.org/devlin/devlin_03_08.html
Magnani L (2004a) Conjectures and manipulations. Computational modeling and the extra-theoretical dimension of scientific discovery. Minds Mach 14:507–537
Magnani L (2004b) Model-based and manipulative abduction in science. Found Sci 9:219–247
Magnani L (2007a) Animal abduction. From mindless organisms to artifactual mediators. In: Magnani L, Li P (eds) Model-based reasoning in science, technology, and medicine. Springer, Berlin, pp 3–37
Magnani L (2007b) Morality in a technological world. Knowledge as duty. Cambridge University Press, Cambridge
Magnani L (2009) Abductive cognition: the epistemological and eco-cognitive dimensions of hypothetical reasoning. Springer, Berlin
Magnani L (2012) Scientific models are not fictions: model-based science as epistemic warfare. In: Magnani L, Li P (eds) Philosophy and cognitive science: western and eastern studies. Springer, Heidelberg, pp 1–38
Mäki U (2009) MISSing the world. Models as isolations and credible surrogate systems. Erkenntnis 70:29–43
Millikan R (2004) On reading signs: some differences between us and the others. In: Kimbrough Oller D, Griebel U (eds) Evolution of communication systems: a comparative approach. The MIT Press, Cambridge, pp 15–29
Mizrahi M (2009) Idealizations and scientific understanding. Philos Stud 160(2):237–252

Morrison M (2009) Fictions, representations, and reality. In: Suárez M (ed) Fictions in science: philosophical essays on modeling and idealization. Routledge, London, pp 110–135

Nersessian NJ (2010) Creating scientific concepts. MIT Press, Boston

Pesic P (1990) Wrestling with proteus: Francis Bacon and the "torture" of nature. Isis 90(1):81–94

Portides DP (2007) The relation between idealization and approximation in scientific model construction. Sci Educ 16:699–724

Raftopoulos A (2001a) Is perception informationally encapsulated? The issue of theory-ladenness of perception. Cogn Sci 25:423–451

Raftopoulos A (2001b) Reentrant pathways and the theory-ladenness of perception. Philos Sci 68:S187–S189. Proceedings of PSA 2000 biennal meeting

Sage J (2004) Truth-reliability and the evolution of human cognitive faculties. Philos Stud 117:95–106

Skelhorn J, Rowland H, Ruxton G (2010) The evolution and ecology of masquerade. Biol J Linn Soc 99(1):1–8

Suárez M (2009) Scientific fictions as rules of inference. In: Suárez M (ed) Fictions in science: philosophical essays on modeling and idealization. Routledge, London, pp 158–178

Suárez M (2010) Fictions, inference, and realism. In: Woods J (ed) Fictions and models: new essays. Philosophia Verlag, Munich, pp 225–245

Sugden R (2009) Credible worlds, capacities and mechanisms. Erkenntnis 70:3–27

Thomson-Jones M (2010) Missing systems and the face value practice. Synthese 172:283–299

Toon A (2010) The ontology of theoretical modelling: models as make believe. Synthese 172:301–315

Tweney RD (2006) Discovering discovery: how Faraday found the first metallic colloid. Perspect Sci 14(1):97–121

Weir AAS, Kacelnik A (2006) A new Caledonian crow (Corvus moneduloides) creatively re-designs tools by bending or unbending aluminium strips. Anim Cogn 9:317–334

Weisberg M (2007) Three kinds of idelizations. J Philos 104(12):639–659

Woods J (ed) (2010) Fictions and models: new essays. Philosophia Verlag, Munich

Woods J, Rosales A (2010a) Unifying the fictional. In: Woods J (ed) Fictions and models: new essays. Philosophia Verlag, Munich, pp 345–388

Woods J, Rosales A (2010b) Virtuous distortion: abstraction and idealization in model-based science. In: Magnani L, Carnielli W, Pizzi C (eds) Model-based reasoning in science and technology. Springer, Heidelberg, pp 3–30

Chapter 4
Models in Action: An Eco-Cognitive Outlook on Experimental Science

Abstract Current scientific practice is often identified with the experimental framework. Yet, what "experimenting" means could be less than perfectly clear. Going beyond the common sense conception of experiment, two broad categories of experiments can be tentatively identified: the generative experiment and the demonstrative experiment. While the former aims at generating new knowledge, new corroborations of hypotheses etc., the latter–which is actually the kind of experiment most laypeople came to terms with in their lives–is designed so that, by being successful, it reverberates knowledge on the experimenters/witnesses, thus instructing them, albeit the experimental outcome was well known before- hand. Prima facie the uninformed observer may not always be able to tell whether an experiment is generative or demonstrative, therefore the existing distinction must rely on something else, namely the framework they are embedded into. The concept of epistemic warfare can be of help in investigating this distinction, also to the scope of showing that it is not a sterile dichotomy but rather a theoretically fruitful continuum, and can help the analysis of epistemically relevant issues such as the repetition/replication of experiments and their potential failure.

4.1 Introduction

The idea of experiment is intuitively connected with the common conception of modern science. Yet, until the second half of the twentieth century, philosophy of science reenacted the ancient bias against craftsmanship and focused chiefly on the theoretical aspects of scientific endeavor. Breaking this tendency, philosophical milestones such as Hacking (1983) and Gooding (1990) claimed the experimental question rightly back in the epistemological feud, but the topic was quickly seized by a different branch of studies, sometimes called social epistemology, sociology or anthropology of science, which focused more on the social dimension of experimental settings (consider Latour 1987; Knorr Cetina 1999).

The aim of this chapter is to make the best of these two approaches (the epistemological care for experimentation, and the social-anthropological outlook), in order to set the framework for an updated and consistent consideration of experiment: that

© Springer International Publishing Switzerland 2015
T. Bertolotti, *Patterns of Rationality*, Studies in Applied Philosophy, Epistemology and Rational Ethics 19, DOI 10.1007/978-3-319-17786-1_4

is, what I mean to provide is an analysis of experimentation able to comprehend both *crucial* experiments carried out in laboratories (and their relationship with theoretical issues such as modeling), and the more modest kind of experimentation we came across with, for instance, during our high school years as scientific laypeople. Experimentation is a particularly pivotal topic for the understanding of science as a whole. Science is a many-headed deity: Hacking (1983) claims in his Introduction that, albeit one decides to deal either with scientific rationality or scientific realism, one topic necessarily ends up defining the other. There are many roads leading to the same castle. This chapter aims therefore to work as a counterweight to the previous one in illuminating the eco-cognitive approach to epistemology, stressing the relevance of the environment and of the cognizant's goals when investigating the nature of scientific rationality.

4.2 Defining the Experiment

I shall begin by sketching out an extensive definition of what can be considered an experiment. In order to be fruitful, the definition of what an experiment is must be neither too broad nor too narrow: I would rather enumerate a list of features that, in my opinion, make up the experiment rather than providing a full definition.

- *Any experiment is characterized by a manipulative dimension.*

Hacking (1983) stressed the importance of *intervention*, and rooted his scientific realism not on our possibility to *know* but on our possibility to *intervene*. This intervention has to be understood in its fully dynamic display: an experiment is not the result of the experiment itself, but the whole process by which this result is achieved—or not. This can be said of thought experiments as well: as contended by Gendler (1998), a thought experiment like the ones conducted by Galileo cannot be reduced to a more or less sound logical argument, because the manipulation and hence the manipulability by an agent are the pivotal feature.[1] This is all the truer if we think of real experiments: you can *tell* somebody how an experiment was conducted (the preparation, the procedure, and the outcome), but the narrative and communicative reconstruction of the experiment, and the description of the involved procedure, are not the experiment itself. Experiments are a manifest example of *manipulative abduction*.

- *All experiments have a transformative nature.*

Experiments can be thought of as mechanical systems: to begin with, we have what is being experimented on; we then add what the experimenters bring to the experiment, that is models, heuristics, techniques, personal hunches and so on. Finally,

[1]Gendler's view on thought experiments was already advocated at the end of the previous chapter, discussing thought experiments as a kind of powerful scientific modeling. Her position on thought experiments is not universal. Some scholars contend that they can in fact be reduced to straightforward arguments (Picha 2011).

manipulation, intervention, the *work* can take place. This work has an output, of any kind. It seems to me that without this *output* there would be no experiment to talk about: the result can be new knowledge, so to say "extracted" with the experimental manipulation. Of course one cannot know that she extracted all the available knowledge from the experiment: just like when you squeeze an orange to have the juice, someone might show you there was more juice left, or that there can be a better way to squeeze it. Yet, as I will explain along the next sections, an experiment does not only generate (absolutely) new knowledge: the output of the experimental transformation might also be a new affordance (of the experimented, for instance), or new commitments (e.g., toward the advance of science). A High Energy Particle Collider experiment will try to produce new knowledge about subatomic particles, whereas a high school physics experiment might help students acquire a new and better understanding of a certain phenomenon, or a renewed commitment towards scientific progress: that, too, is an effect brought about by the kind of transformations enacted by the experiment.

- *All experiments are "situated".*

Any experiment begs for a situation: in this sense, it is a *extremely ecological* act of cognition. It could be argued that experimentation itself projects its situation: experiments are often about the controlled manipulation of a number of variables in order to see "what happens when...", or "if it is the case that x...". The laboratory is the situation *par excellence*, and the next section will focus on it, but it is not the only one: even in a non-scientific setting, but when people wish to make use of a kind of rationality that can be called scientific, the suggestion "Let's make an experiment" entails the setting of a boundary: it is pragmatic (i.e. deciding what matters and what does not matter), but also regards the assigned social roles. Depending on the interactions, the peers of the person who calls for an experiment will be involved as active participants (as in, "Let's make an experiment: what would you do if...?"), or as onlookers/witnesses, expected to provide some opinion about the conclusion and the procedure. In any case, the experiment takes place within a well defined place, which can be more or less physically determined. Hacking refers as "mature" laboratory sciences to those "in which we investigate nature by the use of apparatus in controlled environments, and are able to create phenomena that never or at best seldom occur in a pure state before people have physically excluded all 'irrelevant' factors" (Hacking 1988, p. 507). Let us therefore take a closer look at what we called the boundaries of the experiment, which must be the boundaries of the laboratory.

4.2.1 Setting the Boundaries of the Laboratory

Let us accept our loose intuition about what a laboratory is: informally, we can think that a laboratory is the specific location where scientific experiments take place. But more can be said about the lab: first of all, what are the actual boundaries of the laboratory, understood as the *lieu* of the experiment? With this respect, I invite the reader to make a small recollection and consider what the word "laboratory" makes her think of: she could think of the instrumentation used for experiments, she could think about the instrumentation strategically laid out of workbenches, and perhaps of scientists carrying out experiments on these workbenches. Plus, thinking of the labs she might have attended, she could also think of all this *and* onlookers standing by and witnessing the experiment.

It is clear that the epistemological consequences of where we set the boundaries of the lab are quite significative. In the last case, the onlookers can be students, colleagues, sponsors, visitors at a science exhibition, and so on: I do not believe that including them among the possible targets of the knowledge *transformation* enacted by the experiment necessarily means to shift the investigation from the epistemological plane to a social, anthropological one. Science as an *actual* human endeavor cannot be investigated excluding the human dimension it relies on: this is one of the tenets of the ecological-cognitive outlook. Knorr Cetina wrote that "the power of laboratories (but also their limitations) resides precisely in this 'enculturation' of natural objects. laboratory sciences subject natural conditions to a 'social overhaul' and derive epistemic effects from the new situation" (Knorr Cetina 1999, p. 28).

I believe that the soundness of Knorr Cetina's statement does not exclusively follow from the adoption of an ethnographic outlook on science.[2] Conversely, it is easy to understand that the *enculturation* and the *social overhaul* are mutually implicating in our conception of science, because laboratories are more than a set of instruments, and even more than a set of affordances displayed by those instruments: labs prevent scientists from having to study a natural object "*as it is, [...] where it is [and] when it happens*" (Knorr Cetina 1999, p. 27). In this sense, labs allow the manipulation of the object far better than the natural context would (if it would at all). Even scientific models, as we saw in the previous chapter, play a pivotal role in the economy of the lab and partake of a similar nature, being on the one hand highly manipulative human construals (therefore cultural too), while on the other they are necessarily bound to the natural object: in fact models are fundamental in binding the modeled natural object into a specific phenomenon.[3]

The laboratory includes the experimenters as well, inasmuch as they are not separated from what is being experimented. Albeit Hacking is thankful toward the "large number of studies by philosophers, historians, and ethnographers of experimental science," (Hacking 1988, p. 508) he seems less eager to concede a more significative role to human intervention per se, which is conversely mentioned by Heelan by means of the "instruments, standard procedures, experimental skills, laboratory traditions, and the social context of the research community" (Galison 1988, p. 525). Scientists are not simply interchangeable operators: two teams working on the same raw objects would not apply the same methodologies or necessarily obtain the same

[2]Also Nersessian's outlook on science is often characterized by a particular attention—called "ethnographic"—to the actual dynamics at play in a laboratory (cf. for instance Nersessian 1992, 1995, Nersessian and Patton 2009).

[3]Hacking (1983) contends as well that many phenomena come to happen uniquely as they are *created* in laboratories.

results. Indeed, "not only objects but also scientists are malleable with respect to a spectrum of behavioral possibilities. In the laboratory, scientists are methods of inquiry; they are a part of a field's research strategy and a technical device in the production of knowledge" (Knorr Cetina 1999, p. 29). It seems to me that what could be at stake here is not the dispossession of epistemology by social approaches, but rather the opposite, that is the epistemological flooding of some aspects of scientific endeavor which—by default of better option—have so far been labeled as social but do rather concern an agent-based and factual approach to science, also leaning on an *actually happens rule*.[4]

Why then should we set the limits of the laboratory at the experimenters' level, and not admit the onlookers as well? Why should the "social overhaul" advocated by Knorr Cetina only involve the experimenters? As I will show in the following section, some experiments do not benefit the experimenters at all, in a strict sense, inasmuch as they produce knowledge that had already been acquired, and yet the same experiments cause indisputable epistemic effects on those who *observe* it. For them the experiment still produced a valuable transformation of knowledge, by which they gained a new *understanding* (1) of the phenomenon that was *explained* to them through the experiment, but also (2) of how scientific rationality works. It could be contended, in fact, that the scope of some experiments carried out at schools or science exhibits serve the chief purpose of exemplifying some tenets of scientific method.[5]

We might then conclude that there are indeed many kinds of laboratory, in which different kinds of experiments take place: there are labs for basic research, industrial labs, labs for medical research, and then there are laboratories in schools, science museums, and also the laboratories shown on educational TV programs: most of laypeople are acquainted with the latter kind of laboratory, that is the physics/chemistry/biology lab at high school, or those they see in science museums or on the Discovery Channel. Such acquaintance fuels our *thinking to know* what every lab should be like, which is in fact a *hasty generalization*.

---

[4]This rule was introduced by Gabbay and Woods as a tenet of their new approach to logic, referring to the fact that logic should model how real agents think: one should try to correct the model so it fits the facts, and not try to amend or obliterate facts to make them fit the model (Gabbay and Woods 2001; Woods 2005). In this context, I use it to suggest that philosophy of science should match what science really is, and not arbitrarily cut out aspects of the problem by labeling them as external to the analysis (for instance, "social").

[5]The onlooker's gain of a renewed commitment towards science, be it specific for a particular research/discipline or to scientific endeavor in general, is just as vital for the development of science as the generation of new knowledge through experimentation. Contemporary *knowledge societies* massively rely on the development of science, which in turn relies on the will of citizen to care and spend for it (Magnani 2007): funds are just as vital as genius and intelligence for the survival of science. This view is coherent with Magnani's conception of science as an *epistemic warfare* briefly introduced in the previous chapter, which also includes non-epistemic strategies that are nevertheless crucial for science, such as those for the dissemination of knowledge, the acquisition of funding and so on.

In his classic book *Science in Action*, Latour enacts his anthropological approach to science narrating the epistemological adventure of an anthropologist taking a full immersion in the scientific endeavor. Interestingly, he makes the narrator say: "We came to the laboratory in order to settle our doubts about the paper, but we have been led into a labyrinth" (Latour 1987, p. 67). Specifically, the doubts referred to a reading of endorphin levels, which had to be interpreted through graphs and indicators, yet this bewilderment is common to many onlookers approaching a scientific setting: *we came, we saw, and yet we have not understood anything*. And yet, we saw experiments, at school, at the science museum, on TV, how comes?

In sum, experiments take place in laboratories, and laboratories may include onlookers. Yet not all experiments are geared towards onlookers the same way: to certain experiments anyone can be an onlooker and benefit of the epistemic effects, to others the onlooker is defined by very specific characteristics. As I will argue, this depends mostly on the kind of experiment at stake. If different kinds of the experiment exist, it is legitimate to wonder how many kinds there are, and how we can tell them apart.

4.3 How Many Kinds of Experiment Are There?

A kind of *taxonomy* of experiments is not unusual among philosophers of science, and such differentiations sometimes merge into other connected ones. Gooding, for instance, links the concept of experiment to its *reconstruction*, obtaining six different kinds of reconstruction to be employed in different narratives: namely *cognitive, demonstrative, methodological, rhetorical, didactic, and philosophical* (Gooding 1990, p. 7), each with their peculiar scope. Notwithstanding the utility and soundness of this differentiation, I contend that its root lies at a lower level, and actually underdetermines it. The whole spectrum of experimental activity, as far as natural and model-based sciences are concerned, could in fact be reduced to two major forms of experimentations. One of the advantages of this proposal, could be dubbed a "plea for epistemological austerity", is that every distinction causes some unhappy left-outs: Steinle (1996), for instance, lamented that the "standard view" in the Nineties of the past century would disregard as epistemologically irrelevant those experiments that were aimed at discovery—and not at the test of a clear hypothesis, or at the retrieval of a particular measurement. Such conception would in fact leave out a number of fundamental instances in the history of science: grouping the experiments into two sets, namely "generative" and "demonstrative" experiments could instead cause a lesser number of *homeless* instances.

4.3.1 The Generative Experiment

I could begin by suggesting that what I call the "generative" experiment is the kind of experiment that common sense has acquired, but this would be misleading. I contend, indeed, that the common-sense conception of experiment is somewhat blurred, so that the generative experiment, which is what we *should* think of when we think of a scientific experiment, does not coincide totally with our intuitive conception of experiment.

The *generative* experiment is the experiment whose outcome is *not known* beforehand, and its aim is to manipulate and transform the *experimentandum* (what is being experimented on) into knowledge that is new for everyone. To put it another way, it is the kind of experiment where the *cognitive* target,[6] that is what the experimenters want to obtain, is intrinsic to the experiment (this latter claim might seem a truism, but the next section should prove the opposite).

Most experiments in the history of science can be thought of as *generative* experiments. It is the kind of experiment where you *test* something (a hypothesis, a theory), and is usually comprised within a theoretical framework.

It is also true that "one can conduct an experiment simply out of curiosity to see what will happen" (Hacking 1983, p. 154): not only experiments that are well nested within a particular theoretical frameworks, for instance those aimed at testing a particular hypothesis, or at finding out a particular measurement (think of Millikan's experiment, projected to measure the elementary electric charge), but also entirely "exploratory" experiments are generative. According to Steinle, explanatory experiments do not rely "specific and well-defined procedure, but [include] a whole bundle of different experimental strategies", and their "central epistemic goal is the search for general empirical rules and for appropriate representations by means of which they can be formulated" (Steinle 1996, p. S73).[7]

But also in our everyday life, when we make use of scientific-experimental rationality to put some makeshift model to the test, we recur to generative experiments to gain some new knowledge. I can send myself an email to see if my IMAP server is really experiencing issues, and I can ask a friend to email me as well. I can put a five-dollar bill in a vending machine to test it before butting a twenty-dollar bill, to test if the machinery works properly. Generative experiments are often conducted as parts of model-based activities: I can ask a relative to simulate a social situation to gain better knowledge about some possible consequences of an action of mine, or a man might cast small objects off a table to assess the likelihood of himself surviving

[6] I specify *cognitive* target, as the scope of the experiment, to differentiate it from Hacking's use of the word *target*, by which he refers to a part of the *"materiel"* of the experiment (cf. Hacking 1988, p. 509).

[7] Steinle's aim in describing exploratory experimentation is to allow the appreciation of the epistemological importance of this kind of experiment, while the "standard view" tended to disregard them as part of epistemically irrelevant *discovery* processes. Exploratory experiments are particularly relevant for entering new fields requiring new concepts and new general facts (Steinle 1996). The explanatory experimentation can also be extremely tacit, and consist chiefly of "thinking through doing" (Magnani 2002).

after jumping from a cliff with his car. In those cases, what I gain from the manipulative intervention of the target (that is from the experimentation) is some knowledge I did not possess before.

In a nutshell, the focus in generative experiments should be put on their ability to intrinsically produce new knowledge. This is the kind of experimentation that engages theory (and theories): as suggested by Steinle (1996) and Hacking (1988) among many others, some experiments—which I label as generative—can precede theory inasmuch as they can illuminate new fields of scientific research and provide it with new concepts.[8]

With respect to this kind of experiment, even scientific common sense knows that theories should behave according to the already mentioned *actually happens rule*: experimental observations affect theories. Experiments are where theories can be falsified (Popper 1959), and experiments that do not go as expected can affect the scientific paradigm, taking it to an eventual crisis (Kuhn 1962), or causing scientist to fix the protective belt of the program to keep it progressive (Lakatos 1976). In the next subsections I will show how only the kind of experiment I just sketched out indeed affects theories: I will analyze a wide and yet peculiar class of experiments, that—even though they can be called experiments to their full right—are not expected to affect theories at all.

4.3.2 The Demonstrative Experiment

It is now time to deal with an apparent contradiction: we know that experiments are, so to say, the field artillery of scientific progress, and it is on experimental grounds that new knowledge is either discovered or validated. On the other hand, we also know very well that most experiments we—as laypeople—witnessed (even in decent laboratories) did not add anything to scientific knowledge. It would not be right to arbitrarily exclude them from the category of experiments, because they display all the traits I pointed out in Sect. 4.2, and they also fit with the more demanding description proposed by Hacking (1988).

I am referring to most experiments carried out in schools, exhibitions, museums, and so on. For instance, they can be experiments aimed at demonstrating or *illustrating* a law or a theory, fostering a better understanding. With this respect, at least

[8]Hacking suggests several examples from the actual history of natural science that refute Popper's claims according to which "theory dominates the experimental work from its initial planning up to the finishing touches in the laboratory" (Hacking 1983, p. 155). The debate on the theory-ladenness of experimental facts is often brought to quasi-metaphysical issues: one way to tackle it is to appeal to the intuitive notion of theory (as folk theory). Experiments may precede particular theories, and yet rely on past sub-theories about substances, agency, causation etc. Thus, to say that an experiment precedes theory—and so does the experimental observation that follows such experiment—does not indeed equal saying that the experiment generates new coherent knowledge *ex nihilo*. After all, we could claim that intuitive, hard-wired theory precedes even out every-day observation, even at the lowest levels of the perception of images, sounds etc. (Raftopoulos 2001).

in the Italian school system, theory overwhelmingly precedes experimentation: in chemistry or physics courses, experiments are not even used to stimulate theorization upon the students' minds, but rather as a persuasive proof to show that what was explained in theory *is indeed the case*.[9]

This kind of experiment could be thought of as *deduced* from theory in a strong sense, opposed to the *weak* Popperian sense of experiments *informed* by theory: I mean that the procedure of the experiment is vouched by the theory it means to put in display. Is it a *paetitio principii*? Not really. Consider this example:

1. *Experiment E* (for instance Maxwell's or Faraday's experiments on electromagnetism) is crucial for the establishment of a *Theory T*;
2. *Theory T* is established;
3. *Experiment E** is used at school to prove the adequacy[10] of *Theory T*.

*Experiment E** is a (usually easier) version of *E*, updated according to the theory it means to demonstrate. If its real aim was to *test* the theory, then of course it would be begging the question. But who would expect high-school students to be *actually testing* a theory? Everybody knows that high-school level optics, or electromagnetic physics and so on *do work*. Proving it *n* more times every day, in *n* school laboratories, does not add one bit to the robustness of those theories. *Experiment E** aims at providing students with an actual proof that what they studied (or they are going to study) is really so.

Even if you think about experiments that do not aim at demonstrating a law, but rather at isolating a phenomenon so that it can be shown for some theoretical scope, the defining element is that the experimental *outcome is known beforehand*.[11] Contrarily to the *generative* experiment, in this case the epistemic goal is extrinsic to the experiment itself: it means little to say that the experiment *in se* was successful, because it was planned to be successful. The experiment is successful in its *actual* scope if it operates any *change* within the epistemic configuration of the observer, after she witnessed the positive (staged) outcome of the experiment. That is to say, the experiment is successful if it triggered a new awareness in the observer, for instance a student might be further persuaded about the empirical adequacy of a theory, or a citizen might reconsider the importance of electing a prime minister advocating more funding for scientific research. Or, simply, their aim could be to convey indeed a bit of *local* knowledge about some phenomena but, on the overall, to infuse the belief that science is "interesting", or just "cool."[12]

---

[9]This concept is well exemplified by a sign hung in my chemistry laboratory at high school, which would read something along the lines of "If I listen I forget; if I see I remember; if I do I understand". The experimental dimension is taught as completely subsidiary to abstract theory.

[10]Please understand this word in an intuitive sense, as in "what they taught me about the *Theory T* does indeed happen in real life," and not as laden with implications about the epistemological debate about the truthfulness or acceptability of a scientific theory.

[11]This claim clearly begs for some considerations about the *failure* of an experiment: I will address this issue in Sect. 4.4.2.

[12]This seems to be the case as for the popular blog *I fucking love science*.

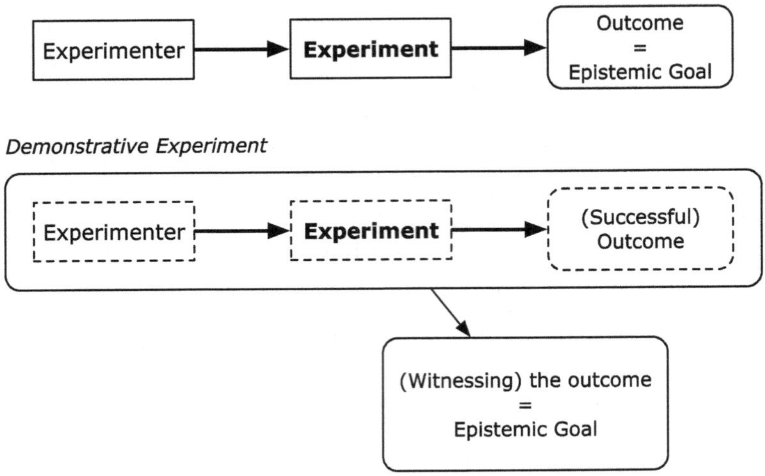

Fig. 4.1 Generative and demonstrative experiments

This class of experiment could be defined as *demonstrative* or *explanatory*, contrasting it with *generative* experiment (Fig. 4.1). Interestingly, one could say that in their scope of disseminating scientific knowledge (for various purposes), demonstrative experiments have become more and more widespread together with the growing impact of science on society. Living science shows in the eighteenth and nineteenth century, analyzed by Raichvarg (2006), provide a clear example of a *demonstrative* experimental framework, which could be seen as the ancestor of modern science exhibitions or scientific shows for general audience on TV. One of the scientists/showmen mentioned by Raichvarg would start his experiments with the following call:

> And if I am here among you, it is because all of you must draw from my demonstrations, the true and natural principles of the forces which are above us, these forces which frighten the ignorant but supply the educated with all the moral pleasures of intelligence (Raichvarg 2006, p. 3).

Raichvarg draws from his analysis a list of characteristics that were typically common to science shows, and still apply to scientific dissemination aimed at general public:

- They reach a wide audience, an audience which could be defined as a public with no scientific training... They come to the fairground for anything but science, but then they meet science face to face...
- The importance of the current events of science, mostly because on a fairground one must astonish everybody to attract everyone and get your pennies back...
- A continuous desire for good pedagogy, together with a continuous desire for wonder, if not for the supernatural! (Raichvarg 2006, p. 4)

These *experimental shows*[13] did not contribute to form scientists, just as contemporary science classes at high school do not mean to train scientists, and science museums do not either (Macdoland and Basu 2007). On the other hand, these forms of dissemination do play a pivotal role in educating people that might undertake or value a scientific career. A living science show, just as a school experiment, may indeed induce in the observer a taste for scientific methodology, or just make her aware of its existence.

A final question concerning the demonstrative experiment might arise: since we are accustomed with experiments carried out at school, often with obsolescent equipment, it seems that the difference between a generative and a demonstrative experiment should be most easily noticed. In my opinion, from a phenomenological point of view, it is not so. We should not be fooled by the *time lag*: if most of our school labs look like museums of past century science it is just because those instruments were once upon a time the cutting edge of generative experimentation (think of Volta's battery and most electromagnetism-relate devices). If we removed this time lag, which is merely contingent, we would be unable to tell one kind of experiment from another, if not by considering the cultural and social framework an experiment is nested in.

Let us make a quick thought experiment (demonstrative in this case, *sic!*): imagine in the near future a highly-funded high school in some advanced country, whose politicians place a great emphasis on education. Just as our high schools have a physics lab, that high school as a High Energy Particle Collider in its basement, and teachers use it to instruct pupils about quantum physics. If the same-old-friendly alien landed on Earth and could not understand human language, and witnessed the experiments carried out in that school, and those carried out at CERN (for instance), it could not be able to tell any difference: what goes on, apparently, is the same. Yet, once our alien managed to set its intergalactic translator to understand our language, it would see at once the difference, since the HEPC at school would be embedded in a pedagogical framework of demonstrative experiments, whose outcomes are already known by the teacher who can therefore lead the pupils along the right path. Time lag, and thus the obsolescence of experimental materials, are not a necessary criterion to tell a generative experiment from a demonstrative one, since *prima facie* they cannot be told one from another, unless considering—as I said—the setting they partake of.

[13]The expression is a bit of an oxymoron, but it means to stress the *staged* dimension of many demonstrative experiments. Concerns about the esthetic dimension of their replication will be addressed in Sect. 4.4.1.

4.4 Consequences of the Distinction

Now that the distinction between generative and demonstrative experiment is in place, I will use it to tackle two epistemological problems, namely the repetition/replication of experiments and their failure. Once again, I will try to match common-sense expectations with the actual scientific practice, past and current.

4.4.1 Differentiating Repetition and Replication of Experiments

In an interesting paper about the conception of experiment repetition in the past centuries, Schickore (2011) sets out stating that "[t]oday it is generally assumed that isolated experimental outcomes1–'one-offs'—are insignificant. Twentieth-century philosophers of science, most notably Popper, made the reproducibility of experimental results the basic methodological requirement for successful experimentation: if an experiment cannot be re-done, it is invalid" (pp. 327–328). Indeed, the possibility of re-doing an experiment became one of the first tenets of contemporary scientific rationality.

Before applying my distinction (between kinds of experiments) to the problem of redoing experiments a brief semantic interlude might be required, which—I think—might let the reader foresee my claim before I make it clear. It is sometimes said that experiments are "repeated", while sometimes they are "replicated". I believe that the two terms can be sensibly separated, each with its own proper meaning.

- **Repeating** an experiment exemplifies the epistemological tenet towards the re-doing of experiments. You *repeat* an experiment when you put the known outcome between brackets and proceed entirely as if it was unknown. The focus of repetition is on *what* outcome will be obtained, and *whether*—changing certain factors—the same outcome will be obtained again.[14]
- **Replicating** an experiment focuses on the replication of the procedure and not only on obtaining the same outcome. You *replicate* an experiment without necessarily putting the outcome between brackets, because what matters is observing *why* a particular procedure yielded such an outcome. Once the reason is found out, it is possible to replicate the experiment with the pragmatic certainty that if the outcome differs from what expected, then a mistake was committed in the procedure.

[14]To make students assimilate this concept, physics teachers often deploy plethoric lists of settings (e.g. here, at the Equator, on mount Everest, on the Moon, on Mars, in a billion years, and so on) where a law (such as "All metal bars expand when heated") must apply for it to be universal. The different settings correspond to a series of real or potential repetitions of one or more experiment concerning the law in question.

I suppose that such semantic characterization foreshadowed quite clearly the rest of my argument. As for generative experiments, I think that *repetition* is the case. Repetition engages the intrinsic epistemic goal of the experiment. Repeating the experiment does not mean necessarily to redo the same experiment over and over. This is what happens every day in schools worldwide, and we know that it has little epistemic value for the progress of science.

> A number of scholars have stressed that scientists rarely try to copy the exact same experiment. Rather, experimenters seek to obtain similar results in different experimental settings, and experimental results are considered valid if multiple determinations of the evidence are possible. The crucial notion here is reproduction by *doing something different* (Schickore 2011, p. 328).

The repetition of an experiment in a generative epistemic context is valuable because it may challenge the previous outcomes of the same experiment, for instance it can interfere with claims of universality (by "doing something different"). Repetition has therefore chiefly epistemic concerns. Repetition can indeed be about the *same* experiment, but in this case it is about looking for *freak factors* of the experimental procedure, and make sure that the result is accurate. Even if scientific truths are notably *provisionally* true, the search for freak factors will end at some point. No branch of science still heats metal bars every day on normal conditions to see if they expand and by what coefficient.[15]

Whereas such use of repetition was already in vogue in Early Modern science, its role was chiefly to corroborate (and make appear *as reliable*) one's own experimental results by the method of the *slight modifications*: Schickore, building his case study on an Italian eighteenth-century microscopist and physiologist, states that "Fontana's methodological thought is particularly interesting because he stressed the importance of *repetition* of his own experiments. The text is packed with claims that experimental trials were repeated 'a hundred times' or even 'a thousand times,' and that thousands of animals were used. Also, the experiments were varied 'in a thousand ways'" (Schickore 2011, p. 328). Only subsequently the stress was placed on the assessment (via repetition) of experimental results obtained by other scientists.[16]

If the repetition is meant to engage the outcome of some other scientist's experiment, then again it can partake of a generative nature. Assessing someone else's experiments is, as a matter of fact, one of the pillars of contemporary scientific practice: a purpose of publishing experimental procedures in peer-reviewed journals is

[15]"Scientists do not repeat the same experiment *ad nauseam*. They perform an experiment a 'sufficient' number of times (whatever that might be), and then perform it no more. The experiment becomes a part of history, to be performed again, if at all, only by science students as an exercise" (Musgrave 1975, p. 248).

[16]This conception was rather absent in early modernity: "Recent methodological frameworks highlight robustness, the importance of multiple determinations of experimental outcomes through a variety of independent procedures. While some parts of Fontana's project could perhaps be reconstructed in hindsight as multiple determinations of experimental results, neither he not Redi [a physician and naturalist at the court of the Grand Duke of Tuscany] explicitly called for independent determinations by different means to make an experimental result more reliable" (Schickore 2011, p. 344).

to offer the experiment to the assessment of fellow scientists (rivals as well), so that other scientists can repeat it and see *if* they obtain the same results. With this respect, an experiment is scientific if it is available for repetition, so that somebody else can repeat it and—perhaps—falsify its previous outcome: it is not necessary, for an experiment to be deemed *scientific*, to obtain necessarily the same outcome upon every different repetition.[17] Also thought experiments, inasmuch as their repetition does not lead *necessarily* to one indisputable result, can be seen as generative in their repetition (Bishop 1999).

As far as demonstrative, or explanatory experiments are concerned, it follows from the initial argument that we should be mostly dealing with *replication*, for a number of reasons. First of all, whereas the redoing of generative experiments has epistemic concerns (since the previous outcome is what has to be challenged), the redoing of demonstrative experiments must face different constrains: indeed, their outcome is already known and their scope is to disseminate knowledge for the benefit of the observers, therefore their peculiar constraints are chiefly *esthetically oriented*.

I do not mean this in a strong sense, *à la* Feyerabend: it is not that experiments carried out in contexts of dissemination are a work of rhetorics. My contention is that the will to reproduce a successful experiment may focus the attention on the reproduction of the *same* procedure, which therefore acquires a ritualized dimension that laminates the epistemic concern. As a matter of fact, being certain about the outcome (be it an experimental result or an experimentally-confirmed theory) causes a shift in the perspective: the objective is not to redo the experiment to see what happens anymore, but to replicate it in the most convincing and understandable way. This can also be said of actual scientific experiments: sometimes, in the reconstruction of a discovery, when things seem to go too smoothly, it may be the case that a more pleasing *demonstrative* experiment was smuggled in place of the original *generative* one.

Interestingly, Hacking reports an annotation of Maxwell's about the work of Ampère which sums up quite neatly the essence of the replicated demonstrative experiment:

> We can scarcely believe that Ampère really discovered the law of action by means of the experiments which he describes. We are led to suspect what, indeed, he tells us himself: that he discovered the law by some process he has not shewn us, and *when he had afterwards built up a perfect demonstration he removed all traces of the scaffolding by which he had raised it* (Hacking 1983, p. 162), added italics.

This methodological reconstruction is akin to the one I put forward in Sect. 4.3.2, by which the demonstrative experiment is somehow deduced from a theory already confirmed as adequate. Consequently, this kind of experimentation (already drawn

[17]Of course, in the latter case, something must be wrong either in one of the procedures, or in the theorization on which the experiment relied. About this issue, see the following Sect. 4.4.2.

from a successful experimental confirmation) is ready for replication without excessive worries about the outcome, but rather about its development: if the experiment is carried out correctly, it will be successful and prove our initial hypothesis.[18]

4.4.2 The Meaning of "Failure"

Repetition and failure are strictly interconnected. As I previously suggested, repetitions of *generative* experiments are aimed at testing the outcome of the experiment (and so at testing the hypothesis, theory or measurement that had been carried out during the experiment):

> Our ability to recognize when data fail to match anticipations is what affords us the opportunity to systematically improve our orientation in direct response to such disharmony. Failing to falsify hypotheses, while rarely allowing their acceptance as true, warrants the exclusion of various discrepancies, errors, or rivals, provided the test had a high probability of uncovering such flaws, if they were present. In those cases, we may infer that the discrepancies, rivals, or errors are ruled out with severity (Mayo and Spanos 2010, p. 18).

If "[a] test 'uncovers' or signals the falsity of *H* by producing outcomes that are discordant with or that fail to 'fit' what is expected were *H* correct" (Mayo 2010, p. 352) then it sparks a procedural loop involving a careful check of the experimental conditions (looking for freak factors), a revision of the hypothesis or ultimately a revision of the model (as described in the previous chapter). Therefore, in case of experimental failure, the existing tension between the experimen*ter* (and her background knowledge) and the experiment*ed* is resolved in favor of the latter, and thus the dignity of the falsifying failure is *respected*. Failure becomes yet another manipulative factor at play in a subsequent experiment. Failures are able to climb back over the experimental framework and crawl inside of general theories from one minimal experimental discrepancy.

> When we falsify a prediction, however "local" it is, we falsify whatever entails that prediction, however general or large-scale. There is, in this respect, no localization of the refuting process. The fact that we may try to find out which part of the refuted whole is to blame is another question the Duhem question (Musgrave 2010, p. 105).

---

[18]Furthermore, Schickore seems to connect the early-modern care for repetition *in se* with a chiefly demonstrative dimension: "References to multiple repetitions have been interpreted as an echo of an Aristotelian conception of experience; as a literary device to bolster an experimental report; as a literary tool to highlight the wealth of the experimenters' patrons; or as an expression of a general commitment to experience that marked the beginning of modern experimental science" (Schickore 2011, p. 329). Such an understanding of repetition clearly embeds it in a demonstrative framework akin to the non-epistemic strategies advocated by Magnani's *epistemic warfare* (see footnote 5). Schickore also hints at how repetition, in Galileo, served as a conceptual wrapper to *run* experimental observations as general facts: "Claiming results that accrued from trials repeated 'a full hundred times' was a way of saying 'things *always* behave this way,' and hoping that the reader would believe it" (Dear 2001, p. 134).

Think of how the inaccurate predictions fostered by Newtonian mechanics about the orbit of Uranus jeopardized the adequacy of Newton's theory in toto: this failure was accepted by Le Verrier, and transformed into new knowledge that managed not only to preserve the adequacy of the theory but also discover a new planet, Neptune.

I suggest, though, that in particular (yet scientific) settings, namely in demonstrative experimental frameworks, Musgrave's claim is wishful or at best it is the object of a mere lip-service. That is, sometimes a "local" falsification does not affect what entailed the falsified prediction at all. Experiments carried out in schools, for instance, can "not work out" for a number of reasons, in a more or less meaningful way (the phenomenon may not occur, or some measurements might be different). What happens in this cases? Nothing at all.

When a demonstrative experiment fails, the general/expected outcome of the experiment is not questioned. This peculiar "experiment *token*" may have failed, but not the "experiment *type*" it stood for (Musgrave 1975, p. 252). Failure is made into something relative to this peculiar occurrence: it is a matter of *here an now*—this particular experiment failed, but by no means it falsified the theory it was meant to prove. This can be supported by a dialectical interplay with the observers, aimed at illuminating and then filling ignorance bubbles with demonstrative *emergency knowledge*: this process is usually introduced by rhetorical questions along the lines of "Okay, you know why the experiment didn't work out?" followed by information—often in-between *ad verecundiam* arguments and plain magical thinking—about the involved instrumentation, secondary phenomena affecting the materiel involved and so on. I label this filling as *magical* because the leading experimenter is saving the expectations of the others by strategically deploying information that was only in her background knowledge: sub-experimenters (for instance pupils, or laypeople visiting a science exhibition) lack the necessary background to make sense of this information, which is therefore offered as self-justifying, or rather justified by the authority of the leading experimenter. There is a significant *appeal to authority* at work in the dissemination of scientific knowledge, even if the latter is presented as immune to authority constraints. Furthermore, it could be said that this *authority overhaul* is *necessary* if only to convey and evoke commitment towards scientific method and its *unconstrained nature*.

One last epistemological effect of this mechanism is worth noting: as already seen in the previous chapter, by constructing his argument against the fictionality of models, Magnani (2012) contrasts a static understanding of science—for instance the one conveyed by textbooks—with the actual understanding of the dynamic nature of scientific endeavor, and states that if they are seen statically then of course models appear as fiction. The demonstrative experimental framework I described raises the stakes. Demonstrative experiments seem to entail the kind of *fictionalism* that sees models as fictions depicting missing systems (Mäki 2009; Thomson-Jones 2010). Why? Consider failure in a demonstrative experiment: the unexpected wrong outcome is injected with emergency knowledge ("I am telling you why the experiment did not work out"), and so the model indeed appears as an awkward fiction (the phenomenon that the model should actualize does not happen). Furthermore, a demonstrative failure turns the observed reality into a fiction as well (a missing system, "which you should have seen in the experiment but you didn't...”), in order to support the cost of the what was to be demonstrated (be it a model, a law, etc.). In case of failure, the tension is resolved in favor of the experimenter and her background knowledge.

What is the final result? Once the observer is faced with a model which underwent a neglected experimental failure (that is, solved through authority-based emergency procedures), she will understand that “there are no actual, concrete systems in the world around us fitting the description it contains” (Thomson-Jones 2010, p. 283). The experimental learning achieves the result of teaching scientific theories as something *necessarily* abstract and incoherent with everyday perceived reality: such a configuration of the experiment awkwardly clashes with Hacking's breakthrough intuition, according to which experiments (and the models they embed) *create* phenomena that might very well not give themselves in everyday reality (Hacking 1983). The constructed/modeled nature of phenomena is a consequence of the experimental framework, and not something that the experiment must cope with as the byproduct of the clash between theory and actual reality.

4.5 Conclusion

The aim of this chapter was to apply the ecological-cognitive outlook onto experimentation, and thus to provide a sensible analysis of the veritable experimental framework in science. As noted in footnotes 5 and 18, this study is coherent with Magnani's conception of “epistemic warfare”, which I introduced in the previous chapter and sees

> [...] scientific enterprise as a complicated struggle for rational knowledge in which it is crucial to distinguish epistemic (for example models) from non epistemic (for example fictions, falsities, propaganda, etc.) weapons. I certainly consider scientific enterprise a complicated epistemic warfare, so that we could plausibly expect to find fictions in this struggle for rational knowledge. Are not fictions typical of any struggle which characterizes the conflict of human coalitions of any kind? During the Seventies of the last century (Feyerabend 1975) clearly stressed how, despite their eventual success, the scientist's claims are often far from being evenly proved, and accompanied by “propaganda [and] psychological tricks in addition to whatever intellectual reasons he has to offer” (Feyerabend 1975, p. 65), like in the case of Galileo. These tricks are very useful and efficient, but one count is the *epistemic* role of reasons scientist takes advantage of, for example scientific models, which directly govern the path to provide a new intelligibility of the target systems at hand, another count is the *extra-epistemic* role of propaganda and rhetoric, which only plays a mere ancillary role in the epistemic warfare. So to say, these last aspects support scientific reasoning providing non-epistemic weapons able for example to persuade others scientists belonging to a rival coalition or to build and strengthen the coalition in question, which supports a specific research program, for example to get funds (Magnani 2012, p. 3).

Magnani's concept was devised arguing about the use and nature of *models* in science, but it can be applied fruitfully to the understanding of other aspects of scientific endeavor. Thinking of generative and demonstrative experiments, it can be said that the former reflect epistemic weaponry, while the latter partake of a non-epistemic nature. Nevertheless, both kinds of experiment are crucial and unremovable for a correct functioning of science: while generative experiment engage the natural framework, and are thus the first-line of scientific and technological progress, demonstrative experiments engage the human framework. Science is a human activity, therefore a fittingly shaped human framework (eager to invest funds, commitments, priorities etc.) is just as essential as the correct exercise of method and rationality.

The distinction I proposed should not be considered a dichotomy, but rather consists in the two poles of a continuum specter covering the experimental dimension. Even if it is possible to find some experiments (like Newtonian mechanics) that are carried out only in patently demonstrative settings, there is not a fixed number of repetitions after which an experiment switches from being generative to demonstrative: Popper had already faced this problem, when dealing with the *diminishing returns* from repeated experiments (Musgrave 1975).[19] On the other hand, the distinction between the two kinds of experiment is sometimes blurred in the actual scientific practice (not in the dissemination to a lay public): as shown by Ampère's example in Sect. 4.4.1 (and other ones in Hacking (1983)), what I called generative experiment has often had a scaffolding role, and once its outcome is assessed, the scaffolding is replaced by a more straightforward and *nicer* experiment informed by the already confirmed theory. Lastly, demonstrative experiments have a minor (if only nominally) role to play as *watchdogs* of the adequacy of well-assessed theories. Said in Lakatosian terms, they provide a further layer to the protective belt of a research program: by repeating *ad nauseam* experiments about basic chemical reaction, light properties, metal bars that expand when heated and so on, we keep assessing the adequacy of fundamental scientific predictions.

It should be noted that even to consider the distinction as two poles of a continuum is slightly problematic because of some anomalies posed by contemporary sciences: in robotics, computer sciences or for instance genetics and cognitive science most experiments can be generative and demonstrative at the same time. A robot, for instance, is at once the product of the manipulative transformation generating new knowledge, and the mediator of dissemination of that same knowledge. This aspect is worth further studying, as is the relationship between the distinction I advocated and thought experiments: thought experiments can be seen at the same time as both generative and demonstrative experiments, depending on the conception of thought experiment rooted in one's background (Gendler 1998; Bishop 1999). If one considers thought experiments as reducible to arguments, then she might think of them as *demonstrative*; conversely if thought experiments are seen as rightful experiments, then no matter how many times a thought experiment is repeated, it could remain *perennially generative*.

Acknowledgments I would like to thank the participants of the MBR'012_Italy conference for the valuable comments on my presentation: many of them contributed to the final form of this chapter. In particular, I would like to thank Viola Schiaffonati for the insight about the anomaly of robotics with respect to my distinction.

[19]See also footnote 15.

References

Bishop M (1999) Why thought experiments are not arguments. Philos Sci 66(4):534–541
Dear P (2001) Revolutionizing the sciences: european knowledge and its ambitions, 1500–1700. Princeton University Press, Princeton
Feyerabend P (1975) Against method. Verso, London
Gabbay DM, Woods J (2001) The new logic. Log J IGPL 9(2):141–174
Galison P (1988) Philosophy in the laboratory. J Philos 85(10):525–527
Gendler TS (1998) Galileo and the indispensability of scientific thought experiment. Br J Philos Sci 49(9):397–424
Gooding D (1990) Experiment and the making of meaning. Kluwer, Dordrecht
Hacking I (1983) Representing and intervening: introductory topics in the philosophy of natural science. Cambridge University Press, Cambridge
Hacking I (1988) On the stability of the laboratory sciences. J Philos 85(10):507–514
Knorr Cetina K (1999) Epistemic cultures. How sciences make knowledge. Harvard University Press, Cambridge
Kuhn TS (1962) The structure of scientific revolutions. University of Chicago Press, Chicago second expanded edition
Lakatos I (1976) Proofs and refutations: the logic of mathematical discovery. Cambridge University Press, Cambridge
Latour J (1987) Science in action: how to follow scientists and engineers through society. Harvard University Press, Cambridge
Macdoland S, Basu P (eds) (2007) Exhibition experiments. Blackwell, Malden
Magnani L (2002) Thinking through doing, external representations in abductive reasoning. In: AISB 2002 symposium on AI and creativity in arts and science. Imperial College, London
Magnani L (2007) Morality in a technological world: knowledge as duty. Cambridge University Press, Cambridge
Magnani L (2012) Scientific models are not fictions: model-based science as epistemic warfare. In: Magnani L, Li P (eds) Philosophy and cognitive science: western & eastern studies. Springer, Berlin, pp 1–38
Mäki U (2009) MISSing the world. Models as isolations and credible surrogate systems. Erkenntnis 70:29–43
Mayo D (2010) Explanation and testing exchanges with Clark Glymour. In: Mayo D, Spanos A (eds) Error and inference: recent exchanges on experimental reasoning, reliability, and the objectivity and rationality of science. Cambridge University Press, Cambridge, pp 351–363
Mayo D, Spanos A (2010) Introduction and background. In: Mayo D, Spanos A (eds) Error and inference: recent exchanges on experimental reasoning, reliability, and the objectivity and rationality of science. Cambridge University Press, Cambridge, pp 1–27
Musgrave A (1975) Popper and 'diminiscing returns from repeated tests'. Australas J Philos 53(3):248–253
Musgrave A (2010) Critical rationalism, explanation, and severe tests. In: Mayo D, Spanos A (eds) Error and inference: recent exchanges on experimental reasoning, reliability, and the objectivity and rationality of science. Cambridge University Press, Cambridge, pp 88–112
Nersessian NJ (1992) How do scientists think? Capturing the dynamics of conceptual change in science. In: Giere RN (ed) Cognitive Models of Science. Minnesota studies in the philosophy of science University of Minnesota Press, Minneapolis, pp 3–44
Nersessian NJ (1995) Should physicists preach what they practice? Constructive modeling in doing and learning physics. Sci Educ 4:203–226
Nersessian NJ, Patton C (2009) Model-based reasoning in interdisciplinary engineering: cases from biomedical engineering research laboratories. In: Meijers AWM (ed) The handbook of the philosophy of technology and engineering sciences. Springer, Berlin, pp 678–718
Picha M (2011) How to reconstruct a thought experiment. Organon F 18(2):154–188
Popper KR (1959) The logic of scientific discovery. Basic Books, Hutchinson

Raftopoulos A (2001) Is perception informationally encapsulated? The issue of theory-ladenness of perception. Cogn Sci 25:423–451
Raichvarg D (2006) Science on the fairgrounds: from black to white magic. Sci Educ 16(6):585–591
Schickore J (2011) The significance of re-doing experiments: a contribution to historically informed methodology. Erkenntnis 75:325–347
Steinle F (1996) Entering new fields: Exploratory uses of experimentation. Philos Sci 64:S65–S74
Thomson-Jones M (2010) Missing systems and the face value practice. Synthese 172:283–299
Woods J (2005) Epistemic bubbles. In: Artemov S, Barringer H, Garcez A, Lamb L, Woods J (eds) We will show them: essays in honour of dov gabbay (volume II). College Publications, London, pp 731–774

Part II
Cognitive Niches and Social Cognition: Using Knowledge as a Tool

This part deals with the *distribution of knowledge* onto the environment operated by human agents (a phenomenon that can be described as "cognitive niche construction"), and also with how this distribution, seen as a collective effort, affects and is affected by *social cognition*. Albeit the interdisciplinary theory of cognitive niche construction robustly accounts for human advancements (including the possibility of science itself), I will tackle the pivotal need to understand the *social mechanisms*, for instance, the simple fact of supporting group living, allowing the coordinated effort of constructing and managing niche construction: this knowledge, I will demonstrate, is also necessary to understand the *crises* impacting on cognitive niches when new *technological* structures (especially when managing communication and decision-making) significantly change basic social dynamics within the niche itself.

Chapter 5
Introducing Cognitive Niches

Abstract This introductory chapter will briefly delineate the theory of cognitive niche construction, on which this part will extensively rely. Cognitive niche construction is a theoretical framework that is proving extremely profitable in bridging philosophy, cognitive science, biology and anthropology studies by offering a real inter-disciplinary ground affording novel approaches to long debated issues in several fields. In brief, it concerns the ability, displayed by human beings and other organisms, to affect their own evolutive processes by *cognitively* shaping their environment in order to modify the selective pressure the latter has on them. Thus, a second hereditary system is introduced, afforded by the ecological persistence of knowledge externalizations serving as cognitive aids.

5.1 Cognition Versus Unpredictability

In order to understand many argumentations in this chapter, it is important to make clear the concept of "cognitive niche." As I will show, this notion is crucial for the establishment of the eco-cognitive outlook.

Eco-cognitive epistemology is tacitly aware that *complexity* and *survival*, at every stage of evolution, are two notions never that far one from the other. The very origin of life on Earth, and the fact that so far there is no strong *evidence* of life having originated anywhere else in the Universe, are considered as the epitome of a random and unpredictable (and hardly reproducible) series of conditions.

The past century, also because of the development of quantum physics, has witnessed a growing awareness about how complexity and unpredictability are dominant constraints in ontogenesis and phylogenesis, that is affecting not only the development of species as we know them, but also of every single individual (Longo and Buiatti 2013). In other words, given an initial set of genetic codes, there is no way to predict their giving rise to a potentially infinite number of new species; similarly, given the genetic code of an embryo, there is no way to reliably predict (or compute) what the phenotypic expression will be in n years, because of the unpredictable effects of a highly unpredictable environment.

© Springer International Publishing Switzerland 2015
T. Bertolotti, *Patterns of Rationality*, Studies in Applied Philosophy, Epistemology and Rational Ethics 19, DOI 10.1007/978-3-319-17786-1_5

The relationship between an organism and its environment is indeed one of coping with uncertainty and unpredictability, and that is what sparked the origin of the multifaceted phenomenon known as cognition, I suggest. Any definition of cognition is conceptually human-centered, and has been only subsequently extended—at least possibly—to animals. This is why I would sat that adopting a barely-essential definition of animal cognition as the one offered by the *Stanford Philosophy Encyclopedia* might serve our scope, and let us rely on something that is not overly biased towards animals, but neither excessively human centered to begin with.

> Cognition is constituted by the processes used to generate adaptive or flexible behavior.[1]

The adaptive behavior implied by cognitive capabilities is a response to the unpredictability of the environment in which the organism must survive. With this respect, the form of inference best characterizing the organism's approach to its surroundings is *abduction*, which in these cases accounts for the attempt (sometimes successful) to elaborate and enact hypotheses about the behavior of relevant objects in an organism's environment.[2] I shall now limit the analysis of abduction to what is requested by our current goal, but it is interesting from the beginning to contrast it with a better known form of inference, that is *deduction*. This very brief example should make the distinction clear, and show why the notion is so relevant for the present task.

- Consider a train running on a rail track. If you know the speed of the train, the path of the tracks and so on, it is relatively straightforward *ceteris paribus*[3] to calculate the position of the train at any given time. In a way, the position of the train can be *deduced* from its speed, its table and the geography of the track. If these are true, that is in accordance with the related state of things in the world, then the train will be where we computed it to be at a given time.
- Consider a lion chasing a gazelle. The lion must anticipate the prey's movement in order to tackle it to the ground and kill it. Still there is no way for the lion, or for any hunter, to *deduce* the position of the escaping prey at any given time. The lion must *perform and enact immediately* a quick appraisal on whether the gazelle will jerk right or left, basing on speed, maybe past experiences, terrain conformation, presence of other lions and so on: the future position of the gazelle is quite unpredictable, therefore an *abductive hypothesis*, enacted at once, is the best the lion can rely on to manage its pursuit.[4]

---

[1]This definition, which is a fortiori applicable to human beings, is not overly forgiving towards animals inasmuch as it at leasts excludes merely instinctual reactions and associations.

[2]I already relied on the wide explanatory power of abduction in Part I. Please refer to the Introduction of this book for an ampler definition and characterization.

[3]That is, assuming there is no accident, no highjacking, no passenger activates the emergency break, no excessive and unjustified delay, etc.

[4]Our perspective does not see any incompatibility between instinct and a certain extent of plasticity, accepting the possibility of instinctual cognition. Certainly the lion has a limited range of options, and does not transparently "decide" what course of action to follow, the feline nevertheless embodied*ly* and opaquely opts for a course of action rather than another (Park 2012; Millikan 2004; Magnani 2007).

Many of the phenomena a cognizing organism has to cope with are much more similar to the second case than to the first. Survival (encompassing both the notions of fitness and welfare) rests on the possibility of continuously appraising unpredictable situations and make the best judgement out of them.

Nevertheless, to make a very long story very short, it is commonly agreed that higher cognitive capabilities are essentially about saving the cognitive effort, and one of the best ways to do this is to realize that unpredictability and similarity are not mutually exclusive and that, even if every biological (and then social) phenomenon is *a priori* unique, it is possible to elaborate certain heuristics in order to exploit the *approximate-cause-effect* relationships nested in biological unpredictability and randomness. This is what Tooby and De Vore described as *accessing* the "cognitive niche."

> At the core of this lies a causal or instrumental intelligence: the ability to *create and maintain cause-effect models* of the world as guides for prejudging which courses of action will lead to which results. Because there are an infinitely large number of possible sequences of behavior (of which the overwhelming majority are maladaptive) "behavioral flexibility" is an insufficient characterization of our innovative adaptive pattern. Our cognitive system is knowledge or information driven, and its models filter potential responses so that newly generated behavioral sequences are appropriate to bring about the desired end. Of course, exploration, trial and error, and feedback are essential to the system, but, by themselves, they are inadequate to construct or maintain it (Tooby and DeVore 1987, p. 210, added emphasis).

If Tooby and De Vore (and subsequently Pinker 2003) understand the cognitive niche as a kind of *stage* of cognition, in their opinion exclusive to human beings, other scholars such as Clark (2005) and Magnani (2009, Chap. 6) have a much more local and objectified vision of a specific cognitive niche as something "constructed:"

> Cognitive niche construction is the process by which organisms modify their environment to affect their evolutionary fitness by introducing structures that facilitate (or sometimes impede) the persistent individuation, the modeling, and the creation of cause-effect relationships within some target domain or domains. These structures may combine with appropriate culturally transmitted practices to enhance problem-solving, and (in the most dramatic cases) they afford potential whole new forms of thought and reason (Bertolotti and Magnani 2015).

5.2 Cognitive Niche Construction: Managing Resources to Contrast Unpredictability and Lessen Complexity

In a nutshell, a cognitive niche consists in a series of externalizations of knowledge into the environment, for instance through material culture, resulting in a modification of the selective pressure that an organism has to face (Odling-Smee et al. 2003; Magnani 2009). The fact of championing cognitive niche construction could be seen as what intrinsically characterizes human beings (which are individuated by the theory as *eco-cognitive engineers*).

Since the dawn of cognition, we have been acting on our surrounding ecologies in oder to make them easier to live in, and we engineered niches ranging from basilar sociality (Dunbar 2004) to material culture (Mithen 1996), through agricultural and hunting abilities. Every single step of development can be framed within the concept of eco-cognitive engineering: we engineer our environment by externalizing and manipulating pieces of knowledge. Otherwise said, humans (like other creatures) do not simply live *in* their environment, but they actively shape and change it while looking for suitable chances. In doing so, they construct *cognitive niches* through which the offerings provided by the environment in terms of cognitive possibilities are appropriately selected and/or manufactured to enhance their fitness as chance seekers (Tooby and DeVore 1987; Pinker 1997, 2003). Lessening the selective pressure means, for our cognitive efforts, to lessen the complexity of the external world by developing simpler models of how the environment work, and to enact them and make the world a less unpredictable place to live in.

A recent book by Odling-Smee et al. (2003) offers a full analysis of the concept of niche construction from a biological and evolutionary perspective. "Niche construction should be regarded, after natural selection, as a second major participant in evolution. [...] Niche construction is a potent evolutionary agent because it introduces feedback into the evolutionary dynamics" (Odling-Smee et al. 2003, p. 2). By modifying their environment and by affecting, and partly controlling, some of the energy and matter flows in their ecosystems, organisms (not only humans) are able to modify some of the natural selection pressure present in their local selective environments, as well as in the selective environments of other organisms. This happens particularly when the same environmental changes are sufficiently recurrent throughout generations and selective change.

In summary, general inheritance (natural selection among organisms influencing which individuals will survive to pass their genes on to the next generation) is usually regarded as the only inheritance system to play a fundamental role in biological evolution; nevertheless, where niche construction plays a role in various generations, this introduces a second general inheritance system (also called *ecological inheritance* by Odling-Smee). In the life of organisms, the first system occurs as a one-time, unique endowment through the process of reproduction (sexual for example); on the contrary, the second system can in principle be performed by any organism towards any other organism ("ecological" but not necessarily "genetic" relatives), at any stage of their lifetime. Organisms adapt to their environments but also adapt to environments as reconstructed by themselves or other organisms.[5] From this perspective, acquired characteristics can play a role in the evolutionary process, even if in a non-Lamarckian way, through their influence on selective environments via cognitive niche construction. Phenotypes construct niches, which then can become new sources of natural selection, possibly responsible for modifying their own genes through ecological inheritance feedback (in this sense phenotypes are not merely the "vehicles" of their genes).

[5]This perspective has generated some controversies, since the extent to which modifications count as niche-construction is not clear, thus entering the evolutionary scene. The main objection regards how far individual or even collective actions can really have ecological effects, whether they are integrated or merely aggregated changes. On this point, (see Sterelny 2005) and the more critical view held by Dawkins (2004). For a reply to these objections, see Laland et al. (2005).

It has to be noted that cultural niche construction alters selection not only at the genetic level, but also at the ontogenetic and cultural levels as well. For example the construction of various artifacts challenges the health of human beings:

> Humans may respond to this novel selection pressure either through cultural evolution, for instance, by constructing hospitals, medicine, and vaccines, or at the ontogenetic level, by developing antibodies that confer some immunity, or through biological evolution, with the selection of resistant genotypes. As cultural niche construction typically offers a more immediate solution to new challenges, we anticipate that cultural niche construction will usually favor further counteractive cultural niche construction, rather than genetic change (Odling-Smee et al. 2003, p. 261).

With a broader explanatory reach than sociobiology and evolutionary psychology, the theory of niche construction simultaneously explains the role of cultural aspects (transmitted ideas), behavior, and ecologically persistence inheritance. Of course niche construction may also depend on learning. It is interesting to note that several species, many vertebrates for example, have evolved a capacity to learn from other individuals and to transmit this knowledge, thereby activating a kind of proto-cultural process which also affects niche construction skills: it seems that in hominids this kind of cultural transmission of acquired niche-constructing traits was ubiquitous, and this explains their success in building, maintaining, and transmitting the various cognitive niches in terms of systems of coalition enforcement. "This demonstrates how cultural processes are not just a product of human genetic evolution, but also a cause of human genetic evolution" (Odling-Smee et al. 2003, p. 27). From this viewpoint the notion of *docility* (Simon 1993) acquires an explanatory role in describing the way human beings manage ecological and social resources to make their own decisions.

Woods (2013) touches a similar problem, related to docility, when, analyzing fallacious reasoning, he stresses the fact that "Whether full or partial, belief states are not chosen. They befall us like measles", in other words, "say so" induces belief (doxastic irresistibility). The problem is related to the effect of what Gabbay and Woods call *ad ignorantiam rule*: "Human agents tend to accept without challenge the utterances and arguments of others except where they know or think they know or suspect that something is amiss" (Gabbay and Woods 2005, p. 27). The individual agent also economizes by unreflective acceptance of anything an interlocutor says or argues for, short of particular reasons to do otherwise, by applying the *ad verecundiam* fallacy. Accordingly, the reasoner accepts her sources' assurances because she is justified in thinking that the source has good reasons for them (the fallacy would be the failure to note that the source does not have good reasons for his assurances). Peirce contended, in a similar way, that it is not true that thoughts are in us because we are in them; "beings like us have a *drive* to accept the say so of others" (Woods 2013): I will focus on this typically human tendency to rely on social cognition in Chaps. 7 and 8.

It is noteworthy that all these information resources do not only come from other human beings. This would clearly be an oversimplification. Indeed, the information and resources that we continuously exploit are—so to speak—*human-readable*. Both information production and transfer are dependent on various *mediating structures*, which are the result of more or less powerful cognitive delegations, namely, niche construction activities. Of course, it is hard to develop and articulate a rich culture as humans did, and still do, without effective mediating systems (writing, artifacts, material culture, etc.). Hence, I can say that, first of all, docility is more generally concerned with the tendency to lean on various *ecological* resources, released through cognitive niche construction. Secondly, social learning cannot be seriously considered without referring to the agency of those mediating structures, whose efficiency in storing and transmitting information far exceeds, from many perspectives, that—direct and non-mediated method—of human beings. To this respect, Chap. 9 will deal with the handling of information resources by non-biological mediators that are nonetheless able to exhibit forms of "cognition."

It is well-known that, from the point of view of physics, organisms are far-from-equilibrium systems relative to their physical or abiotic surroundings.[6] Apparently they violate the second law of thermodynamics because they stay alive, the law stating that net entropy always increases and that complex and concentrated stores of energy necessarily break down. It is said that they are open, dissipative systems (Prigogine and Stengers 1984), which maintain their status far from equilibrium by constantly exchanging energy and matter with their local environments. Odling-Smee, Laland and Feldman quote Schrödinger, contending that an organism has to "feed upon negative entropy [...] continually sucking orderliness from its environment" (Schrödinger 1992, p. 73). Cognitive niche construction is a way that an organism (which is always smartly and plastically "active," looking for profitable resources, and aiming at enhancing fitness) has to stay alive without violating the second law: indeed it "cannot" violate it. In this sense cognitive niche construction can be considered *necessary*: "To gain the resources they need and to dispose their detritus, organisms cannot just respond to their environments [...] to convert energy in dissipated energy" (p. 168).

Evolution is strictly intertwined with this process and so it has consequences not only for organisms but also for environments. Sometimes the thermodynamic costs are negligible, like in the heat loss caused by photosynthesis that is returned to the universe, "which is in effect infinite," (p. 169), sometimes they are not, in this case *abiota* of the environment have no capacity to contrast the niche-constructing activities of organisms (like for example, the atmosphere, which is in a new physical state of extreme disequilibrium in relation to exploitation of the Earth's limited resources). The only no-costs exception is when organisms die—and lose their far-from-equilibrium status). In this case the dead bodies are returned to the local environment in the form of dead organic matter (DOM), still a kind of niche construction, so to say, also called "ghost niche construction" (Odling-Smee et al. 2003, p. 170). Of course *biota* alone can actively resist most thermodynamic costs imposed

[6]It is important to note recent research—based on Schrödinger's focusing on energy, matter and thermodynamic imbalances provided by the environment—draws the attention to the fact that all organisms, including bacteria, are able to perform elementary *cognitive functions* because they "sense" the environment and process internal information for "thriving on latent information embedded in the complexity of their environment" (Ben Jacob et al. 2006, p. 496). Indeed Schrödinger maintained that life requires the consumption of negative entropy, i.e. the use of thermodynamic imbalances in the environment.

on them by other niche-constructing organisms, often performing counteractive niche-constructing activities.[7]

References

Ben Jacob E, Shapira Y, Tauber AI (2006) Seeking the foundation of cognition in bacteria. From Schrödinger's negative entropy to latent information. Physica A 359:495–524
Bertolotti T, Magnani L (2015) Bridging the gap between niche construction, cognitive niche, and cognitive niche construction, (Forthcoming)
Clark A (2005) Word, niche and super-niche: how language makes minds matter more. Theoria 54:255–268
Dawkins R (2004) Extended phenotype—but not extended. A reply to Laland Turner and Jablonka. Biol Philos 19:377–397
Dunbar R (2004) Gossip in an evolutionary perspective. Rev Gen Psychol 8(2):100–110
Gabbay DM, Woods J (2005) The reach of abduction. A practical logic of cognitive systems, vol 2. North-Holland, Amsterdam
Laland KN, Odling-Smee FJ, Feldman MW (2005) On the breath and significance of niche construction: a reply to Griffiths, Okasha and Sterelny. Biol Philos 20:37–55
Longo G, Buiatti M (2013) Randomness and multilevel interactions in biology. Theory Biosci 132:139–158
Magnani L (2007) Animal abduction. From mindless organisms to artifactual mediators. In: Magnani L, Li P (eds) Model-based reasoning in science, technology, and medicine. Springer, Berlin, pp 3–37
Magnani L (2009) Abductive cognition: the epistemological and eco-cognitive dimensions of hypothetical reasoning. Springer, Berlin
Millikan R (2004) On reading signs: some differences between us and the others. In: Kimbrough Oller D, Griebel U (eds) Evolution of communication systems: a comparative approach. The MIT Press, Cambridge, pp 15–29
Mithen S (1996) The prehistory of the mind. A search for the origins of art, religion, and science. Thames and Hudson, London
Odling-Smee FJ, Laland KN, Feldman MW (2003) Niche construction. The neglected process in evolution. Princeton University Press, Princeton
Park W (2012) Abduction and estimation in animals. Found Sci 17(4):321–337
Pinker S (1997) How the mind works. W. W. Norton, New York
Pinker S (2003) Language as an adaptation to the cognitive niche. In: Christiansen MH, Kirby S (eds) Language evolution.Oxford University Press, Oxford, pp 16–37,http://www.isrl.uiuc.edu/amag/langev/paper/pinker03LanguageAs.html
Prigogine I, Stengers I (1984) Order out of chaos man's new dialogue with nature. Bantam, London
Schrödinger E (1992) What is life? With "mind and matter" and "autobiographical sketches". Cambridge University Press, Cambridge (originally published in 1944)
Simon H (1993) Altruism and economics. Am Econ Rev 83(2):156–161
Sterelny K (2005) Made by each other: organisms and their environment. Biol Philos 20:21–36
Tooby J, DeVore I (1987) The reconstruction of hominid behavioral evolution through strategic modeling. In: Kinzey WG (ed) Primate models of hominid behavior. Suny Press, Albany, pp 183–237
Woods J (2013) Errors of reasoning. Naturalizing the logic of inference. College Publications, London, (Forthcoming)

[7]The relationship between biota and abiota in the management of cognitive niches will be reviewed in Chap. 9.

Chapter 6
Curating the Richness of Cognitive Niches

Abstract This chapter focuses on the curation of cognitive niches understood as the curation of eco-cognitive chances: as chances can be faked, it seems intuitive to think that the inhibition of chance-faking contexts is a good activity of chance curation. Yet, could this activity sometimes be counterproductive? The question will be answered positively considering the case of bullshit as a case of fake chances, but also as a fertile ground for learning and developing intuitions. Ultimately, this chapter will argue that the peculiar context, that is the cognitive niche supporting the (potentially) fake chances, is the discriminating factor: indeed, a rich cognitive niche may benefit from certain kind of fake chances—which should therefore not be inhibited—whereas a poorer niche might not benefit from this situation, and therefore the preclusion of fake chances is an act of chance curation in those contexts.

6.1 Introduction

My research did not focus on the biological arguments in favor of niche construction theory, but rather took it as a given and elaborated on a number of related issues that had been disregarded so far, mainly related to the management of cognitive niches and the underlying cognitive dynamics. It seems therefore proper to commence with the analysis of a phenomenon that is strictly connected with the management of cognitive niches, that is chance curation.

Providing an introduction to cognitive niche theory, I stressed the relevance of two strictly related notions: chances and affordances, denoting niche construction as the activity of selecting the best ecological (and cultural) chances, that is those able to resonate with the agent's ecological, cognitive and cultural endowments, and thus become *affordances*—an affordance is in fact a relationship that is situated in a nexus between the ecological (external) dimension and the individual (cognitive) one.

Parts of this chapter were originally published in L. Magnani, T. Bertolotti (2013). Selecting chance curation strategies: Is chance curation related to the richness of a cognitive niche? *International Journal of Knowledge and Systems Science* vol 4(1) (pp. 50–61). Reprinted by permission of the publisher.

© Springer International Publishing Switzerland 2015
T. Bertolotti, *Patterns of Rationality*, Studies in Applied Philosophy, Epistemology and Rational Ethics 19, DOI 10.1007/978-3-319-17786-1_6

The activity of niche construction, carried out through the selection of chances and the discovery/construction of appropriate affordances, must not be seen as an exclusively *poietic* activity, one of "adding" to the environment. In many cases, what is at stake is the opportunity of editing in a negative sense, in removing or downplaying, an existent ecological or cognitive trait: as defined by Oshawa and McBurney (2003), a chance is a new event or situation conveying both an opportunity and a risk in the future. Recently, a number of contributions have acknowledged the abductive dimension of seeking chances with relation to science (Magnani 2005; Magnani and Bardone 2008; Magnani 2009; Abe 2009). As maintained by Magnani and Bardone (2008) and Abe (2009), the process of chance detection (and creation) is resulting from an inferential process—mainly abductive—in which the agent exploits latent clues and signs signaling or informing the presence of an action opportunity (Magnani and Bardone 2008). In this case, as argued by Magnani (2009) the abductive inferential dimension has to be considered beyond its sentential/computational one.

According to Peirce, an inference is a form of sign activity in which the word "sign" encompasses several types of sign, for instance, symbol, feeling, image, conception, and other representation (Peirce 1931–1958, 5.283). Moreover, the process of inferring—and so the activity of chance seeking and extracting—is carried out in a distributed and hybrid way. As already said, this ecological-cognitive approach considers cognitive systems in terms of their environmental situatedness: instead of being used to build a comprehensive inner model of its surroundings, the agent's perceptual capacities are seen as simply used to obtain "what-ever" specific pieces of information are necessary for its behavior in the world. The agent constantly "adjusts" its vantage point, updating and refining its procedures, in order to uncover a piece of information. This resorts to the need of specifying how to efficiently examine and explore and to the need of "interpreting" an object of a certain type. It is a process of attentive and controlled perceptual exploration through which the agent is able to collect the necessary information: a purposefully moving through what is being examined, actively picking up information rather than passively transducing (Thomas 1999). In this sense, humans like other creatures are ecological engineers, because they do not simply live their environment, but they actively shape and change it looking for suitable chances, epistemic for example, as in the case of the production of modelings through abductive abstractions, previously analyzed in Chap. 3.

Generally speaking, the activity of chance-seeking as a plastic behavior is administered at the eco-cognitive level through the construction and maintenance of cognitive niches. The various cognitive niches humans live in are responsible for delivering those clues and signs informing about a (environmental) chance. So, the mediating activity of inferring as sign activity takes place (and is enhanced) because of the presence of the so-called eco-cognitive inheritance system relating to cognitive niche construction. That is, humans can benefit from the various eco-cognitive innovations as forms of environmental modifications brought about and preserved by the previous generations. Indeed, many chance-seeking capacities are not wired by evolution, but enter one's behavioral repertoire because they are secured not at the genetic level, but at the eco-cognitive one—in the cognitive niches. The second important point, already mentioned, is that humans as chance extractors act like eco-cognitive engineers. Accordingly, they take part in the process of extracting chances by performing smart manipulation in order to turn an external constraint into a part of their extended cognitive system.

Chances are provided by the continuous eco-cognitive activity of humans as chance extractors, but all human agents do not stand at the same level: some individuals, in fact, do not only explore and make use of the chances available in their ecologies, but also *take care* of them. If chance discovery (and chance utilization) are, as I contend, inferential tasks, then the quality of these inferences heavily depends on the quality of available chances. As suggested by Abe (2010), chance discovery is coupled by an activity of *chance curation*: deriving the word from the museum jargon, Abe remarks that the "main task of curator is a curatorial task, which is multifaceted. curator comes from a Latin word 'cura' which means cure [and care]. Then originally it was used for a person who takes care of a cultural heritage" (p. 794). The word was subsequently applied to the person who takes care of data in IT settings:

> They use "data curation" because they think data have value. Not only for keeping data but also usability of data for the public, they use the word "curation". Actually, most of data are neither art works nor archaeological artifacts. However, it is important to view data from the aspect of what should be preserved (p. 795).

Following this extension, Abe defines "curation" as the "task to offer users opportunities to discover chances". This will be our starting point. If indeed "curation" may describe the intentional offering of chance discovery to users, then the concept can in fact be further expanded to comprehend the activity of those who, within a cognitive niche, try to reduce the occurrence and the appeal of fake chances (Magnani and Bardone 2010). Thus, to make a case in a non-computational environment, a parent overlooking her children's playground is indeed a curator, and she will be likely to guide them towards chances affording safe playing, while she will be on the lookout for potentially dangerous or fake chances—such as bullies, ill-intentioned adults, unsafe facilities and so on.

The example I just proposed was quite straightforward, but the next case I am going to tackle will not be that immediate. My aim is to point out the non-monotonic nature of curation activities: that is, any kind of chance (however potentially dangerous) cannot generate a one-size-fits-all response in curation because, depending on the particular cognitive niche offering that chance, it could be either potentially harmful, or potentially beneficial to users—in this latter case not to be inhibited.

6.2 Chance Curation and Bullshit

In the first part of this section, I will show why bullshit, as described by Frankfurt, is a kind of potentially dangerous chance-faking, and therefore an efficient curation activity should be aimed at contrasting bullshit. In the second part, I will juxtapose a different perspective on the same topic which instead depicts bullshitting as a potentially benefiting phenomenon that fosters the production of new ideas and contributes to learning processes.[1]

6.2.1 Why Bullshit Equals a Faked Chance

The first expert I would like to call in our discussion of chance curation is Harry Frankfurt, who brings to the matter his notion of *bullshit*. The Oxford American Dictionary defines the word as follows: "stupid or untrue talk or writing; nonsense", while the verb *to bullshit* has a solely negative connotation, as it means "to talk nonsense to (someone) in an attempt to deceive them". The notion of bullshit introduced by Frankfurt (2005) may thus help describe a fundamental feature of human beings: they often exhibit a carelessness about truth—a trait favoring deception and fraud—via a systematic undervaluing of truth (Misak 2008), at least when we consider truth as connected to concepts such as reason, evidence, experience, inquiry, reliability, and standard or rational belief—for example in agreement with a kind of scientific mentality. According to Frankfurt, there is an important distinction to make between a bullshitter and a liar. The difference between the two is that the liar has a general concern about truth, while the bullshitter has none. And this is just because, in order to *intentionally* tell a lie, the liar has to know what the truth is. Although the liar fails to be cooperative with respect to a certain state of things in the world, he is indeed cooperative with respect to his attitude towards truth: she is *intentionally* creating a fake chance, potentially dangerous for the victim, but profitable for the liar. Just like a hunter setting a trap: the hunter skillfully masters a fake chance, dangerous for the prey but advantageous for himself: similarly, the fabrication of a lie may require a great deal of knowledge and it is mindful. It requires the guidance of truth. In this perspective, what is relevant about lying is that there is always a *reason why* somebody may not be telling the truth: lies and deceit can be therefore detected.

With this respect, people recur to both verbal and non-verbal cues enabling them to detect potentially deceiving situations (Vrij 2008). Even conceding that sometimes people fear the consequences of knowing the truth, and therefore they prefer not to investigate, it does not mean that they would not succeed in asserting it. Quite the opposite: I *fear* the truth when I know what it might be and I know it is within my reach. So, given the fact that a liar is committed to the truth-value of one's belief, lying may contribute to hiding a chance, but at the same time it might be revealing it, as one detects the deceiving intention of the liar: a trap (faked chance) may reveal the intention and—furthermore—the presence (real chance) of the hunter.

According to Frankfurt, the case of bullshit is different, as the bullshitter is supposed to lack any concern or commitment to the truth-value of what he says. What turns out to be extremely puzzling is not the content, but the bullshitter's attitude. For instance, as I have already pointed out above, a liar voluntarily gets a thing wrong.

[1] This point is all the more interesting considering that Frankfurt's proposal was indeed received by *curators* of certain kind of cognitive niches as a curation manual (Perla and Carifio 2007).

But in doing so he conveys a certain commitment to the truth-value of what he claims, and so he allows the chance to be debunked. A bullshitter does not care about it. As just mentioned, a liar has a deceptive intention that can be detected. Whereas the case of bullshitter is different: normally, when a person believes P, she intends to believe P. And that intention becomes meaningful to other people. In the case of bullshitter, he believes without a real intention to believe what he believes. So, in Frankfurt's vision, what really defines a bullshitter is his attitude towards truth: he fails to be committed towards truth. He simply does not care whether what he says is true or false, accurate or inaccurate.

The illustration of bullshit I have presented so far allows us to argue that bullshitters are basically unwilling chance-fakers (thus all the more dangerous, according to Frankfurt). Why are they so? The deceiving character of chance-faking is related to the fact that they act as if a chance was present, when it is not. Roughly speaking, what comes out of the bullshitter's mouth is *hot air* or *vapor*, meaning that the informative content transmitted is *nil*.

The deceiving character of chance-faking is particularly evident in the case of those cognitive processes involving a collaborative dimension, for instance, the case of second-hand knowledge, that is, the possibility of passing a chance on to another person.

As argued by Simon, we exhibit a constant tendency to lean on what other people say. This disposition to rely on social channels in problem-solving activities is what he called *docility* (Simon 1993). People tremendously benefit from aids and resources provided by their fellows. This has a major cognitive advantage: that almost anybody can trust other people and so have at disposal chances that, first of all, he has never personally experienced, and, secondly, that are already available to pick up. That is one of the most important assets describing cognitive economy, that is, the need to reach a sort of trade-off between the accuracy of a decision and the limited time one is bounded to. Indeed, trust is not informatively empty: one decides to trust another person, because she has reasons to do so. There are a number of clues we make use of in order to consider a particular source of information (a person, for instance) as trustworthy or not. For instance, people usually tend to trust people that exhibit some authority (Jackson 2008). What happens then to a bullshitter?

I said that a bullshitter does not really intend what he says to believe in. He does not have any concern about the source of what he chooses to believe in. It just happens to him to believe. If so, then information transmission becomes highly *noisy*. This is another fundamental difference with lying. As already illustrated, a lie is not informatively empty, because people have various mechanisms for detecting lies. Our lying detector is based on our ability for mind-reading. Basically, we can guess that a person might lie, because we know that we can lie in the first place. Would we say the same about bullshitters? Do we have an analogous sort of bullshit detector enabling us to see when one is faking a chance? However trivial this question might be, the answer seems to be that we have nothing like that.

Following Frankfurt, it is possible to claim that a bullshitter is defined by the kind of attitude he has about truth: he exhibits no commitment regarding what he came to believe in. My suggestion is that we can infer that he is bullshitting only because we are already familiar (or expert) with what the bullshitter is talking about. The cues that are meaningful to us are only related to what he is talking about. But, as one can easily note, in the case of second-hand knowledge this is precisely what is missing. This would be a kind of vicious circle, as we would need what we lack (knowledge) in order to detect bullshitting: the detection of bullshitting is therefore extremely contextual.

I contend that bullshitting as chance-faking can be seen as a particular kind of semantic attack or, more generally, an example of cognitive hacking. By the term semantic attack, I refer to all those situations characterized by a more or less fraudulent/violent move aiming at hacking a chance potentially available to a person or a group of person. A chance might be an idea, a word, a piece of news, a statement, or an explanation. In the current terminology, a semantic attack is concerning with the manipulation of the meaning a person assign to something that he is going to use in his decision-making process.

Interestingly, Thompson (2007) has recently argued that a semantic attack, and therefore—I could add—the notion of bullshitting as chance-faking, may be a threat for all those situations in which our cognitive performances are mediated by various technological cognitive artifacts as, for instance, computers or artificial agents. According to Thompson there are three categories of threats that can put a computer or a network of computers in danger. The first category groups those physical attacks aimed at physically destroying hardware, connections, etc. Then, we have syntactic attacks that regard, for instance, viruses and worms. They are thought to destroy software or alter its normal functioning. Semantic attacks belong to the third category of threat. They do not aim at destroying hardware or software, but they manipulate the perception of reality. In doing so semantic attacks distort the decision-making process, whether it is performed by an autonomous agent or by the human user of a computer system. On some occasion particularly violent semantic attacks may even lead to information warfare. Bullshitting as chance-faking can be considered as a form of semantic attack. As already pointed out, bullshit appears to be a good chance when it might not be. And even when there is no fraudulent intent from the bullshitter to deceive, bullshit actually manipulates the way a cognitive agent or a decision-maker interprets her own reality. With this respect, the link between this kind of linguistic strategic exchange and the kind of dynamics I had analyzed in the previous chapter, dealing with camouflage, is quite apparent: camouflage *explicitly* messed with an agent's perceptual faculties, making her interpret the surrounding reality in a way that is not factual.

The threat posed by bullshit in Frankfurt's view can be further understood by this comparison: if camouflage is a strategic use of deception, it could get extremely counter-productive if agents in the same party (for instance soldiers in the same platoon), because of indiscriminate use of camouflage, failed to recognize each other and therefore could not connect, or even worse, attack and harm each other. Similarly, in Sect. 2.3 of this book, I had extended the strategic use of camouflage to the dialectical and rhetorical use of fallacies and figures of speech: the claim I had made in the following section make sense only if such deceptions are deployed intentionally. It is easy to imagine that a context where fallacies, loaded metaphors, innuendos and figures of speech are used in an *empty* fashion would soon turn the dialogue or address in an undifferentiated land of confusion.

6.2.2 Why a Bullshit Can Be a Valuable Opportunity

Now that all the curation-oriented minds are set against bullshit, I shall call in other experts who will provide us with a different outlook on bullshit. R.J. Perla and J. Carifio devised an argument interestingly contrasting Frankfurt's. The essence of their claim is that:

> Bullshit is not always bad or subversive to the truth; rather it is often a highly dynamic and necessary matrix for the development of expressive, creative, critical, and higher order thinking and representation that give birth to the truth or/and new truths (Perla and Carifio 2007, p. 122).

Otherwise said, and as I will show more specifically, bullshit is a provider of chances that are not necessarily fake—at worse they are random—but always related to a context and that might be revealed as real and advantageous chances. Claiming that Frankfurt sets off from a purely theoretical and prescriptive ground, refusing any compromise with psycho-cognitive studies, Perla and Carifio root their argument in empirical research.

> [...] Basic research in language acquisition and development has shown (and continues to show) convincingly that the use of words, concepts, and conceptual relations is a highly emotive process that is extremely difficult to develop, and that imitation, modeling and "talking above oneself" or "beyond one's comfort zone" or experiences is necessary to develop increasing knowledge and skill in a particular area [...]. *This later point is especially true for highly complex instructional areas such as mathematics, science, philosophy, and other highly abstract and technical fields of study* (p. 123, added italics).

The authors deploy a number of thought-provoking issues, to the scope of showing, so to say, the value of bullshit *in particular contexts*: it must be stressed that Perla and Carifio's claim does not maintain that all bullshit is positive and should be encouraged. Conversely, bullshit might (more or less casually) concoct a series of viable chances in "highly complex instructional areas such as mathematics, science, philosophy, and other highly abstract and technical fields of study." Otherwise said, the utility of bullshit depends on the cognitive niche in which it is taking place. Practices like *monkey see, monkey do* can indeed be valuable chance-discovery mechanisms to guide the acquisition of new context-related affordances by one who enters a new cognitive environment for the first time.

We could say that on the one hand bullshit has a strong linguistic connotation, but on the other hand its use could be expanded to comprehend activities performed by an agent who lacks the proper know-how and is not interested in acquiring it: for instance, weekenders who see a tennis court and decide to have a go, occasional museum visitors, and so on. I am not referring to *amateurs* as opposed to *professionals*, because amateurs have a commitment towards the know-how of the activity (at worst, they may fake this commitment and be satisfied with their amateur level, but they *know* and care about what the actual know-how is).

Bullshit, and the uncommitted relationship it entertains with truth, can in fact be easily spotted in relationship with many learning processes, from the lowest to the most complicate ones. A far less than complete list could include:

- Infants, when learning to speak, often talk *non-sense*. First with words that have no meaning, mostly repeating and imitating sounds (blabbering), then composing sentences that, in a dialogical intercourse, often lack semantic appropriateness, let alone any commitment to truth;
- Children, when learning to draw, start by casual doodling;
- A beginner taking up a musical instrument can try to obtain random sounds from it;
- Pupils often play randomly with their calculators;
- Philosophy students, in their early years, use words and arguments in a way that is clearly beyond their comfort zones.

One could come up with many more examples in which a learning process benefits from elements, which could be seen as bullshit, as chance-discovery tools. Of course, there are many other cases where this does not happen: whereas a pianist-to-be is often allowed to play some notes aimlessly, one who is to earn her driving license is not encouraged to *play around* with the car, nor is a person who is taking a shooting course. It depends on the particular cognitive niche that the activity partakes of: driving evokes a rather peculiar cognitive niche, where much knowledge is at play, but the material part is relevant as well, as it consists of heavy and expensive equipment (cars) that may harm the niche operators (drivers, passengers and people passing by).

To sum up, it could be argued that bullshit, instead of being only negative, does sometimes offer something to the cognitive niche. Even better, it can be a tool, which cognitive agents construct and rely on as a *repository* of cheap-and-easy chances to explore the functioning of the niche. Thus, Perla and Carifio exploit the biological etymology of the notion of bullshit to state that:

> Just as excrement provides the raw materials for sustaining life, conceptual and linguistical bullshit (which Frankfurt associates with "hot air") provides the opportunity to exercise one's developing rhetorical style, imagination, storytelling ability, humor, and creativity (Csikszentmihalyi 1996). Perhaps more importantly, bullshit provides a vehicle for cognitive and social engagement and the opportunity to develop more "accurate," cogent and informed ideas and views via discussions with more knowledge people (Schunk 2004), however "more knowledgeable" people are defined. This last point is where Frankfurt's interpretation of bullshit has both oppressive and intimidating undertones and consequences (pp. 123–124).

Interestingly enough, from the previous subsection we could have drawn the idea that bullshit and H. Simon's concept of *docility* (Simon 1993) are mutually exclusive: a bullshitter is not docile and the docile person does not bullshit. This is acceptable if the exchange is considered purely under the perspective of informational content. From a pragmatic point of view, things are slightly different (and this why Perla and Carifio regard Frankfurt as a fascist): the *chance* to bullshit—in particular settings—may inform the agents about the possibility for "cognitive and social engagement."

Bullshit can then be effectively understood as a mechanism for revelation of chances: it is *not* a chance for learning (as a book would be), but it is a chance inasmuch as it signals (and fosters) the presence of a particularly rich environment where a particular kind of learning can take place. Even in such situations, bullshitting does not equal being docile, but instead it is a way of saying "here you can be docile." This kind of tag, in a learning environment, can be coupled with "here you can make mistakes" *in order to learn and practice*. Docility and the possibility of error (not of intentional lie) are therefore connected.

Similarly, *fallacious* reasoning can in some cases be *redefined* and considered as a good way of reasoning. "Fallacies" are usually seen as errors that people make. Nevertheless, these errors occur in ways of reasoning and acting that from some perspectives are good and from others are bad. Informal fallacies (such as the inductive ones like "hasty generalization") and formal fallacies (such as abduction), seen from the viewpoint of agent-based reasoning, present themselves as ways of reasoning that, in spite of the fact that they can be seen as fallacious from ideal perspectives, are fruitful for the cognitive agent: there is, so to say, a way of being rational through fallacies. The peculiarity of the so-called fallacies, seen in the perspective of agent-based reasoning, is that mistakes that are actually committed do not seem to be mistakes to those who commit them. In some sense we can say that they are ways of reasoning that are *felt* as truth preserving by the reasoner but are *not considered* truth preserving by the logicians! From the point of view of classical logic a fallacy is a bad argument that in general *looks* good. From the point of view of agent-based reasoning, a fallacy is not an argument that looks good but is bad, but an argument that is bad in some aspects and good in some others. Among the positive aspects of fallacies one should include their heuristic value, that is their ability to quickly individuate new chances and new affordances.[2]

> The point that should be stressed here is not that one epistemic context (discovery or justification) or cognitive operation (metaphoric or hypothetico-deductive) trumps the other, but that metaphoric operations and the context of discovery often provide the initial insight or inspiration that leads to views, theories, and ideas that can then be subjected to the context of justification and hypothetico-deductive operations (p. 129).

To conclude this case, it can be said that if we explicitly overemphasize Perla and Carifio's contention, then we can stress the mimetic nature of bullshit, and try to define it as the *potential discovery of chances through mimetic behavior*. After all, bullshitters do often imitate somebody's behavior, actions, words, commitments and so on. Thus transposed, bullshit traces back to the bio-pedagogical *topos* of replication (replicating somebody else's behavior is a most basic chance discovery heuristic), with the exception that a bullshitter is perceived as an imitator that does not care to learn. Even granting that this perception corresponds to the truth, and is not just superimposed by an authority who *knows more* (and, pace Frankfurt, this

---

[2] A full illustration of the various negative and positive cognitive role of fallacies is given in Magnani (2009, Chap. 7). Fallacies are a highly relevant topic in eco-cognitive epistemology and therefore mentioned several times along this book: in particular, they had a crucial theoretical role when exploring linguistic camouflage and those strategies aimed at its debunking (Sect. 2.4).

negative and conflictual judgment could be just the epiphenomenon of a moral clash, as Magnani would contend (Magnani 2011), would it be sufficient to label bullshit as a disruptor and hider of chances?

If, in an eco-cognitive perspective, bullshitting is indeed about acting carelessly *but mimetically* upon the locally available chances, and thus carelessly producing new ones (that might be actual or fake), then it might be appropriate to consider a famous quotation by René Girard, the great scholar of imitation (and its less desirable and violent outcomes): "mimetic desire, even when bad, is intrinsically good, in the sense that far from being merely imitative in a small sense, it's the opening out of oneself" (Adams and Girard 1993, p. 24). The philosophical shift is only apparent: Girard was clearly not speaking about chance curation, but maybe he was. His contention, and some extracts of Perla and Carifio's, are strikingly similar: the former says that mimetic desire, that is imitation, "is intrinsically good" because it produces "the opening out of oneself"; the latter argue that bullshit provides a "vehicle for cognitive and social engagement and the opportunity to develop more 'accurate,' cogent and informed ideas and views via discussions with more knowledge people." If we do not concentrate on the different philosophical jargon, we can see that the core concept does not change.

6.3 Chance Curation as a Consequence of the Richness of a Cognitive Niche

The two perspectives on bullshit I just analyzed offer very distinct conclusions as far as bullshit is considered under the lens of chance discovery and the curation of a cognitive niche. Following Frankfurt, we can say that bullshit impairs chance discovery attitudes, whereas relying on Perla and Carifio's analysis bullshit seems to impregnate the discourse with chances potentially worth discovering. What about curation, then? Should an effort of chance curation (carried out by an individual, an institution or a group) encourage or downplay bullshit mechanisms? It interesting to see how a similar question should be (and has actually been) asked about religion: with the respect of chance curation, religious beliefs are just as puzzling. Let me briefly focus on how should chance curation deal with religion, in order to get some inspiration about the way to solve the current issue about bullshit.

As pointed out by Atran (2002) and Bulbulia (2009), the main issue with religious beliefs is their *counterfactual* nature—and thus the flawed pragmatic behavior guidelines that those beliefs could disseminate in the cognitive niche: a host of crooked chances. Atran contends that "[...] to take what is materially false to be true (e.g. people think and laugh and cry and hurt and have sex after they die and their body disintegrate) and to take what is materially true to be false (e.g., people just die and disintegrate and that's that) does not appear to be a reasonable evolutionary strategy" (Atran 2002, p. 5). Conversely, Bulbulia suggests that religious beliefs could be modeled as characterized by a specific cognitive marker activating or defusing practical inferences about those very same beliefs (Bulbulia 2009). Even if this problem is worth a more elaborate study, it can be argued that—as far as chance discovery is concerned—religion seems to foster a heap of potentially dangerous chances that are perceived by the religious agent but are in reality fake (*counterfactual*) chances.

An opposition, similar to the one I analyzed about bullshit, arises here: this is why I found it proper to look at a religion for solving our *quérelle* about bullshit. On the one hand, scholars such as Dennett and Dawkins rely on the counterfactuality of religion (and thus the dangerous fake chances it offers) as a main reason for the need to explain it *away* (Dennett 2006; Dawkins 2006); on the other hand the sociobiological tradition, best condensed in Wilson's *Darwin's Cathedral* (2002), would reply that the production of counterfactual chances is not necessarily harmful for the survival of groups, because even if it can be costly for the single individual and impair her practical inferences, it does benefit the group as a whole (and thus the individual as well): this can happen thanks to the fact that those counterfactual chances, albeit ecologically awkward, do indeed embody powerful social and moral affordances (allowing humans to climb to the next level of natural selection, that is the group level).[3]

Knowing that bullshit is not the sole case where chance curation is not a univocal activity, I can now try to tackle the core of our problem. The big question would be: "provided that bullshit, religion and many other activities sometimes foster the production of fake chances, while other times are powerful tools for the discovery of real chances, how can you tell one situation from the other—and act consequently?" In other words, one could wonder why Perla and Carifio depict Frankfurt as an epistemic fascist, while the latter would probably consider Perla and Carifio as pernicious epistemic anarchists. In my eco-cognitive perspective, neither of those views is completely right: the reason resides in the most ancient philosophical answer: *it depends*. On what? On the cognitive niche.

This idea is tacitly embedded in Perla and Carifio's view, in fact they mostly refer to *learning* situations being favored by bullshit. Learning-oriented cognitive niches are, by conception, extremely rich in disseminated chances (otherwise they would not be about learning): some of these chances might be tacit, or structural, and in such an abundance bullshit might indeed act as a positive element, that is as a careless (and hence effortless) way of creating and discovering hidden chances. Connecting to the last contention made in the previous section, that is conceiving bullshit as the mimetic repetition of unchecked previously acquired information, it could be easily imagined that this kind of behavior—immersed in a friendly and knowledgeable environment—might easily suggest new epistemic chances for instance during a brainstorm between scholars.

---

[3]Sustainers of the *New Atheists*, sociobiologists and cognitive scientists of religion will probably deem the last two paragraphs to be quite an oversimplification. It is probably so, but as I stated the intention here is not to exhaustively eviscerate the matter (which will be the object of a dedicated study in the near future), but just to point out how—as for chance discovery and chance curation—religion does pose a similar riddle to bullshit. A more complete analysis of religion, from an epistemological and pragmatic perspective, will be the object of Part III.

As it should be clear by now, bullshit works in the "positive" way described by Perla and Carifio *only if* the cognitive niche is (at least potentially) rich. Then indeed it can display a heuristic nature, because the richness of the niche can act as *gravity* and precipitate the most interesting information contained in bullshit from their status of "hot air." That is, the niche can assess which careless chances individuated by bullshitters do indeed correspond to actual chances. It must be noted that there are many higher creative heuristic methods that, even in a rich cognitive niche, are at play in chance discovery: for instance analogies, metaphorical reasoning, conceptual binding, simulation and so on. One should take note that, when saying that bullshit can play a positive role in a rich cognitive niche, the key concept is *rich cognitive niche*: what Perla and Carifio fail to underline is that bullshitting can be a positive chance discovery tool *because of* and *only in* cognitive niches that are already virtuous: in such cases bullshit can become a plus factor, which the chance curator should not downplay.

Conversely, Frankfurt's dismissal of bullshit is indeed appropriate in a cognitively poor niche, where chances are scarce and sparse—or in a cognitive niche characterized by high sensitivity to errors. Consider medical discourse: it is clearly not a poor cognitive niche, but bullshit could lead to loss or harm of human lives, therefore higher creative heuristics are allowed, but not bullshit. Intelligence environments as well favor ignorance over bullshit: hot air could lead to dangerous and unwilling escalations of conflict.[4] Therefore, in poor or precarious cognitive niches Perla and Carifio's account of bullshit does not fit and remains much of an idealization, and chance curation activity—in order to preserve the few available chances or the high quality of chances needed in certain environments such as the medical one—should rightfully contrast bullshit without being considered authoritarian or "fascist".

Along this line, as I briefly sketched out the case of religion at the beginning of this section, it could be argued that even in the discussion the case of chance curation is similar. Religion can offer a surprising repository of existential, social and moral chances but the efficacy depends on the richness and robustness of the cognitive niches it is rooted in. This is why religiosity fostered (or did not impair) socio-historical processes such as the flourishing of ancient Mediterranean civilizations, the arts in the Renaissance, the funding of American colonies by the Pilgrim Fathers etc. on the one hand, but on the other hand religion and superstition contribute to contemporary plagues such as the diffusion of AIDS in Africa, abuse and mistreatment of women in some developing (but not only) geographies…

Similarly to what was the case with bullshit, a "platonic" chance curator should judge the impact of religion differently according to the cognitive niche the believer is set into: a rich cognitive niche can not only absorb the less desirable effects brought about by religiosity but benefits from some effects traditionally connected to religion (social cohesion, mood improvement deriving from faith and prayer, better acquisition of moral character etc.), while a more precarious state of the cognitive niche might require (for sake of chance curation) a downplay and containment of religious activity.

[4] As stated earlier, bullshiting is not to be confused with intentional lying.

6.4 Conclusion

Summarizing, in this chapter 1 focused on the notion of chance curation, and on how it should deal with phenomena that can be seen as ambiguous with respect to chance discovery, such as "bullshitting" and religiosity. Such activities are interesting because some studies let them be perceived as chiefly disruptive of eco-cognitive chances, while others praise them as benefiting the retrieval and construction of local chances.

My claim was that the ambivalence of such phenomena is actually an epiphenomenon of the cognitive niche they are nested into: a rich cognitive niche might indeed take advantage of the carless creation of chances that might be either fake or real, reducing the harm coming from fake chances; conversely a poor cognitive niche—individuating a situation with scarcity of chances—is most likely damaged by ambivalent phenomena such as bullshit or blind religiosity. The activity of chance curators should therefore not only consider the impact of a given phenomenon on chance discovery, but should pay particular attention at the overall state of the cognitive niche, so that well-meaning chance curation does not turn into chance disruption.

In the two following chapters, I will turn my speculations onto another topic that has often been the unquestioned victim of chance curation activity for the sake of the cognitive niche: gossip. The intellectual and philosophical tradition has excluded gossip from any theoretical value, both as an activity and as an object of study: Perla and Carifio claim that to outright rejection of the potential benefits of bullshitting is viable only by ignoring contemporary cognitive and psychological studies, and I argue that the philosophical rehabilitation of gossip underwent the same theoretical path. If cognitive niches must rely on social groups for maintenance purposes, and gossip—as traditionally considered—disrupts such things as social harmony and good will, then efforts in chance curation and in building moral character must indeed suppress instances and inclinations towards gossip. Yet, as I will show in the next two chapters (first from the point of view of social epistemology in the following chapter, then from that of philosophy of technology in Chap. 8), recent cognitive and paleo-anthropologic studies sparked a new philosophical evaluation of gossip, letting it emerge as a basic operator in cognitive niche construction, and not necessarily as a disruptor.

References

Abe A (2009) Cognitive chance discovery. In: Stephanidis C (ed) Universal access in HCI, part I, HCII2009. LNCS, vol 5614. Springer, Berlin, pp 315–323

Abe A (2010) Curation in chance discovery. In: 2010 IEEE international conference on data mining workshops. IEEE, pp 793–799

Adams R, Girard R (1993) Violence, difference, sacrifice: a conversation with René Girard. Relig Lit 25:9–33

Atran S (2002) In gods we trust: the evolutionary landscape of religion. Oxford University Press, Cambridge

Bulbulia J (2009) Religiosity as mental time-travel. In: Schloss J, Murray MJ (eds) The believing primate. Oxford University Press, Oxford, pp 44–75
Csikszentmihalyi M (1996) Creativity, flow and the psychology of discovery and invention. Harper-Collins, New York
Dawkins R (ed) (2006) The god delusion. Transworld Publishers, London
Dennett D (2006) Breaking the spell. Viking, New York
Frankfurt H (2005) On bullshit. Princeton University Press, Princeton
Jackson S (2008) Black box arguments. Argumentation 22:437–446
Magnani L (2005) Chance discovery and the disembodiment of mind. In: Oehlmann R, Abe A, Ohsawa Y (eds) Proceedings of the workshop on chance discovery: from data interaction to scenario creation, international conference on machine learning (ICML 2005), pp 53–59
Magnani L (2009) Abductive cognition: the epistemological and eco-cognitive dimensions of hypothetical reasoning. Springer, Berlin
Magnani L (2011) Understanding violence. Morality, religion and violence intertwined: a philosophical stance. Springer, Berlin
Magnani L, Bardone E (2008) Sharing representations and creating chances through cognitive niche construction. The role of affordances and abduction. In: Iwata S, Oshawa Y, Tsumoto S, Zhong N, Shi Y, Magnani L (eds) Communications and discoveries from multidisciplinary data. Springer, Berlin, pp 3–40
Magnani L, Bardone E (2010) Faking chance. Cognitive niche impoverishment. In: Setchi R, Jordanov I, Howlett RJ, Jain LC et al (eds) KES 2010, part III. Springer, Berlin
Misak C (2008) Pragmatism and solidarity, bullshit, and other deformities of truth. Midwest Stud Philoso 32:111–121
Oshawa Y, McBurney P (eds) (2003) Chance discovery. Springer, Berlin
Peirce CS (1931–1958) Collected papers of charles sanders peirce. Harvard University Press, Cambridge, vols. 1–6, Hartshorne, C. and Weiss, P., eds.; vols. 7–8, Burks, A. W., ed
Perla RJ, Carifio J (2007) Psychological, philosophical, and educational criticisms of Harry Frankfurt's concept of and views about"bullshit"in human discourse, discussions, and exchanges. Interchange 38(2):119–136
Schunk D (2004) Learning theories: an educational perspective. Prentice-Hall, New York
Simon H (1993) Altruism and economics. Am Econ Rev 83(2):156–161
Thomas HJ (1999) Are theories of imagery theories of imagination? An active perception approach to conscious mental content. Cogn Sci 23(2):207–245
Thompson P (2007) Deception as a semantic attack. In: Kott A, McEneaney W (eds) Adversarial reasoning: computational approaches to reading the opponent's mind. Chapman & Hall/CRC, pp 125–144
Vrij A (2008) Detecting lies and deceit pitfalls and opportunities. Wiley, New York
Wilson DS (2002) Darwin's Cathedral. Chicago University Press, Chicago and London

Chapter 7
Gossip as Multi-level Abduction: The Inferential Ground of Linguistic Niche Construction

Abstract As presented in the introductory chapter of this part, cognitive niche construction is strictly connected with the need to lessen the unpredictability characterizing most human endeavors: as posited by niche construction theorists, this gives rise to a new kind of pressure resulting from the modified environment. If living in groups and increased sociality are, for instance, examples of cognitive niche construction aimed at improving fitness and welfare, they produce a new series of drawbacks liked to the unpredictability of human behavior. Hence, a new kind of niche construction must reduce this further unpredictability by making human behavior more predictable and controllable. It is interesting to connect recent studies concerning the coevolution of language and enculturation (Castro et al., Biol Philos 19:712–737 2004) with the emergence of assessors and curators overlooking the maintenance of a give niche, ecological at first, then more and more cognitive. It is in this perspective, I suggest, that it could be interesting to frame, and speculate on, the recent re-evaluation of gossip. Dunbar's famous hypothesis (Dunbar, Rev Gen Psychol 8(2):100–110 2004) that gossip developed as an evolutionary assorting device (creating bonds but boundaries as well) can be understood as the selection and formation of the fundamental ground for supporting a cognitive niche: language could in fact mediate, in an unprecedented way, the diffusion and elaboration of information about peers involved in the perpetration of the niche. Thus, language can effectively be considered as a super-niche (Clark, Theoria 54:255–268 2005), and as projecting a zero-level cognitive niche which scaffolds all subsequent niches, just because it is able to organize and maintain the human groups necessary for niche construction.

7.1 Introduction

As presented in the introductory chapter of this part, cognitive niche construction is strictly connected with the need to lessen the unpredictability characterizing most human endeavors: as posited by niche construction theorists, this gives rise to a new kind of pressure resulting from the modified environment. If living in groups and increased sociality are, for instance, examples of cognitive niche construction aimed

© Springer International Publishing Switzerland 2015
T. Bertolotti, *Patterns of Rationality*, Studies in Applied Philosophy, Epistemology and Rational Ethics 19, DOI 10.1007/978-3-319-17786-1_7

at improving fitness and welfare, they produce a new series of drawbacks liked to the unpredictability of human behavior. Hence, a new kind of niche construction must reduce this further unpredictability by making human behavior more predictable and controllable. It is interesting to connect recent studies concerning the coevolution of language and enculturation (Castro et al. 2004) with the emergence of assessors and curators overlooking the maintenance of a give niche, ecological at first, then more and more cognitive.

It is in this perspective, I suggest, that it could be interesting to frame, and speculate on, the recent re-evaluation of gossip. Dunbar's famous hypothesis (2004) according to which gossip developed as an evolutionary assorting device (creating bonds but boundaries as well)[1] can be understood as the selection and formation of the fundamental ground for supporting a cognitive niche: language could in fact mediate, in an unprecedented way, the diffusion and elaboration of information about peers involved in the perpetration of the niche. Thus, language can effectively be considered as a super-niche (Clark 2005), and as projecting a zero-level cognitive niche which scaffolds all subsequent niches, just because it is able to organize and maintain the human groups necessary for niche construction.

7.2 Gossip and Group Cognition: An Epistemological Shift

This chapter frames the analysis of gossip, and gossip-based knowledge, within the debate concerning applied (Coady 2012) and social epistemology (Goldman 2010). This seems all the more fitting as David Coady connects the origins of applied epistemology to the tradition of applied ethics (p. 1 and ff.): indeed, gossip is a topic whose philosophical relevance is chiefly pertaining to applied ethics. Many contributions to what I hold to be the most comprehensive philosophical work about gossip, *Good Gossip* (Goodman and Ben-Ze'ev 1994), are indeed essays by moral philosophers concerned with the defense of such a widespread human habit, often surrounded by a negative reputation—by which the moral bias (gossip is bad and harmful) turns into an epistemic one (to rely on gossip is a poor method of enhancing one's knowledge).

While I will soon confront the definitional hardships of gossip, it should suffice to say from the beginning that I will consider *gossip*, in its amplest conception, as a kind of social discourse, apparently idle or purposeless, between two or more speakers that are at least comfortably acquainted, focusing about some third party—usually absent—they are (at least) acquainted with, and usually displaying a broadly evaluative/moral dimension.

[1]Dunbar's hypothesis will be introduced in this chapter, and analyzed in further detail in the next chapter (Sect. 8.3.1).

Philosophy has already dealt with this *moral* nature of gossip (the first section of the aforementioned *Good Gossip* is essentially a treatise of applied ethics): the main issue was to decide whether gossip is decent, acceptable, or not, and I will make occasional references to the issue of morality (since, coherently with the recent trends, I acknowledge that a discourse on gossip may not *totally* prescind from a discourse on morality).[2] Yet, a further cluster of philosophical problems about gossip remains to be tackled, and it concerns *epistemology* and the analysis of the *epistemic value embedded by gossip*. The source of the interest is the common sense ambivalence towards gossip: on the one hand gossip is despised as rumor-mongering, deceitful and inspired by ill feelings (and therefore one is often encouraged not to give too much weight to gossip), but on the other hand everyone has relied—to different extents—on gossip in the course of their human experience. Gossip is such a fundamental source of knowledge concerning our fellow human beings that it begs indeed for a serious epistemological analysis focusing on the *kind* of knowledge conveyed by gossip, how it actually manages to *share* any knowledge, and its further *knowledge-forming* capabilities. This, together with the need to individuate and develop the most fitting epistemological framework, is the task I set before the reader and myself in this chapter.

Gossip has also been the object of a number of different, non-philosophical studies in the past 50 years, rehabilitating it not only as something worth being studied, but also as a pivotal informational and social structure of human cognition. The revaluation of gossip is indebted towards two main disciplines: anthropology on the one hand, cognitive/evolutionary studies on the other hand. Anthropologists focused on gossip as a means of social regulation (see for instance Gluckman 1963 and Yerkovich 1977), but the mechanism often maintained a sense of otherness—both a geographical and chronological distance: gossip could be indeed more than mere idle talk, but that was true as far as *other* people, at a different stage of history, or in another land or society, were concerned—keeping gossip as an anthropological topic seemed to shield "decent" Westerners from the implications (sometimes hard to accept) of these studies. Conversely, evolutionary psychology and sociobiology immensely boosted gossip's reputation (no pun intended) by showing its relevance as far as it concerns the dawn of language and sociality (Dunbar 2004; Wilson et al. 2002): gossip was presented as a *natural* tool for the enforcement of a determinate order within the group. Dunbar, specifically, maintains that gossip evolved along the lines of grooming to allow hominids the possibility to cope with life in large social groups: in his view, the origin of gossip is strictly connected with the development of language itself: this view (together with the sociobiological perspective of Wilson et al. (2002) that will soon be taken in further consideration) stressed the pragmatic-moral origin of gossip as a collective device that—by circulating social information—monitors, reports and punishes deviants and free riders.

[2]Recently, the ethical discourse about gossip has interestedly been woven with evolutionary studies yielding results also in the field of ethics of technology (Bertolotti 2011; Bertolotti and Magnani 2013).

A few words of caution, though, should accompany the reception of gossip studies, both evolutionary and philosophically oriented. One should fundamentally avoid, as I will suggest in the following chapter, to commit a *naturalistic fallacy* and identify the evolutionary (hence *natural*) origin of gossip with its intrinsic "goodness." Similarly, if in this chapter I will succeed at persuading the reader about the peculiar epistemic character of gossip, I do not mean to endorse an overoptimistic image of gossip, forgetful of its many negative aspects. For instance, the fact that gossip is indeed an extraordinary enforcement tool as described by Dunbar (2004) does not entail that it cannot be (or has not been) used to enforce irrational or abusive norms and regulations, more or less tacitly.[3] It must be considered, though, that whereas I deal with a topic that has been much explored by applied ethics in a normative and prescriptive dimension (consistently with Coady's assertion that applied epistemology usually has a normative interest), the task I mean to accomplish is chiefly *descriptive*: I mean to spell out the inferential functioning of gossip, and the epistemic structures it supports without—for the time being—adopting a normative outlook on the subject, which might instead be undertaken in a future study when the descriptive stage is deemed sufficiently accomplished.

In brief, the aim of this chapter will be to provide a novel and complementary philosophical outlook on gossip, set within a recent tradition broadly definable as *applied* or *social epistemology*, and that would be consistent as far as it concerns individuals and groups on the one hand, knowledge content and types of inference on the other. Hence, my argumentation will be subdivided as follows: in this section, I will try to settle the relationship between gossip and group, arguing that gossip is a collaborative epistemic synergy; in Sect. 7.3 I will see how gossip can be defined an inferential process underpinning a kind of inquiry; finally Sect. 7.4 will analyze the peculiar kind of inference at stake at individual and group-level, namely different kinds of abduction. In the conclusion (Sect. 7.5) I will wrap it all up and suggest some likely developments of my studies.[4]

---

[3]I will come back, along my argumentation, to the issue of the negative aspects of gossip when dealing with the weak pragmatic mode of corroboration for gossip-based knowledge in Sect. 7.3.3.

[4]Some more words of caution should conclude this introductory matter, stating how my analysis is not intended as totally competitive with other theories concerning gossip, elaborated for instance in evolutionary and psycho/cognitive paradigms (on which anyway I partially rely). For instance Yerkovich (1977), Baumeister et al. (2004), Dunbar (2004), Daniels Jr. (2012)—to mention a few heterogeneous traditions—stress how gossip has a role that transcends the informational one and is often aimed at consolidating the bonds and commitments between gossipers at the indirect expense of those being gossiped about. The point worth noting here, is that gossip (a bit like language), is such a complex and multilayered phenomenon that it can afford (if not plainly need) different levels of explanation that are not necessarily mutually exclusive: ultimately, deciding the priorities between these explanations (i.e., did gossip originate as an informational or a bonding mechanism?) faces the risk of becoming a "chicken and the egg" question which I will not tackle in my analysis.

7.2.1 Definitional Issues

Gossip is a complex object of speculation first of all because it is hard to break it down to a basic unit, and it is therefore easier (and more useful) to frame it as a functional dynamic phenomenon. The very word "gossip" may refer to multiple things, namely the activity of gossiping, the single or cluster of exchanged information, and the agent who engages in gossip: yet, when one thinks of gossip, she intuitively refers to the activity of gossiping, which involves more than one people engaging in the exchange of a peculiar kind of information, and not to an abstract piece of gossip, or two ideal subjects engaging in a discourse called gossip. This is why the perspective needed to study gossip from an epistemological point of view is necessarily *applied* and *social*. Indeed, virtually any literature focusing on gossip has associated it to the presence of a determinate group, even to the point of being labeled as a *group-serving behavior* (Wilson and Dugatkin 1997): albeit I will make use of findings that can be related to evolutionary psychology (Baumeister et al. 2004; Dunbar 2004), my interest will focus on an epistemological analysis of gossip and thus only accidentally and locally challenge concurrent hypotheses about the existence of a "group mind" put forward by different sociobiological studies (Wilson and Dugatkin 1997).[5]

Considering its peculiar nature, it is most hard to provide a univocal definition of gossip that is not already laden with theoretical implicatures (as the one I just mentioned, referring to the relationship between gossip and groups, that is functional to my analysis and consistent with the common sense conception of gossip). In order to have a better working hypothesis for my epistemological analysis of gossip, though, it is interesting to proceed *negatively* and add something about what gossip is not: it is important to distinguish gossip from another kind of discourse it is often confused with, that is *rumor*. Consider this annotation:

> This is one difference between rumor and another form of communication with which it is often confused, gossip. Gossip may well be first-hand. By contrast, no first-hand account of an event can be a rumor, though it may later become one (Coady 2012, p. 87).

Distinguishing gossip from rumor will be crucial at the final stage of this chapter, when, in order to understand some implications of gossip at group-level, I will rely

[5]The evaluative dimension of gossip, not concerning gossip in a moral perspective but rather about the moral content of gossip, is the best example of how gossip is eminently group-laden: from the relevance of the shared information (Yerkovich 1977) to the accuracy of a report on someone's behavior (Ayim 1994; Wilson et al. 2002; Beersma and Van Kleef 2011), to the moral judgement of such behavior (Gluckman 1963; Taylor 1994; Dunbar 2004; Magnani 2011), sometimes to the point of triggering a pragmatic reaction concerning the individual on which the gossip focuses. It is worth noticing that one should not understand this reference, and all subsequent ones, to *morality* and *morality evaluation* in a sense that makes this chapter an essay of moral philosophy: I am not interested in providing a moral appraisal of gossip, rather in expanding the well-established thesis (for instance by Gluckman 1963, Yerkovich 1977, Taylor 1994, Daniels Jr. 2012) that gossip is concerned with *expressing* moral judgements. In addition, I mean to stress how the attention to a naturalized dimension of morality (Magnani 2011; Boehm 2002) acknowledges the emergence of a proto-moral attitude, which continuously divides the (social) world in what should be approved and what should be disapproved (and eventually sanctioned): this does not require the explicit engagement of complex and explicit moral frameworks.

on some considerations about the functioning of rumor offered by Gelfert (2013) and Coady (2012).

Slightly elaborating on Coady's aforementioned annotation about the difference between gossip and rumor, it seems consistent with anyone's experience of gossip to say that a peculiarity of gossip resides in the abundance of first-hand testimony: as we will see, albeit different kinds of knowledge are selected and used in a gossiping exchange, usually the prop for the exchange itself is ideal when it refers to the first-hand testimony of one of the participants, namely relating to what happened to them, or they saw, heard and so on. When referring to rumor, we will explore the possibility of gossip becoming *as* rumor, what epistemic processes underpin this possibility and what is its significance for group-level dynamics.

My extrinsic working hypothesis should make the case that gossip is not just a group serving behavior, but even from an epistemological perspective its informational analysis cannot be separated from concepts such as *group*, whose members share first hand testimonies and elaborate on it, cumulating into the *knowledge base* that a group shares, and so on. It is therefore extremely difficult to isolate a base unit of gossip that is not laden with assumptions about its existence conditions: even if you can think of an isolated exchange of gossip, it makes theoretical sense as "gossip" only if it is inserted within a broader collective epistemic effort.

Indeed, the thesis I will argue in favor of in this chapter is that gossip is an *inferential activity* (i.e. moving from some premises to some conclusions), displaying a *collaborative nature* within the group of gossipers, which ultimately manages to reach a kind of *collective self-appraisal*. In order to propagate this view of gossip, I will refer to a series of stereotypical gossiping interactions such as the following one:

> Jason: "Guess what! Petra must be having an affair!"
>
> Lynda: "No way! How do you know?"
>
> Jason: "I saw her this morning at the station, holding passionately a man that was not her husband…"
>
> John: "Oh no, that was her brother. She muttered something about not being able to come to the corporate picnic yesterday because he was visiting…"
>
> Patricia: "That's impossible. My aunt and Petra's mother were very close friends, Petra is an only child."

Before proceeding with my actual analysis, though, it is fit to clear the field from an objection that the reader is likely to put forward, and at the same time provide a sensible insight concerning the definitional issues gossip has to face—by highlighting what I think to be yet another peculiarity of gossip. More precisely, I do not think that the reader will fail to recognize as gossip the examples I provide, but one might argue that I am committing an unjustified generalization because many instances of gossip do not display the collaborative functioning I individuate, but are rather characterized by a number of passive hearers.

Quite often, gossip does not *prima facie seem* to instantiate the collaborative epistemic effort I will stress in the chapter, and it is rather about one single arguer who does all the job, telling someone else the whole information and possibly the moral evaluation. Following my example, we can imagine Jason telling this to another acquaintance belonging to the same group: "I had no idea Petra was such a hussy! I saw her holding passionately a man at the station, John thought it was some relative of hers, but no! Besides, Patty said she's been acting weird lately…" In this case, Jason is providing his friend (explicitly) with all the information: is he thus making the conclusion obvious, and the inferential and collaborative part superfluous? Probably not (especially if we assume gossip to be yet another process of *coalition enforcement* (Rohwer 2007; Boehm 2002), as proposed by evolutionary studies) and this lets us elaborate on yet another interesting difference between gossip and rumor-mongering. Coady rightly states this characteristic of rumor, while assessing its epistemic value:

> All else being equal, a person is more likely to pass on a rumor, and more likely to pass it on with a high estimate of plausibility, if he or she thinks it is true (Coady 2012, p. 94).

This is still generally true for gossip,[6] but one more feature is present in the latter case. As far as rumor is concerned, unless one deems herself expert in the topic of a rumor, or has other good reasons for doing so,[7] she will not take a stand to reject a rumor that she judges false—she will merely not pass it on. Conversely, in gossip everyone is an *expert* about one's acquaintances, and a self-proclaimed moral expert in human affairs (how often do we proclaim ourselves unable to proffer a personal moral say-so on a given situation?): if one is not an accidental bystander, even her silence and passivity will be a form of engagement and collaboration, as gossip makes no (moral nor epistemological) room for indifference. This should make sense with everyone's real life epistemic behavior: if you receive a forwarded email about the latest scam, threat, cover-up, etc. you forward it if you think it is at least probably true, but otherwise chances are you will simply delete the email, while you will not confront the sender and explain to her why the news she forwarded is at best inaccurate. Conversely, if you are sitting at the pub and a friend of yours starts badmouthing another friend of yours, unless you agree with the accuser you are hardly likely to refrain from contributing to the discussion defending the accused friend: if you abstain from doing anything, the party will *ceteris paribus* assume that you agree with the criticism.

This view is consistent with what was suggested by cross-disciplinary milestones in gossip studies, such as Dunbar's evolutionary hypotheses (Dunbar 2004) and Yerkovich's fieldwork (Yerkovich 1977), in their observations of how one is seldom a passive hearer in gossip: at worse, the most passive participant to the exchange is not exerting her right to disagree with the information suggested, or to dismiss it as irrelevant. Therefore, even if a participant does not contribute with information, just for the fact that she is necessarily agreeing or disagreeing, and deciding whether to rebroadcast in her turn the information, makes her an *active* part of the epistemic process, possibly appraising previously issued inferences.

According to my perspective, a new understanding of gossip can be achieved by raising the stakes and trying to frame it within the toolboxes of applied and

---

[6]Refer Sect. 7.4.2, where I will elaborate on the epistemic unfeasibility of gossip if it mostly consisted of spreading malicious distortions and fake news.

[7]For instance, pragmatic or strategic reasons, that yet are not linked to the object of rumor by a relationship of relevance.

social epistemology: that is to say, I will approach gossip as it relates to groups understanding them as peculiar "knowers," composed by a community of individuals that *epistemically* behave as parts of a group.

7.2.2 Group Mind or Epistemic Synergy?

Socio-epistemological studies on gossip could seem to have an earlier hypothesis, developed by sociobiology, advocating that gossip, as a group-serving behavior, is a clue and an effect of the presence of something such a *Group Mind*, informing the behavior of the group itself. In a nutshell, the existence of the group mind is a sociobiological theory relating to the acceptance of groups as a further level of natural selection: groups as *superorganisms* exhibit their own degree of fitness (which is then mediated upon organisms), but in order to perform at their best there must be a "group mind" coordinating the activities (Wilson and Dugatkin 1997; Wilson et al. 2002). This hypothesis is clearly modeled upon the socio-cognitive organization of eusocial animals such as ants, bees, termites, etc.

Nevertheless, I think there is room for a more convincing hypothesis concerning the relationship between gossip and its effects on the cognitive and epistemic performance of a group. A useful notion comes at play here, allowing us to perform a shift from the cognitive to the epistemological ground: the concept of "synergy" (Kelso 2009). I suggest that groups as epistemic agents are not best captured by the Group Mind Hypothesis (or GMH) inasmuch as they should be considered as instantiating "soft" cognitive systems. Coherently with the terminology employed by Anderson et al. (2012) to define cognitive systems, colonies of eusocial animals can be said to be organized by a group mind inasmuch as their cognitive dynamics are "component-dominant" and not "interaction-dominant." That is to say that the group mind in a beehive, or in a colony of termites, relies on the cognition of single individuals whose contribution is determined by their genetic inheritance.[8] In this sense eusocial animals do not display "soft" cognitive systems, because components cannot be easily detached and reattached. Conversely, "other systems, such as flocks of birds, are more fluidly put together. It the latter case, it doesn't matter which particular birds are part of the flock—any old bird will do—and each bird is capable of taking up each position in the flock" (Anderson et al. 2012, p. 716). In this case, what the group's behavior seems to instantiate is a synergy: "a functional grouping of structural elements (molecules, genes, neurons, muscles, [individuals], etc.) which, together with their supporting metabolic networks, are *temporarily* constrained to act as a single coherent unit" (Kelso 2009, p. 83).

---

[8]A warrior ant will exhibit a given behavior and a given cognition, and thus contribute to the cognitive performance of the hive, in a way that is necessarily different from that of a queen, and of a worker—in fact ants exhibit a high level of "social resilience" in the division of labour: if moved, they will try to come back to the precedent position (Sendova-Franks and Franks 1994).

How does the synergy concept help my analysis of gossip? Because it allows to hypothesize, from an epistemological point of view, that certain groups can be regarded as displaying *epistemic synergies* in their distribution of epistemic labour. What does this mean? I can say, drawing from the antecedent discourse, that an epistemic synergy is a "soft-assembled *epistemic* system." More specifically, it obtains when a number of individual epistemic agents join their efforts and temporally act as one collective epistemic agent: this also means that some prerogatives that are usually comprised and carried out by one epistemic agent—i.e. knowledge gathering, inferencing, assessing, and so on—can be dislocated among the parts of the *synergetic group* so that individuals, instead of being full epistemic agents, may concentrate on empowered epistemic faculties. The notion of synergy also accounts for the lack of persistence of this structural epistemic configuration, allowing individuals to break apart, or to fluidly exchange epistemic roles, without compromising the structure of the epistemic synergy.[9] Furthermore, synergies are disposable structures imposing only epistemic constraints on the group members: as I will suggest in Sect. 7.2.4, one can be part of as many synergetic groups as she wants provided that she *knows* how to be part of them.[10]

When evaluating the opportunity to accept my hypothesis over other, better established ones such as the Group Mind Hypothesis, it is important to bear in mind that the latter is not an epistemological hypothesis but an evolutionary and biological one: therefore, any will to corroborate it or to discard it would be eccentric to the aim of this chapter. Indeed, my hypothesis stands at a different level and this is why I think that the epistemology of cognitive group phenomena can be rendered using the concept of *epistemic synergy* in a more convincing and theoretically parsimonious way than by the GMH. The latter is particularly cumbersome from an ontological perspective inasmuch as it often requires the *actual* existence of groups as levels of selection: if groups are endowed with the same kind of existence normally attributed to cells and multi-cellular organisms, then the further hypothesis about the *actuality* of the group mind becomes necessary. Seeing human groups in their cognitive and epistemic performances *as instantiating synergies* accounts for their temporary and often contingent character, and for the fact that in many cases pluralities of human agents (and also other social, but not eusocial, animals) indeed behave *as one*.

Furthermore, the notion of epistemic synergy saves us from being locked in yet another sum/parts dilemma. The synergy allows in fact the emergence of an ontologically and epistemologically *deflated* collective knower. This should leave us one

---

[9] It is the exact opposite of a traditional computational model: a single machine is a hard-assembled system, and most components cannot switch roles, and when they do, as in the case of virtual devices, it requires a heavy software manipulation. Indeed, hard storage cannot behave as CPU, CPU cannot become a Random Access Memory, a Mobile Storage cannot become a Sound Drive, and so on. Conversely, the synergetic group could be computationally imagined as a network of machines, each capable of working stand-alone, joining their efforts so that one acts principally as a processor unit, another one manages the graphic output, other ones the storage of data, and so on, and where any machine can potentially be assigned to any role.

[10] The intuitive idea that those groups should not be in conflict or ecologically compete is a pragmatic, and not an epistemological concern.

step further away from group-mind theories: while the latter impose a non-deflated commitment about the existence of groups, I mean that the groups exist only as *projected* epistemic subjects, the result of a "soft-assembled epistemic system." The group can be regarded as an epistemic agent only as the projection of the epistemic agency of individual group members, cognizing and inferencing through the mediation of the Knowledge Bases they share in virtue of being part of the group, and whose maintenance is heavily indebted towards gossip as I am about to argue.

7.2.3 Groups Are the Subject of Gossip

If—epistemologically speaking—gossip makes sense only as the cognitive effort of a group (understood as supporting an epistemic synergy), then it might be appropriate to enquire whether groups can be seen as the (collective) subject of gossip. Here the meaning of subject is the grammatical one, that is to say, it is groups that perform the action of gossiping.

One could argue that this is not the case, because clearly it is individuals who gossip about each other. Yet, as I suggested earlier, and as I am going to argue all along this chapter, gossip is ultimately about updating and sharing the collective Knowledge Base of the group, composed of shared information about people and shared evaluation (also moral) of the information.

The pragmatic group-serving effects brought about by gossip, namely the enforcement of norms and rules due to the possible report of misbehaviors and free riders (Dunbar 2004; Beersma and Van Kleef 2011), but also the easiness and effectiveness at tearing apart positive reputations, are an effect of the epistemic structure of gossip.[11] The idea I will develop in this section is that, through gossip, individuals enrich the Knowledge Base of the group they belong to (KB^G), and in return have their individual Knowledge Bases (KB^I) updated by the information gained by other members. The notion of *Group Knowledge Base* is strictly connected to the functioning of the *epistemic synergy* introduced in the preceding section. Defining the epistemic synergy as a "soft-assembled epistemic system" held together by epistemic constraints, I claimed that group members could join, leave or change roles in the synergy simply if and by meeting such epistemic constraints.[12] One of these constraints is sharing the same knowledge about social matters of the group: such knowledge, even if clearly located in the minds of—and in the discourses between—group members, can be abstracted to a single Knowledge Base to whom all of the members partake and, as I will argue, is updated through gossip.

[11]Once again, it is proper to remember that efficacious epistemic systems do not necessarily provoke positive social outcomes.

[12]As we will see, these constraints, as far as gossip is concerned, amount to knowing gossip and how to gossip, and knowing that coincides with being a member of the group. cf. the following section.

Beersma and Van Kleef (2011) experimentally proved that the presence of a witness, likely to issue a judgment and hence gossip, does indeed reduce the local occurrence of misbehavior. The reason seems to reside indeed in the epistemic processes enacted by gossip. Let us remember that for gossip to be effective, it must be pervasive. That is to say, if we take *m* to be a misbehavior, that is an anomaly deviating from the standard approved behavior (for instance, cheating) carried on while another member of the group (for instance, Individual 1) is present, the resulting situation could be schematized as follows:

1. $m \in KB_1^I$: Anomaly *m* is witnessed by Individual 1, and stored in her Knowledge Base.
2. If $m \in KB_1^I \rightarrow m \in KB^G$: If *m* is stored in Individual 1's KB, and if no more up-to-date instance is present in KB^G, then *m* will be transferred to KB^G.
3. If $m \in KB^G \rightarrow$ *possibly* $m \in KB_2^I, \ldots KB_n^I$: Once received by KB^G, anomaly *m* can possibly be transferred to all other connecting Individual Knowledge Bases.

The efficacy of gossip as an information device resides in the interplay between (2) and (3), which could be metaphorically seen as establishing a *potential difference* allowing the formation of a flux. Let us explain why. The device functions only if a member of the group may, basing on her first and second-hand experience, believe (2) to be reasonably adequate. (2) therefore must meet a dual condition: on the one hand it must be quite *accurate*, on the other hand it must be *thought so*. The interplay between (2) and (3) is epistemologically *self-enforcing*: if (2) is accurate, and known to be accurate, then it is also believed, but if (2) is believed, then it affects (3) even in those cases in which (2) does not happen.

From an agent-based epistemological perspective (Woods 2005; Magnani 2005) the distinction between the actuality of (2) and the mere *belief that (2) is accurate* is meaningless: for the deterrence to be efficacious, it is sufficient that an individual *believes* that her misbehavior/anomaly will be recorded and reported, thus entering the common knowledge of her group.

However, from the epistemological perspective, the fact that (2) is *believed* to be actual is relevant as well, inasmuch as it shows how the Group Knowledge Base reflected in individual ones comprises also the *rules* of gossip (consistently with Gluckman 1963). Of course KB^G displays an *epistemic* reality as a *postulated knower*,[13] not an ontological one (as the Group, and the Group Mind, in the sociobiological *Group Mind Hypothesis* introduced at the beginning of this chapter), and it exists only as long as individual members keep updating and sharing their personal *KB*. Indeed, the various individual $KB_1^I, \ldots KB_n^I$ contribute to forming KB^G in virtue of the possibility that KB^G exists. That is to say, gossip as an epistemic synergy exists because an information is known by individuals *in virtue* of the fact that it is known by the group.

Coming back to the process we were analyzing, the possibility in item (3) is relevant as well. It signifies that a member of the group might not have received

[13]This aspect will be further developed in Sect. 7.4.2.

information m because of constraints due to the fact that the whole process of information diffusion is taking place in a framework of bounded rationality, where time and pragmatic possibilities play a pivotal role. Seeing it from a member's perspective, it is as if I could be *sure* that the group—postulated as an abstract *knower* embodying the KB^G—knows about my misbehavior/anomaly m, but it might be the case that a fellow of mine has not received the update *yet*. Gossip, though, remains effective because any member, *in virtue* of being part of the group (and thus embodying the synergy that produces the KB^G) is *entitled* to knowing m, and therefore she could be reasonably expected to know m: this state could be defined as an *epistemic prerogative* shared by all members of the group.[14]

As I will argue in Sect. 7.4.2, this epistemic *asymmetry* between the perceived *necessary character* of KB^G ("The group knows everything") and the *possibility* of $KB^I_1, \ldots KB^I_n$ ("Single members may *not* know everything") partly roots the strategic success of gossip at group level and its poor epistemic value at individual level: the knowledge of the group as embodying an epistemic synergy is superior than that of each single individual, but is actualized only by its projection in the single, Individual Knowledge Bases, in all of their being limited and constrained.

I suggest that the relationship between KB^G and the multiple KB^I can be conceptualized an *epistemic osmosis*. By this expression, I refer to the way some knowledge is circulated according to its *gradient of relevance and update*: an update in KB^I_1 will be transferred to KB^G if the former has not received it yet, and once KB^G is updated, the new information will reverberate (through subsequent gossiping interactions) to all the remaining Individual Knowledge Bases: this step is crucial, and it means that by sharing some information as gossip to a fellow group member *equals* inserting the gossip in the Knowledge Base of the Group, according to the process I have just described (Fig. 7.1).

The left part of Fig. 7.1 represents a schema of gossip circulation. Individual 1 introduces the bit of information which is then circulated among group members throughout the mechanisms and constraints typical of gossip (which I will analyze in the subsequent sections), and the characters with the blackened head symbolized the group members updated with the new gossip. The ovals labeled *t 0*, *t 1* and *t 2* stand for the diachronic element involved in the actual circulation of gossip. Nevertheless, such diachronic element is immaterial to understand the circulation of gossip: whereas it is clear that gossip does travel, it is unthinkable to calculate "the speed of gossip" because it depends on a number of contextual factors. When trying to engage *strategically* a group of gossipers, the tricky part is indeed managing to know whether one individual has received the gossip *yet*.[15] It is therefore simpler to conceptualize the diffusion of gossip as an epistemic osmosis taking place between the single individuals and the Knowledge Base their share. When gossiping, a person is not sharing a piece of information with A, knowing it will then reach B and C, subsequently D and E and F etc. in a precise timely fashion: conversely, from an

[14]Section 7.4.1 deals with related issues while outlining the actual inferences enacted by gossipers.

[15]To make a trivial example, a free rider must try to find the individuals who have not received any information about his behavior in order to cheat them.

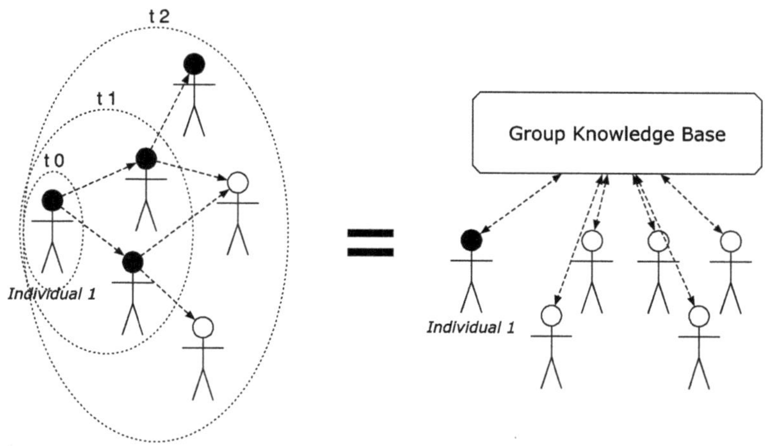

Fig. 7.1 The diffusion of gossip conceptualized as epistemic osmosis (*right*)

epistemological perspective, by sharing a gossip with *A* she is potentially committing that gossip into the Knowledge Base of the Group itself, to be received by those who find it relevant and interesting (see Fig. 7.1, right part). At the same time by gossiping with different people, the same person continuously receives information from the same Knowledge Base.

Summing up, in this section I contended that only groups, considered as *epistemic synergies*, can be the subject of gossip. Single individuals are the subjects of individual instances of gossip, but they acquire meaning as gossip only in reference and in virtue of their membership to the group. Therefore, from the epistemological perspective, it can be argued that groups gossip, through individuals.

7.2.4 Groups are the Condition and the Output of Gossip

To further explore the relationships between gossip and groups, we will not clash with the findings and hypotheses of earlier studies on gossip arguing about the role of gossip as a manager of identity and a tool to set the boundaries of a group (Gluckman 1963; Dunbar 2004; Daniels Jr. 2012): it should be noted, though, that the group-managing utility of gossip depends indeed on the epistemic structure by which it manages social knowledge.

It is rather trivial to say that groups are the *condition* of gossip. Even better, we should say that cooperation is the condition for gossip. Groups are required in order to promote degrees of regularity in cooperation patterns and, as the intellectual potential of individuals grow, the stabilizing element within a group can be identified as knowledge. Without the need for a group as a collaborative entity, there would be no need for the emergence of gossip as an epistemic synergy shaping the group at the informational level. It is renown that human beings cooperate by forming groups that transcend the level of kinship, and thus of genetic membership (Gintis et al. 2005; Magnani 2011). The feeling of belonging to a group that is more "abstract" than family has to be mediated by something else, namely by *knowledge*—intended as human producible and readable information. This knowledge can be mediated by external artifacts,[16] which thus contribute to group membership, but these are externalizations of precedent knowledge (which can be defined as "cognitive niches" (Godrey-Smith 2002; Odling-Smee et al. 2003; Laland and Brown 2006): the other face of cooperation—seen as an assortative, "positive" behavior—is the moral enforcement (achieved through the threat of punishment) of the knowledge supporting and determining the coalition (Magnani 2011).

It seems indeed legitimate to claim that a group is held together by a highly informational process (at the same time supporting—and triggered by—moral inferences): first of all, in order to *know who* are the other members of the group, or *what* are the conditions for membership.[17] Linde (2001) seems to be pointing to the same phenomenon making the case for the utility of narratives in shaping *social tacit knowledge*, which constitutes ultimately the identity that a group and its member can relate to: according to Linde, social tacit knowledge is a shared repository that "tacitly" hosts the moral, social and pragmatic guidelines of a given community (be it a firm, an association, and so on).

> When the notion of tacit knowledge is used within the field of knowledge management, it is frequently used to describe any form of non-quantifiable knowledge, particularly the knowledge about social interactions, social practices, and most generally, how a group or an institution gets things done (p. 160).

Consistently with my claim in the previous section about the epistemic asymmetry between the *perceived necessity* of KB^G and the *possibility* of $KB^I_1, \ldots KB^I_n$, Linde claims that the social tacit knowledge, embodying the group identity, is different than the various *personal* instances of tacit knowledge (that is to say, how much a single individual can retain from her exposition to the loaded narratives shared within the group). Of course, the social tacit knowledge exists only as a projection of the various personal ones, and is abstracted from the different circulating narratives about *who did what*. Linde also acknowledges that the circulating narratives are far more useful

---

[16]Linde (2001) explores the role of occasions and peculiar institutional artifacts as tools regulating storytelling with a social purpose.

[17]It is intuitive that, in order to proclaim herself a member of a group, an individual should at least know one of these two types of information, so that she can either name the other members (in a kind of membership by association, "I am in the same group as Paul, Mary and Peter.") or name the objective/commitments of the group (as in "I am a member of the group of those who believe in Jesus Christ," or "of those who think overfishing should be stopped," and so on): it makes little sense to imagine that an individual proclaims herself member of a group and does not know who the other members are or what are the conditions for membership. Of course this has to do with the problem of spontaneous assorting (Wilson and Dugatkin 1997), and not with groups that are imposed upon unwilling or passive individuals by external forces.

at enforcing the identity of the group (and thus at fostering feelings of membership) than, for instance, corporate mission statements (p. 162).[18]

At this point, groups could be reasonably seen as the output of gossip dynamics, at least as far as knowledge-dominant, assortative interactions are concerned. Gossip, furthermore, is a self-explicative process that brings in itself the rules of its functioning: thus, membership is achieved through a slow epistemic fine-tuning that shapes the way a group processes information. Linde contends that "part of becoming a member of any institution, formal or informal, is learning to tell the stories of that institution, and learning to tell one's stories in a way coherent with those of that group. Part of what one needs to know to be a member is what the stories of the group are, what events in the past are judged to have relevance on the present, what values the stories exemplify, and when it is appropriate to tell them" (p. 163). Of course her interest transcends gossip and focuses more on workplace narratives contributing to company identity, but her analysis is strikingly similar to Gluckman's pioneering account of gossip (Gluckman 1963), where—reporting about the Makah tribe in North America—he claimed that one of the achievements regulating the membership to the Makah was being able to "gossip as a Makah". Gluckman's approach is often labeled as "functional" rather than information-based, but from my perspective such definition does a wrong to the concept of membership, which is epistemically dependent on the ability to share a certain kind of information in a given way.

Conclusively, we could state that the relationship between groups and gossip is not only clear and corroborated at psychological and anthropological level, but also from an epistemological point of view gossip is a *collaborative activity*. Such a perspective, whose exact inferential functioning will be explained in the next two sections, could be said to offer a compatible but alternative approach to Wilson and Dunbar's insights about the usefulness of gossip at group level (Wilson et al. 2002; Dunbar 2004), by providing the epistemological reasons as far as the levels of shared information and knowledge are concerned.

7.3 The Inferential Structure of Gossip-Based Knowledge

In the previous section, I have modeled gossip as a kind of epistemic osmosis between the Knowledge Bases of different individuals that are members of a group, aimed at maintaining a group Knowledge Base that is projected as a kind of "repository" of the knowledge (eminently social) of the group. As we will see, while for every member gossiping is knowing about the group, for the group instantiating an epistemic synergy gossip is about *self-knowledge*.

[18]Such a perspective on gossip is not new, but Linde's studies corroborate from another angle Baumeister, Zhang and Vohs' contention that the benefits of gossip reside in its being a tool for cultural learning about the moral and social norms of the group (Baumeister et al. 2004).

The aim of this section, and of the following one, is to have a clearer view of the epistemic value of gossip, analyzed through an epistemological understanding of the inferential grounding of gossip, and to grasp what kind of knowledge is a gossip-based knowledge. The final claim will be that one of the main (and overlooked) functions of gossip is to allow groups members to perform a constant self-appraisal of themselves as a group. I will claim that this kind of global appraisal can be modeled as an emergent system of different kinds of abductions.

7.3.1 *Gossip as Inquiry*

It can be said that my perspective was brilliantly foreshadowed by Peircean scholar Maryann Ayim, in her article *Knowledge Through the Grapevine: Gossip as Inquiry* (Ayim 1994). Ayim relies on the characterization of *science* depicted by Charles Sanders Peirce: "the pursuit of those who are devoured by a desire to find things out" (Peirce 1931–1958, 1:8).

> Gossip's model captures several aspects of Peirce's notion of a community of investigators. Describing what he sees as the causes of "the triumph of modern science," Peirce speaks specifically of the scientists' "unreserved discussion with one another, …each being fully informed about the work of his neighbour, and availing himself of that neighbour's results; and thus in storming the stronghold of truth one mounts upon the shoulders of another who has to ordinary apprehension failed, but has in truth succeeded by virtue of the lessons of his failure. This is the veritable essence of science" (Peirce 1931–1958, 7:51). […] If Peirce is right that the unreserved discussions with one another are a cornerstone in the triumph of modern science, then gossip, by its very nature, would appear to be an ideal vehicle for the acquisition of knowledge. Gossips certainly avail themselves of their neighbours' results, discussing unreservedly and sharing results constitute the very essence of gossip (Ayim 1994, p. 87).

Ayim observes—forerunning social epistemology—that the epistemic ground and the social one are deeply intertwined in gossip, just as they are in science. After all, part of the success of scientific endeavor rests in the asymmetry between what a scientist may individually know, and what is known in the whole body of scientific literature. What we refer to as "Science" could be very well defined as the product of an epistemic synergy which, just as the Group Knowledge Base, does not exist *in se*, but in the minds of individual scientists and in the knowledge externalizations (in books, universities, lectures) produced by scholars and scientists. Here the main difference is that gossip (usually) does not rely on material repositories that gossipers can access.[19]

The analogy, though, is not only about the presence of a community—which was already noted as crucial in the previous section—but also about the inferential component as well. The root, Ayim contends, has to be traced back to the Peircean notion of *abduction*: "The good scientist, as described in the work of Charles Sanders

[19]Things change when such a repository exists, as in the case of gossip mediated by Social Networking websites (Bertolotti 2011).

Peirce, will be likely to start with a hunch, or retroduction, [or *abduction*] as Peirce calls it, a tentative hypothesis appealing because of its great explanatory capacity" (p. 89).

7.3.2 Gossip as Abduction

Ayim does not further investigate the abductive nature of gossip as its inferential basis, but there is more to be said on the matter. Let us consider the famous Peircean schema of abduction (Peirce 1931–1958, 5.189):

1. The surprising fact C is observed.
2. But if A were true, C would be a matter of course.
3. Hence there is reason to suspect that A is true.

As I have already noted, and as we will discuss further on, gossip is often thought to have an "evaluative" component. For gossip to be evaluative one often thinks that it should embed a clear moral judgement of the information. What if the evaluative component of gossip should rather refer to the action of making an appraisal rather than stressing the judgmental feature? This is not to underplay the moral role of gossip, which must be necessarily considered, but to investigate the inferential ground affording the moral evaluation itself. Let us consider the following example of gossip, bearing in mind the Peircean schema of abduction.

> Jason: "Guess what! Petra must be having an affair!"
>
> Lynda: "No way! How do you know?"
>
> Jason: "I saw her this morning at the station, holding passionately a man that was not her husband…"

The *evaluation as appraisal* is clearly present in this very likely example of gossip:

1. Petra was holding passionately a man (who is not her husband) at the station. *(The surprising fact C is observed.)*
2. But if Petra was having an affair, then she would be holding passionately a man who is not her husband. *(But if A were true, C would be a matter of course.)*
3. Hence Petra must be having an affair. *(Hence there is reason to suspect that A is true.)*

The result of this abductive appraisal is now offered to the Knowledge Base of the Group, and it will be repeated among those who, in the same group, know Petra. What about the evaluation? In this case the evaluation is an appraisal of the state of things in the group's (social) world. The necessity of an explicit moral evaluation for gossip to be *morally evaluative* is further reduced by the existence of "thick concepts," that is concepts that are imbued with a moral content and cannot be accepted neutrally (Putnam 2002): words such as *affair*, *adultery* and so on, even if expressed without explicitly blaming the subject, do evoke a set of moral rules that is enforced (or favored) within the group. If we think that the fact that Petra is holding passionately a man who is not her husband is worthy of being told, it is because it is unusual, and by reporting it we also—more or less tacitly—convey the opinion that we would not want to be in the shoes of Petra's husband.

Ayim, optimistically but not without a certain reason, stresses how this hypothetical (*qua* abductive) nature of gossip should make it a viable way for achieving truth on social matters. The fact that (abductive) gossip can be withdrawn is compared to how scientists hold scientific truth—that is, provisionally.

> On this analysis of the scientific process, gossip may appear to be even more analogous to science in its procedure for arriving at the truth, with gossipers, ever ready to attribute no security whatever (CP 6.470) to the beliefs and claims of others, subjecting them to the harshest of critical analyses, adopting such claims and beliefs "only ...on probation" (7.202), insisting on the stringent tests that Peirce saw as a vital component of scientific progress, and standing "ready to abandon one or all as soon as experience opposes them" (1.635). The difference between science and gossip lies not in their procedure, then, but in the type of subject matter that will characteristically interest them (Ayim 1994, p. 90).[20]

Here, Ayim seems to see only one side of the coin. What she refers to is that gossip, as a dynamic—usually quite fast-paced—information exchange, can warrant for an extreme openness towards the constant renegotiation of what is held to be true. Coherently with this, we can very well imagine that the precedent interaction is enriched by two additional information, stressing the *nonmonotonic* dimension of gossiping collective inferences.

> Jason: "Guess what! Petra must be having an affair!"
>
> Lynda: "No way! How do you know?"
>
> Jason: "I saw her this morning at the station, holding passionately a man that was not her husband..."
>
> John: "Oh no, that was her brother. She muttered something about not being able to come to the corporate picnic yesterday because he was visiting..."
>
> Patricia: "That's impossible. My aunt and Petra's mother were very close friends, Petra is an only child."

The contributions of John and Patricia are crucial for establishing the *provisional truth* that will be then received by KB^G. If John's was conclusive, then there would be no rumor that Petra is having an affair. It seems more reasonable that a married woman hugs her brother at the station than a lover, if the former is in town. The matter would be settled were it not for Patricia's remark, which refutes John's defense. In this sense, the abductive nature of gossip can indeed be seen as a *collective inference to the best explanation*: performed by many agents instead of one (hence collective), it is an abduction involving the formation of a set of hypotheses that are accepted as plausible (at the beginning of our case, Petra could have been holding either her lover or her brother), and then evaluated (also by resorting to moral knowledge) so that the *most plausible* is accepted, that is the one able to "explain" more. That is the *best* explanation, even if—as usual when dealing with abduction—we have to note

[20]Brackets refer to paragraphs in Peirce (1931–1958).

that the adjective "best" has to be taken in a Pickwickian sense: actually abduction never reaches the status of best hypothesis, we have to intend the word "best" in a contextual and provisional way.[21] Indeed,

> [The] evaluation has a multi-dimensional and comparative character. Following Peirce the economics of abduction are driven in turn by three common factors: the cost of testing (1.120), the intrinsic appeal of the hypothesis, e.g., its simplicity, (5.60 and 6.532), where simplicity seems to be a matter of naturalness (2.740); and the consequences that a hypothesis might have for future research, especially if the hypothesis proposed were to break down (7.220) (Magnani 2009, p. 20).

But what about the other side of the coin? I believe that Ayim's comparison of gossip and science is excessively biased in favor of gossip.[22] It is right, and extremely revealing from an epistemological perspective, to take scientific inquiry—as depicted by Peirce—to be an effective model of how gossip works. Yet, once the model has been successfully employed, it is opportune to achieve the best understanding by "climbing down the stairway of abstraction," as Nancy Cartwright would say.[23]

---

[21] Magnani (2013) explains that abduction (also when intended as an inference to the best explanation in the "classical" sense I have indicated above) represents a kind of reasoning that is constitutively provisional, and it is possible to withdraw previous abductive results (even if empirically confirmed, that is appropriately considered "best explanations") in presence of new information. From the logical point of view this means that abduction represents a kind of nonmonotonic reasoning, and in this perspective we can even say that abduction interprets the "spirit" of modern science, where truths are never stable and absolute. Peirce also emphasized the "marvelous self-correcting property of reason" in general (Peirce 1931–1958, 5.579). So to say, abduction incarnates the human perennial search of new truths and the human Socratic awareness of a basic ignorance which can only be attenuated/mitigated. In sum, in this perspective abduction always preserves ignorance because it reminds us we can reach truths that can always be withdrawn; ignorance removal is at the same time constitutively related to ignorance regaining.

[22] This is not meant to be a critique of Ayim's work, which brilliantly revealed the similar orientation towards inquiry. Ayim's positive bias might come from the feminist reading that she offers of gossip, as she presents it as the—often—sole available form of inquiry allowed to subjugated human categories (women, immigrants, etc.), and it should not be forgotten that the essay was published in a collection entitled *Good Gossip*, whose aim was to vindicate gossip from the excessive and unreasoned traditionally attributed negativities.

[23] A first objection is trivial, and should therefore be limited to a footnote: considering the non-monotonic nature of gossiping abductions, we can easily imagine that—in the conversation about Petra's possible affair—if the conversation stopped with Jason's observation, or with John's defense, or with Patricia's rebuttal, the final conclusion would been very different. This often happens in gossip interactions: being *idle* and unfocused, gossip happens as it can, and those who detain the most valuable information might not be present and might not have shared it yet. In this sense,
(Footnote 23 continued)
it can compare to science in a very amateurish way: a scientist *could not be excused* for joining a discussion on a field she's an expert within, without having read of the latest groundbreaking advancement because it was published "just" a month earlier.

7.3.3 About the (Lack of?) Corroboration of Hypotheses in Gossip

Ayim contends that gossiping appraisal is also characterized by the subsequent test "against the real world" until the gossipy theory passes "the test of evidence" (Ayim 1994, p. 90). Such view, albeit being legitimated and motivated by Peirce's discourse and (at least *prima facie*) applicable to the standard image of scientific process, to diagnostic reasoning and other forms of sentential inference, is proved partial by recent developments in abduction studies (which rely, in turn, to alternative—but not incompatible—accounts of abduction).

The point at stake is that the result of an abductive inference, even if not put to the test (inductively, for instance), can be accepted and made operative just the same: I already debated this issue when dealing with the *poietic* nature of scientific models in the first part of this book (Sect. 3.3) Contrasting the traditional conception of abduction, in which results are always corroborated to some degree (which legitimates them as "best"), recent discussions on abduction (Magnani 2013; Woods 2013) have argued that, in certain occasions, abductive results are accepted *just so*, in the sense that they do not undergo any inductive testing (as in the case of a medical diagnosis) but are enacted at once, sometimes tacitly. This is the case, particularly relevant for gossip, of moral inferences: abducing the moral judgement (usually about the goodness of another agent's acting) immediately opens the way for the pragmatic enforcement of the judgement itself.

In gossip, one way of corroborating the abduced hypothesis would be to confront the *gossipee*. In the case I was just examining, though, none of the speakers would be eager to ask Petra whether she is actually having an affair or not.[24] The other way would be to *enact* the hypothesis. That is usually the pragmatic self-corroboration of gossip: as remarked by Gabriele Taylor, "the manner in which a person is represented in thought, whether justly or generously, in terms of clichés or caricatures, is bound to affect at least to some degree one's behavior towards her, if not necessarily what one does, at least how one does it" (Taylor 1994, p. 44). Is this a corroboration of the hypothesis? Even if it is a kind of corroboration, can it be said to display some analogous element with respect to scientific praxis, and its relationship with empirical evidence? The matter can get tricky here. It is absolutely not a corroboration if one envisions an epistemologically virtuous, textbook-like image of science, as the one that Ayim derives from a certain reading of Peirce, where hypotheses are *sacramentally* held as *plausibly* true and everything is subject to rigorous testing and open to revision, if necessary. The prescriptive-*qua*-abductive dimension of gossip is likely to foster its morally violent aspects: the exchange of social information—intertwined with tacitly (or *thickly*) expressed judgments—elicits the sometimes unwarranted confidence experienced by gossipers, for whom what was originally a tentative hypothesis becomes in turn an unquestionable truth for the gossiping group (in turn adoptable as a practical guideline). This is to say

[24]They could not be sure about the truthfulness of Petra's answer, even if they asked her.

that gossip, understood as a comprehensive, social-oriented knowledge management tool may, because of its very abductive nature, produce beliefs that instead of being checked against the reality of things, are just put into action, thus achieving a just-so narration of the social group. The output of gossip could be *akin* to the truth (in a pragmatic sense) about social matters just because it performs the very explanation it provides.

Concluding this section, it should be argued that by adopting the concept of abduction, one describes gossip in a twofold way that, in my opinion, fittingly captures its essence and its ambiguity. While assessing the epistemic value of gossip, we could say that:

- As far as its positive value is concerned, defining gossip as an abductive activity stresses its epistemologically creative dimension: it is sensible with the claim that gossip is a cognitive activity deeply rooted in one's social ecology, and therefore it contributes to understanding the difference between gossiping (i.e. sharing and elaborating information about one's social acquaintances and friends) and spreading false information.
- As far as its negative value is concerned, labeling gossip as an abductive activity spells out the constitutively tentative and provisional character of the knowledge it provides. This would not be so bad in an abstract consideration, but real agents need to operate basing also on whatever information gossip conveys, and they usually assume to be pragmatically true what is less than epistemologically likely, and the outcome of such situation is often grim.

Analyzing the (often lacking) mode of corroboration afforded by gossip allows us to connect the reason for the questionable epistemic value with the traditionally well-known negative aspects of gossip. I have argued that gossip is ultimately about abductive inferences. Evolutionary studies have maintained that gossip has a group serving effect and has therefore been selected by cultural evolution. This could mean, and Ayim's essay seems to agree, that gossip is a viable way for achieving a certain kind of social truth, at least as long as the social domain is concerned. Such a view seems to be coherent with the Peircean motto that "abduction is akin to truth." The note that breaks this pleasant harmony is nothing less than Peirce's "pragmatic" conception of truth as "the opinion which is fated to be agreed to by all who investigate." What could emerge from this situation is that gossip's "truth" is ultimately a kind of self-imposed truth, a *poietic* inference that appraises the social situation but at the same time it significantly *determinates* it.[25]

[25]This could be reverberated by social dynamics such as "scapegoating," described by Girard (1986): gossip plays a crucial role in individuating deviants who might bear the responsibility of a crisis just because they display what Girard calls "victimary signs," signs of differentiation in times of undifferentiation and chaos. Similarly, if gossip plays a substantial role at poetically projecting the group's self-appraisal, then there is little wonder in discovering that, after the scapegoating takes place, the removal (or killing) of an innocent victim does indeed create a climate of general

7.4 Epistemic Synergy and Multi-level Abduction

In the two previous sections I pointed out some essential characteristics of gossip defined as a *collaborative* and *inferential* epistemic synergy, updating the knowledge base of the group through abductive inferences. In the first section, I contended that gossip is an informational behavior that should be predicated about groups. Such groups are not simple agglomerations of individuals but correspond to groups as circles of gossip, in a mutual dependence between the group (seen as instantiating the epistemic synergy) and the activity of gossip itself. In the second section I started exploring the inferential dimension of gossip, which was best described as an abductive one. In this final section I am going to assess this at a higher level of specification, clearing the way to a possible understanding of gossip-based knowledge. The questions I mean to answer in this section are "What is inferentially specific about a gossiping exchange?", "How many levels of inference are there?" and if there are more than one "Are those inferential levels similar?"

7.4.1 Abducing Gossip at Individual Level

A common issue in the various analyses of gossip is whether it should be considered at individual or at group level. As already stated, I favored a group-centered perspective, but by defining groups as instantiating epistemic synergies I am arguing that the result, albeit to be understood at group level, is strictly dependent on functions happening at lower levels.

As a matter of fact, nobody has ever seen a group gossiping. Groups exist as forests do: for a forest to disappear, much of the trees must disappear first. Therefore, it is appropriate to seek the fundamental unit of gossip at the individual level. In order to acquire a fuller understanding of what I am about to disclose, a little introduction is needed.

Many of the essays in the already quoted collection *Good Gossip* (Goodman and Ben-Ze'ev 1994), but other studies as well (Yerkovich 1977; Dunbar 2004; Foster 2004), stress that gossip has two main characteristics: it is a *pleasurable* and *idle* activity (it is important to keep noticing that, apart from cases of pathological narcissism, usually it is pleasurable to gossip and not to be gossiped about). I am about to show that gossip is pleasurable *because* it is an idle inferential/abductive activity. A quick (but fundamental) disambiguation on how the adjective "idle" should be appreciated: Goodman and Ben-Ze'ev (1994), Dunbar (2004) already note that the idleness must not be understood in the Heideggerian pejorative connotation, but

(Footnote 25 continued)
distention and wellbeing, as if the crisis had actually been solved. Girard's theory could be seen as a fitting example of gossip as a purveyor of *military intelligence* as described in the previous footnote. I will refer to Girard's theory when analyzing the sacrificial mindset at the end of Part III, Chap. 13: in particular, the aforementioned link between scapegoating and the pragmatic construction of truth will be echoed in the description of heroic self-sacrifice in Sect.13.2.3.

rather in a pragmatic perspective. Gossip is an activity that people can engage in their spare time, for a few minutes as they wait for the bus, and also *while* they are carrying some other activity (Dunbar 2004). Human beings are natural-born problem solvers: from the lowest levels of perception all the way up to scientific endeavor, human beings are making up spontaneous or theoretical abductions.[26] As we know from everyday experience, we feel a certain intrinsic pleasure when we solve a problem (psychologists would say that the *reward system* in our brain is activated), and we experience further pleasure when someone else gratifies us over our cognitive accomplishment. As I pointed out in the previous section, gossip is an abductive appraisal. It is therefore enjoyable *because* the enjoyment comes from being able to make an appraisal.[27] Having been able to discover Petra's secret affair is rewarding per se, but being acknowledged by Linda, Patricia, and John as the one who framed her first is even more rewarding. This is true both if the gossiping session is indeed a collaborative effort, or if one illustrates to her acquaintances a brilliant reconstruction, or eye-witnessing, expecting a confirmation rather than a true contribution. Consider the utterance I examined earlier on as an example:

> Jason: "I saw Petra this morning at the station, holding passionately a man that was not her boyfriend...She's having an affair!"

What is Jason doing? In his spare time, while chatting with his mates on the train, or over a cup of coffee during a break, he is solving a problem, or what appears to him as an anomaly. In this case, the gossip is that *Petra is having an affair*. Is this a fact? No, it is Jason's appraisal of a series of clues, namely that he saw her at the station that very morning, that she was holding a man, with a certain *élan*, and that such man was not her husband. In this case, the abduction can be very easily framed through Brunswick's *lens model* as the inference of an inaccessible distal social event from a series of proximal, accessible clues (Fig. 7.2).

Of course the proximate clues can point to several distal events. As I suggested in the aforementioned simulated conversation, it could be her brother that she was holding, but then again, more gossip suggests that this is not the case (Fig. 7.3).

---

[26] Much of Part I of this book dealt indeed with this aspects.

[27] Linking the pleasurability of gossip to its inferential ground, to the ability of the speaker to *connect the dots* with respect to the clues she presented, is not radically new. If we could on the one hand trace it back to Keat's poetical epigraph, according to which "Beauty is truth, truth beauty"—and hence the acquisition of truth should provoke an aesthetic pleasure—on the other hand there is a more recent and provocative hypothesis to contemplate. Harvard anthropologist Frank Marlowe suggested an evolutionary, inferential explanation for men's fascination with ampler female breasts, and connected it with a woman's "residual reproductive value" (Marlowe 1998). Whereas smaller breasts undergo less changes during a woman's lifespan, larger breasts tend to sag more conspicuously with advancing age—hence, large breasts, whether sagging or not, could provide a valuable clue as to the woman's age, and hence to her likelihood of being still fertile and able to carry on a successful pregnancy. I do not mean to overlook, but neither to engage, the questionable sexism of Marlowe's hypothesis: what I find interesting is the connection he draws between something that is pleasurable, in this case *sexually/aesthetically*, and the fact that the same thing affords a correct abductive appraisal, which is not that distant from our speculation about the enjoyability of gossip.

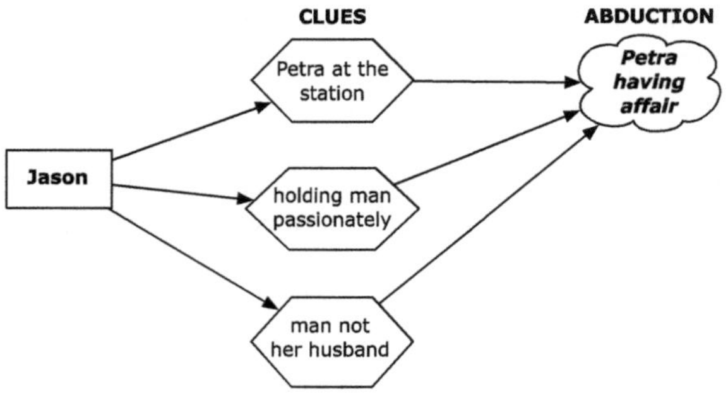

Fig. 7.2 Gossip as the inference of a distal social event

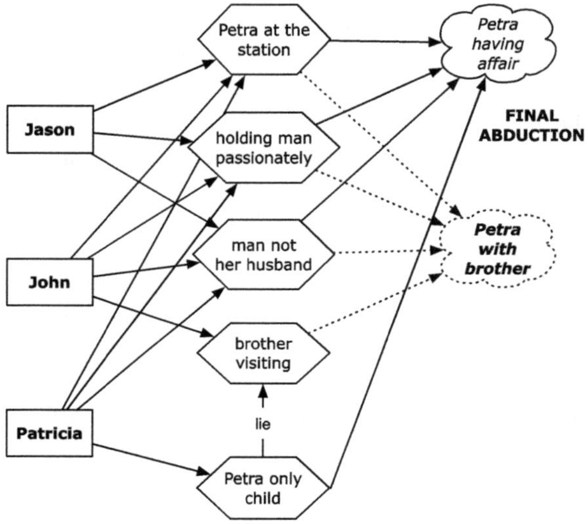

Fig. 7.3 New clues point out to the abduction of a different social event

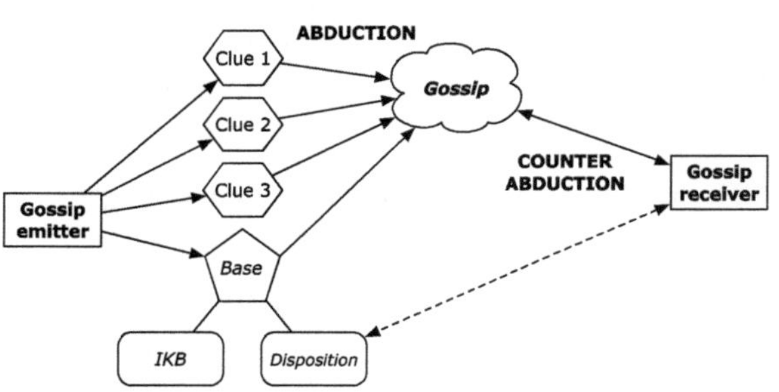

Fig. 7.4 A model of how gossip is abductively produced and received

Let us analyze this first step in greater detail: the production of a gossip equals to making an abduction upon a series of contextual clues, and what can be called the *Base* (see Fig. 7.4). The constitution of the base reflects a dual nature: one part coincides with the individual's best-updated Knowledge Base about the group, the other with her *Disposition*. The Disposition is composed of the agent's feelings, her mood, her personal cognitive endowments (for instance, being good at recognizing faces rather than remembering names) and so on: in sum, it is what affects the abductive capacities of an individual in virtue of being that precise individual and not any other one in the group. Conversely the Individual Knowledge Base, which as I said in Sect. 7.2.3 mirrors the Group Knowledge Base, affects the abductive capacities of an individual in virtue of being part of the group itself.

Nevertheless, if the model so far could resume gossiping information exchange, it would be all too easy. Gossip is in fact not a solipsistic informational activity but it is engaged between two or more parties, gossiping together. The party emitting the information is mirrored by one or more receiving parties, whose reception is not passive but consists of an abductive counter-appraisal.

This process is crucial if we keep in mind the initial definition of gossip as a "soft-assembled epistemic system," in which the prerogatives that are usually comprised and carried out by one epistemic agent—i.e. knowledge gathering, inferencing, assessing, and so on—can be dislocated among the parts of the *synergy* (cf. Sect. 1.2). Albeit making sense of clues and providing a general explanation is the most patent example of abductive inference, also matching an explanation against one's background is an abductive activity, leading to the acceptance or the rejection of the proffered explanation. Thus, as shown when dealing with the definitional issues of gossip (cf. Sect. 7.2.1), a "passive" hearer may not refrain from participating to the abductive interaction even by being a simple gatekeeper, evaluating the retransmission of the information she received.

The effectiveness of the informational exchange rests significantly on the counter-appraisal: thus, even apparently passive participants act as gatekeepers, whose aim is to assess as best as possible the abduction that lead the first speaker to gossip by matching her statement against the receiver's own *Individual Knowledge Base* (which mirrors at its best update the Group Knowledge Base), and against the best informed (abductive) guess about the speaker's Disposition.[28] If this mechanism is accurate, it would partly explain the social efficacy of gossip and why, according to many of its commentators, it is not necessarily all for evil—even more, it is *necessarily not* all for evil. As stated by Collins, in her "feminist" defense of gossip,

> If all gossip aims at denigrating its subject, whom will it persuade? My malicious intent will usually only succeed where those whom I seek to persuade do not see my aim. If all those who gossip admit to the aim of denigration, their co-gossips have no reason to *interpret* their claim about *S*, the subject, as telling them anything about *S*, except that the malicious gossip does not like her (Collins 1994, p. 108, italics added).

When engaging in gossip, therefore, partners are far from exchanging simple information, but rather pass on clustered information, whose composing parts (maybe not present, but abductively traceable as the mold that generated them) may heavily affect one another and the gossips themselves. Reading Collins' analysis within the present formalization, it can be argued that:

1. The receiver may operate on her Individual Knowledge Base and on the received gossip. Since the Individual Knowledge Base mirrors the informational model of

[28]Gossip is often regarded as an exchange where the participants also aim at consolidating their relationship by talking or badmouthing about somebody else (Yerkovich 1977; Gluckman 1963 endorsed this view as well). When that is the case, it could be said that the participants' aim is still inferential, but the *primary* objective is to assess each other's Disposition towards a certain person.

social interactions as well, she will already have an idea of the emitter's *Disposition*: we must not forget that the possibility of gossip is entailed by participants recognizing each other as members of a same social group. Therefore, the receiver can abduce the core of the gossip (that is the clues and the evaluation offered by the first speaker) basing on her best-updated knowledge.

2. If the counter-appraisal reveals that the gossip is significantly incompatible with the receiver's *IKB*, and neither the first speaker nor other participants are able to make a case explaining the discrepancy, then the receiver's counter-abduction does not focus on gossip anymore but on the cluster of the emitter's Disposition embedded in the gossip, and formulate a new abductive hypothesis about the Disposition of her conversational partner, which might in turn become a subsequent topic of gossip.

This explains why, in some occasions, it is possible to obtain valuable knowledge also in gossip situations: an accurate and updated knowledge of one's interlocutors can guide the receiver through fine-tuning the heuristics for extracting the relevant knowledge from a cluster of gossip. It would also explain why to rely on gossip delivers better epistemic and strategic pay-offs to those who are the most immersed in the gossiping interactions: this way they have the most updated *KB*s, so they can perform the best inferences about the quality of the received information and extract the relevant core. Conversely, one who gossips once in a while is less likely to be able to efficaciously manage the clusters of information she receives, and is hence incapacitated to tell the relevant core of gossip from the originating disposition, for instance.

If this is true, then the argumentation embeds a tacit element that should be made explicit since it will be pivotal to fully understand the following and conclusive step: that is, the presence of a temporal dimension. Gossip, indeed, is heavily dependent on the timeliness of most human endeavors. What you hold to be gossip may turn into a fact (for you) if, over enough time, you may be able to assert the information independently from the source of the gossip. Similarly, the flow of time is a necessary frame for the construction of a Group Knowledge Base, and the possibility that it is mutually updated by Individual Knowledge Bases. My final contention is that timeliness can be of help to understand the development of gossip at group level, by matching the latter to a closely related epistemological phenomenon, that is rumor.

7.4.2 Abducing on Gossip at Group Level: Confronting Gossip with Rumor-Based Knowledge

Whereas we dealt so far with the kinds of abductive inferences taking place between single gossiping agents, I stated through the chapter that the actual actors of gossip are *groups as embodying epistemic synergies*, and that gossip could be conceived as an abductive appraisal performed by the group on itself, through the *emergence* of a series of abductive inferences: the notion of emergence conveys a strong spatiotemporal connotation (Holland 1997). While we dealt with the inferential building blocks of gossip in the previous section, I will now explicit my final claim and show the kind of abductive processing performed by the gossiping group itself. Though the analysis must move from the single occurrences of gossip, I have stated since the beginning that gossip makes sense only if understood within a continuous and reiterating pattern of similar exchanges: to paraphrase Aristotle, just as "one swallow does not a summer make, nor one fine day," so one socially-oriented conversation does not make gossip.

My first step was, in fact, to consider the gossiping group as embodying the epistemic synergy (that is, gossip). Synergies—as soft assembled, function-dominant epistemic systems—depend heavily on the operations carried out by their parts. Synergies are difficult to individuate because it is hard to tell the reach of the single components within the system. Translated in the case of gossip, this difficulty reflects the theoretical impasse in defining gossip as a strategy that matters chiefly for the group, but chiefly for the individual as well. This impasse, though, reflects an epistemological one as well: as introduced in Sects. 7.2.3 and 7.2.4, gossip individuates a situation in which the "group" knows everything but all of its members know something (ranging from *a lot* to *very little*). But how can we still say something like "the group knows" if we denied an ontological reality to the group? And how can something deprived of ontological reality display a gnoseological one?

The answer rests once again in the resolution of the tension originating between the epistemic effort of the individual parts of the synergy, and that of those individuals when elaborating on the outputs of the epistemic synergy they enact—a process that is necessarily set within the flowing of time. We have already seen in Sect. 7.2.4 that groups are the necessary condition for gossip to take place, and the parsimonious notion of group I sketched out is compatible with another kind of social epistemic activity, analyzed by Gelfert as "rumor-based belief" (Gelfert 2013). Gelfert recognizes the affinity between gossip and rumor, but claims that: (a) rumor is not necessarily focusing on social-related information, and (b) "certain instances of first-hand testimony may well count as gossip, whereas rumor, by necessity, is based on hearsay that cannot—at least not momentarily—be confirmed by independent evidence" (p. 8). Coady's differentiation of rumor from gossip seems to be compatible:

> This is one difference between rumor and another form of communication with which it is often confused, gossip. Gossip may well be first-hand. By contrast, no first-hand account of an event can be a rumor, though it may later become one (Coady 2012, p. 87).[29]

Rumors as well circulate in groups, but these are not as closed as gossiping groups: since rumors are often context-dependent, they circulate among groups defined by the same context (such as, groups of investors, users of a particular electronic brand, advocates of a nutritional habit and so on). A characteristic of rumor stressed out by Gelfert, though, can be stimulating for my investigation of the group-level inferential dimension of gossip:

---

[29]Gelfert's and Coady's characterizations of gossip and rumor seem to present two sides of the same coin, as the former places the stress on evidence, while the latter on testimony.

> From the perspective of the recipient, mere existence of independent evidence is irrelevant, if it is out of reach—perhaps because it is temporarily unavailable, kept secret, or unintelligible to anyone lacking significant expertise. Rumor-mongering, thus, does not arise from a *suspension* of "secure standards of evidence", but instead indicates a collective need for corroborating (or disconfirming) evidence to be made available. [...] What rumors typically require is at least the presumption that relevant evidence *can be had*—and, often enough, is in fact enjoyed by epistemically privileged outsiders (Gelfert 2013, pp. 8 and 9, original italics).

Can groups, with respect to the gossiping interactions I described so far, be seen as *recipients of rumor*? At group level the difference between rumor and gossip becomes inapparent because gossiping interactions act as a filter, embodied by participants acting as inferrers and gatekeepers, allowing only certain information to reach the Group Knowledge Base. Interestingly, Coady offers us a view on the diffusion of rumor that illuminates how progressively, after a certain stage, we can say that the difference between gossip and rumor is inapparent.

> [...] real rumors are often spread among people who know each other. This means that those who hear a rumor may be in a position to make a rational judgment about whether or not to believe it, on the basis of their knowledge of their informant's reliability, as well as the reliability of their informant's informant, their informant's informant's informant, and so on (to the extent that they know who all these informants are) (Coady 2012, p. 90).

The process individuated by Coady is indeed reminiscent of the abductive appraisal of gossip (described in Sect. 7.4.1) performed by matching the new gossip on the speaker's assumed *Disposition* and the receiver's best updated Knowledge Base. What is clearer in rumor, that is the absence of first-hand testimony, is implicit in gossip when analyzed as a group-level interaction, inasmuch as after being absorbed in the Group Knowledge Base, and corroborated by a sufficient number of gossiping interactions, the first hand testimony that is crucial for the beginning of gossip becomes eventually irrelevant.

Thus, single instances of gossip act as *authoritative sources* of information for the group's Knowledge Base, because of the reciprocal and collaborative assessments I described in the last section. In this sense, framing my approach in Gelfert's lexicon, it can be said that groups have an *epistemic dependence* on the information exchanged in gossiping interactions.

Remaining faithful to our understanding of groups as instantiating epistemic synergies, it is interesting to see the hiatus between the kind of epistemic processes displayed by individuals and those displayed by the same individuals when operating on the refined output of a synergetic effort (such as gossiping), that is the knowledge recognized as belonging to the Group Knowledge Base. Whereas, as I just explained, abductive inferences at individual level can make use of personal fallacies (such as *ad hominem*, bad company, poisoning the well etc.) to assess the interlocutor's *disposition* (Bardone and Magnani 2010), the group-level inferences will rely far less on the same inferences since they have *already* been performed and the information has already been assessed. In this sense, it could be argued that the first stage of gossip is meant to abductively *select* which information must make part of the Group Knowledge Base, while the second stage—performed synergistically at group level—engages and performs other inferences on the Group Knowledge Base itself.

Gelfert seems to avail this interpretation, stressing the importance of *timeliness* in his appraisal of the acceptability of rumor-based knowledge. His reasoning, in facts, orbits around a kind of "If *p* were true, I would have heard it *by now*": he concludes that "if a rumor is told by a source that one has reason to believe is part of a by and large trustworthy social network, the 'filter effect' [...] may well outweigh any argument from the lack of coverage-reliability of official sources of information" (p. 22). Timeliness, that is to take into account a dynamic conception of information sharing (be it rumor or gossip) is essential to this view, just as—in the current analysis of gossip—I stressed that the aim of gossip is to make sure that an individual's Knowledge Base is as updated as possible with respect to the Group Knowledge Base: you cannot conceive something's being *updated* without referring to its actually chronological development. It is therefore particularly interesting to read Gelfert's argument if we consider the group (embodying an epistemic synergy) as the epistemic subject of rumor-based belief:

> One might worry that there may come a point when incoming testimonial reports may themselves be justifiedly dismissed on the ground that, if they were true, one would have heard about them (earlier) [...but, after] all, if H rejects α's report that p on the basis that, if p were true, she would have heard it by now, she overlooks the fact that her hearing p from α may very well be the first time that she "legitimately" encounters the report in question—where what is "legitimate" depends on the time lag that is objectively appropriate to H's position in society, degree of separation from the original source of the information, membership in relevant epistemic communities and so forth (pp. 19 and 20).

The benefits of gossip, which have been fascinating sociobiology, can indeed be framed properly through an epistemological modeling. As shown in the precedent sections, individuals pay the abductive costs of gossip, thus allowing—for the benefit of an abstract subject (the group)—a crystallization of information that is "decently warranted," in the epistemological sense that it has been more corroborated than falsified *so far*. The epistemic asymmetry between the consistency of the IKB with respect to the GKB was already introduced in Sect. 7.2.3, but there seems to be an asymmetry concerning the inferential level as well: albeit gossip maintains a chiefly abductive and hypothetical dimension, two broad inferential areas can be individuated, at least from a theoretical perspective. The inferences performed at group level (that is, on the information belonging to the Group Knowledge Base) are not anymore chiefly *simple* fallacies regarding the person, but rather more plausible abductions performed on *rich* clues such as the persistence and diffusion of information and its corroboration by independent sources, to reach decently plausible conclusions ("This is the third time that somebody, for different reasons, tells me that Petra could be having an affair: I'm starting to believe you might be right..."): it is as if a group member was able to tell the difference between information that consist in the personal elaborated addition to her Individual Knowledge Base, thus concerned by a certain epistemic regime, and information that belongs to her Individual Knowledge Base *in virtue of* belonging to the Group Knowledge Base. The epistemic value of the information acquired through gossip, therefore, depends on how easily an agent can access to the Group Knowledge Base and update her own Knowledge Base: it is strategically effective to rely on gossip only if the agent can operationalize in a different way those beliefs that are abductively produced at individual level (still to be forwarded, for them to be assessed, to the Group Knowledge Base) and those that are abductively produced *by* the epistemic community relying on different clues. Groups, as instantiating epistemic synergies (gossip, in this case), behave *smartly* if their members can make the distinction I just sketched out, and can mostly rely on best-updated Knowledge Bases.

7.5 Conclusion

The aim of this chapter was to provide a new outlook on gossip that, setting off from anthropological and psychological studies, could be defined as *epistemological*. The matter at stake was to identify the informational and inferential dynamics of gossip, both as far as individual speakers and as far as the group they belong to were concerned, and to understand the formation of gossip-based knowledge.

The first section explored the tension between seeing gossip as an individual-centered and a group-centered (and group-serving) activity: that tension was mitigated, as the model I used represented gossip as a collaborative *epistemic synergy* in which the goal for individuals is to share and update a Knowledge Base of social information, adherent to a synergetic Group Knowledge Base, projected as embodying the knowledge of the group itself. In the second and third sections, I investigated how these Knowledge Bases were formed and consolidated: following earlier insights, I suggested that the best frame for capturing such knowledge-assessment and formation dynamics is *abductive reasoning*. By referring to abduction I could make sense, in a collaborative dimension, of the hypotheses-generating nature of gossip: this accounted on the one hand for gossip being a kind of (often successful) strife for the truth, while on the other hand for the scarce epistemic value of the single instances, as often captured by common sense. The third section, in fact, differentiated the kind of abductive inferences at play *between gossipers*—in which the appraisal is dominated by informational heuristics that could be defined as qualitative fallacies, mostly dependent on the person—from those enacted at *group level*—more similar to the quantitative handling of rumor by means of fallacious heuristics operating on notions such as the persistence and the distributed corroboration (or non-falsification) of information.

My approach theoretically grounds the intuitive notion that relying on gossip is a *sometimes good, sometimes bad* strategy. If gossip can be described as a synergetic epistemic osmosis between Individual and Group Knowledge Bases, then it is clear that the "group," as a deflated epistemic subject projected by the synergy, has the best Knowledge Base from which correct inferences can be drawn: the success of individual inferences operated on gossip depend on how updated an individual Knowledge Base is, with respect to the Group Knowledge Base. It follows, from the analysis carried out in the last section, that one is more likely to perform successful inferences based on gossip the more she is exposed, and participates, to gossiping interactions, because her Individual Knowledge Base will match as best as possible the Group one.

This last claim, no matter how trivial it may appear, could stimulate future research in social and applied epistemology—as an empirical research question as well: could gossip be seen as a form of epistemic altruism—thus incorporating the likelihood of an *epistemic sacrifice*?[30] After all, if gossiping aims indeed at offering to the group a best-updated Knowledge Base, then the cost of information assessment is entirely absolved by gossipers themselves, who might receive and share (willingly or not) poor, inaccurate, or toxic knowledge. Inaccuracies in the information will be found out by gossips themselves—at times jeopardizing their own epistemic welfare (and the actions that follow from it)—who will then include or exclude the information from future gossiping inferences, thus brining the Group Knowledge Base to a further level of refinement.

It is perhaps proper to conclude by stating that this chapter did not mean to provide the *ultimate* explanation of gossip, but rather to provide an analysis of how epistemic subjects are concerned by gossip, and how this impacts (and is affected by) gossip's knowledge-forming capacities. This is not, nor it is meant to be in contrast with parallel ethical, psychological and cognitive explanations of gossip, making the case for the more negative and dangerous effects of gossip, namely the enforcement of unreasonable norms and regulation, the diffusion of prejudices and stereotypes which can lead to mobbing or worse, or the simple consolidation of the relationship between two gossipers at the expense of some unaware victim.[31] Just like appreciating the brilliant engineering behind a deadly pistol should not make us unaware of the problematic and negative uses that pistol could be employed in, to acknowledge the perfectly decent epistemic functioning of gossip, which partially explains its evolutionary success, should not necessarily make us blissfully forgetful of the potentially devastating effects of gossip on everyday human groups.

Such a conceptualization of gossip may indeed show the fruitfulness of the approach I am following in this book, and help individuate the recurrence of eco-cognitive inferential patterns in most differentiated, but still human, cognitive domains. If the center of human cognitive activities is about being able to cope with the external world, then one should not be surprised to see, for instance, the recurring of abductive patterns in strategic ecologic cognition, in science and now in social cognition: abduction is a sense-making inference, allowing the cognizant to the *bigger picture*.

As we saw in the previous chapter, eco-cognitive inferences are first of all about making a *model* of one's surroundings, then—provided that model provides an initial survival-value—it is just about *fine-tuning* it with feedback from the actual ecology (or reality, used in a naïf sense): no model is final, and abductive inferences serve both in providing the initial model, through our cognitive capacities and our cognitive hard-wired and cultural endowments, and in the progressive improvement or rebuttal of the model. Yet, our survival does not depend uniquely from our relationship with

[30]Sacrificial cognition will be analyzed in Part III, Chap. 13.

[31]The next chapter will deal with these aspects relating to gossip.

the physical and animal world. The notion of "ecological dominance" (Flinn et al. 2005) suggests how, at a certain point of our evolutionary development, human beings became the primary threat to each other, overcoming the menaces posed by any other animal: this shift can be associated to the passage from ecological niches to cognitive niches. As proved by the history of mankind since the rise of ancient empires, it is not numbers who decided the fate of wars and battles, but technological and cultural developments: in other words, what matters is the advancement of cognitive niches. As cognitive niches are supported and maintained by human beings, the first level of cognitive niche construction has to do with solving problems about which individuals can be trusted or not, and to what extent.

Indeed, whereas for the development of an ecological niche the apparent biological endowments of an agent would usually suffice, the maintenance of a cognitive niche is usually the work of an explicitly-minded group: for this scope, being able to read the intentions so to infer the trustability of an agent is as fundamental as knowing the presence of ecological chances and risks in one's environment. This is why, although the domain changes, the eco-cognitive inference underlaying gossip are not that different from those I illustrated in Part I (Chap. 2), for the detection of nearby agents. Gossip should be rightly analyzed as a fundamental tool preparing for niche construction, and a cognitive niche of its own, aiming at *making sense* of one's human environment by relying on the sharing of apparent traits, from which broader (and strategically useful) categorizations can be inferred.

In the next chapter I will carry on my analysis of gossip, merging the study of cognitive niches and the kinds of inference characterizing them together with an approach that is typical of philosophy of technology. Adopting a philosophical methodology that will characterize Chap. 11 as well, I shall focus on the "translation" of a given behavior (and the related inferential patterns) from one cognitive niche to the other, namely a high technological one which offers dramatic improvements meant to enhance cognitive performances. This will also allow me to deal with the ill reputation of gossip as a mediator of violence, a trait I overlooked—for clarity purposes—in the current chapter.

References

Anderson ML, Richardson MJ, Chemero A (2012) Eroding the boundaries of cognition: implications of embodiment. Top Cogn Sci 4:717–730

Ayim M (1994) Knowledge through the grapevine: gossip as inquiry. In: Goodman RF, Ben-Ze'ev A (eds) Good gossip. University Press of Kansas, Lawrence, pp 85–99

Bardone E, Magnani L (2010) The appeal of gossiping fallacies and its eco-logical roots. Pragmat Cogn 18(2):365–396

Baumeister RF, Zhang L, Vohs KD (2004) Gossip as cultural learning. Rev Gen Psychol 8(2):111–121

Beersma B, Van Kleef GA (2011) How the grapevine keeps you in line: gossip increases contributions to the group. Soc Psychol Pers Sci 2(6):642–649

Bertolotti T (2011) Facebook has it: the irresistible violence of social cognition in the age of social networking. Int J Technoethics 2(4):71–83

Bertolotti T, Magnani L (2013) A philosophical and evolutionary approach to cyber-bullying: social networks and the disruption of sub-moralities. Ethics Inf Technol 15(4):285–299
Boehm C (2002) Variance reduction and the evolution of social control. Paper presented at annual workshop on the co-evolution of behaviors and institutions, Santa Fe
Castro L, Medina A, Toro M (2004) Hominid cultural transmission and the evolution of language. Biol Philos 19:712–737
Clark A (2005) Word, niche and super-niche: how language makes minds matter more. Theoria 54:255–268
Coady D (2012) What to believe now: applying epistemology to contemporary issues. Blackwell, New York
Collins L (1994) Gossip: a feminist defense. In: Goodman RF, Ben-Ze'ev A (eds) Good gossip. University Press of Kansas, Lawrence, pp 106–114
Daniels JW Jr (2012) Gossip in the new testament. Biblical Theol Bull: J Bible Cult 42(4):204–2013
Dunbar R (2004) Gossip in an evolutionary perspective. Rev Gen Psychol 8(2):100–110
Flinn MV, Geary DC, Ward CV (2005) Ecological dominance, social competition, and coalitionary arms races: why humans evolved extraordinary intelligence. Evol Hum Behav 26:10–46
Foster EK (2004) Research on gossip: taxonomy, methods and future directions. Rev Gen Psychol 8(2):78–99
Gelfert A (2013) Coverage-reliability, epistemic dependence, and the problem of rumor-based belief. Philosophia 41(3):763–786
Gintis H, Bowles S, Boyd RT, Fehr E (eds) (2005) Moral sentiments and material interests: the foundations of cooperation in economic life. The MIT Press, Cambridge
Girard R (1986) The Scapegoat [1982]. Johns Hopkins University Press, Baltimore
Gluckman M (1963) Papers in honor of Melville J. Herskovits: gossip and scandal. Am Econ Rev 4(3):307–316
Godrey-Smith P (2002) Environmental complexity and the evolution of cognition. In: Sternberg R, Kaufman K (eds) The evolution of intelligence. Lawrence Erlbaum Associates, Mawhah, pp 233–249
Goldman AI (2010) Introduction. In: Haddock A, Millar A, Pritchard D (eds) Social epistemology. Oxford University Press, Oxford, pp 1–28
Goodman RF, Ben-Ze'ev A (eds) (1994) Good gossip. University Press of Kansas, Lawrence
Holland J (1997) Emergence: from chaos to order. Oxford University Press, Oxford
Kelso JAS (2009) Synergies: atoms of brain and behavior. In: Sternad D (ed) Progress in motor control. Springer, Berlin, pp 83–91
Laland KN, Brown GR (2006) Niche construction, human behavior, and the adaptive-lag hypothesis. Evol Anthropol 15:95–104
Linde C (2001) Narrative and social tacit knowledge. J Knowl Manag 5(2):160–171
Magnani L (2005) Abduction and cognition in human and logical agents. In: Artemov S, Barringer H, Garcez A, Lamb L, Woods J (eds) We will show them: essays in honour of Dov Gabbay, vol II. College Publications, London, pp 225–258
Magnani L (2009) Abductive cognition: the epistemological and eco-cognitive dimensions of hypothetical reasoning. Springer, Berlin
Magnani L (2011) Understanding violence. Morality, religion, and violence intertwined: a philosophical stance. Springer, Berlin
Magnani L (2013) Is abduction ignorance-preserving? Conventions, models, and fictions in science. Log J IGPL 21(6):882–914
Marlowe F (1998) The nubility hypothesis. Hum Nat 9(3):263–271
Odling-Smee F, Laland K, Feldman M (2003) Niche construction. A neglected process in evolution. Princeton University Press, New York
Peirce CS (1931–1958) In: Collected papers of Charles Sanders Peirce. Harvard University Press, Cambridge, vols 1–6 Hartshorne C, Weiss P (eds) vols 7–8, Burks AW (ed)
Putnam H (2002) The collapse of the fact/value dichotomy and other essays. Harvard University Press, Boston

Rohwer Y (2007) Hierarchy maintenance, coalition formation, and the origin of altruistic punishment. Philos Sci 74:802–812

Sendova-Franks AB, Franks NR (1994) Social resilience in individual worker ants and its role in division of labour. Proc R Soc Lond Ser B: Biol Sci 256(1347):305–309

Taylor G (1994) Gossip as moral talk. In: Goodman RF, Ben-Ze'ev A (eds) Good gossip. University Press of Kansas, Lawrence, pp 34–46

Wilson D, Wilczynski C, Wells A, Weiser L (2002) Gossip and other aspects of language as group-level adaptations. In: Hayes C, Huber L (eds) The evolution of cognition. The MIT Press, Cambridge, pp 347–365

Wilson DS, Dugatkin LA (1997) Group selection and assortative interactions. Am Nat 149:336–351

Woods J (2005) Epistemic bubbles. In: Artemov S, Barringer H, Garcez A, Lamb L, Woods J (eds) We will show them: essays in honour of Dov Gabbay vol II. College Publications, London, pp 731–774

Woods J (2013) Errors of reasoning. Naturalizing the logic of inference. College Publications, London (forthcoming)

Yerkovich S (1977) Gossiping as a way of speaking. J Commun 27:192–196

Chapter 8
Niche Construction Through Gossip and Mobbing: The Mediation of Violence in Technocognitive Niches

Abstract As hinted in the previous chapter, gossip is empowered by a "pragmatic" notion of truth, according to which truth, at least in social matters, is the opinion held by the majority of authoritative sources. The power of gossip relies in the immediate pragmatic enactment of the predicated truth, as acknowledged by (Taylor, Good gossip pp. 34–46 1994). Magnani, in Understanding violence. Morality, religion, and violence intertwined: a philosophical stance (2011) opened a new perspective on the philosophical study of violence, showing how even the maintenance of cognitive niches involves coalitions supporting axiological positions, likely to trigger violence, either structural or individualized. Even if topics such as mobbing and bullying deriving from gossip might seem void of philosophical relevance, cultural studies such as the Girardian tradition (Girard, The scapegoat [1982] 1986; Violence and the sacred [1972], 1977; Job: The victim of his people [1985] 1987) show how a mechanism called *mimetic rivalry*, rooted in envy and fear against differences, informs an historically ever-present motif: scapegoating, and other related methods for resolving conflicts and crises. In this chapter, I will take advantage of the emergence of a contemporary phenomenon, that is mobbing in Social Networking websites, as a case illuminating two issues. First, the violent element embedded even in the most innocuous gossiping appraisal; second, the kinds of sidetracking that might happen when a long-established cognitive behavior—such as gossip—is so-to-say *translated* into a different cognitive niche, which seems to afford a new, potentiated version: should we say that we are facing the same behavior?

8.1 Introduction

As hinted in the previous chapter, gossip is empowered by a "pragmatic" notion of truth, according to which truth, at least in social matters, is the opinion held by the majority of authoritative sources. The power of gossip relies in the immediate pragmatic enactment of the predicated truth, as acknowledged by Taylor (1994). Magnani (2011) opened a new perspective on the philosophical study of violence,

© Springer International Publishing Switzerland 2015

T. Bertolotti, *Patterns of Rationality*, Studies in Applied Philosophy, Epistemology and Rational Ethics 19, DOI 10.1007/978-3-319-17786-1_8

showing how even the maintenance of cognitive niches involves coalitions supporting axiological positions, likely to trigger violence, either structural or individualized. Even if topics such as mobbing and bullying deriving from gossip might seem void of philosophical relevance, cultural studies such as the Girardian tradition (Girard 1977, 1986, 1987) show how a mechanism called *mimetic rivalry*, rooted in envy and fear against differences, informs an historically ever-present motif: scapegoating, and other related methods for resolving conflicts and crises.

In this chapter, I will take advantage of the emergence of a contemporary phenomenon, that is mobbing in Social Networking websites, as a case illuminating two issues. First, the violent element embedded even in the most innocuous gossiping appraisal; second, the kinds of sidetracking that might happen when a long-established cognitive behavior—such as gossip—is so-to-say *translated* into a different cognitive niche, which seems to afford a new, potentiated version: should we say that we are facing the same behavior?

8.2 Internet Actualizing Gossip's Ill Repute

Cyber-bullying (that is mobbing carried out through means related to the Internet, that is via Instant-Messages, Social Networks and so on), and other issues related to violence being committed online in prosocial cognitive niches, are beginning to constitute an emergency worldwide. The effects of pervasive cyber-bullying, at the level of school-system and individual students, are akin to those of ordinary bullying. Yet, recent events, climaxing in teenagers or young adults taking their lives after episodes of cyber-bullying, put forward the urgency to elaborate and enforce specific policies about this phenomenon.[1] Institutions are particularly sensitive to the problem especially as far as teenagers are concerned inasmuch as, in cases of inter-teen episodes, they cannot easily appeal to ordinary justice (i.e. threaten to sue). This is not to say that episodes are necessarily dealt with *extra lege*, in many jurisdictions parents can sue on behalf of their minor children: the matter at stake is that, for this to happen, a clear image of the episode must penetrate up to the parents' awareness—or the caregivers'—but in order to report the episodes the youth must have a decently clear conception of her active and passive rights (so to acknowledge that her rights have been violated).

[1]These cases received peculiar attention by the media, but new ones are arising every day: May 2010, a UK boy hangs himself after being "plagued" by online bullying. September 2010, a US college student commits suicide after his roommate spies on him and divulges his homosexuality on Twitter, offering public viewings of his encounters. This event sparked the debate on cyber-bullying in the US. June 2011, in Italy, elementary school teachers abusively mock a pupil on their easily accessible Facebook profiles—as a result, no action is taken against the teachers but the pupil changes school. January 2012, a New York girl commits suicide by walking in front of a bus after being bullied in real life and on Facebook. April 2012, in Georgia (US), a teen sues over a Facebook bullying episode achieved via a fake mocking profile of the girl herself. The case exposed insufficient State legislation about the matter.

In order to develop the most suitable policies, institutions should not be satisfied with a statistical and sociological approaches the phenomenon, but should seek a deep ethical and philosophical understanding—also referring to the evolutionary past of human beings. Lest some misunderstanding arise, it is important to state that the outlook on morality I adopt in this chapter is the same as in the previous one: a naturalistic and "agent-based" one. I do not mean to analyze the violence sparked by gossip in techno-cognitive niches as a moral issue, comparing it to a more or less consolidated moral perspective, but rather to show how it is nested within the moral perception and moral judgements performed by individual, real agents. The aim of this chapter is to show a way to fill a theoretical gap, offering some answers (and some questions too) that can illuminate future policy-oriented research and reflection.

One of the key elements to understand the violent societal changes brought about by Social Networking Websites, most often in the form of cyber-bullying, is the notion of sub-morality. This concept, introduced by Magnani (2011) in his investigation of the relationship between morality and violence, partakes of a more fluid and naturalistic conception of morality. In brief, not only human beings deal with fully-recognized moralities, such as the Christian morality, the Capitalistic one, Utilitarianism and so on, and with the violent conflicts that big moralities engage with each other: the lives of human beings are also affected by sub-moralities, that are smaller axiological systems that can regard one individual or a group (usually small). Sub-moralities are linked to the notion of *over-moralization*, namely the tendency to attribute extreme moral value to things that—according to standard moralities—should be at best *indifferent*. Therefore, one could profess herself a Christian as for her broad moral engagement, and yet endorse separate sub-moralities—for instance preaching male superiority, or the wrongness of dressing unfashionably, of being fat, and so on. Sub-moralities, disregarded by traditional moral studies, have a high potential to provoke everyday conflictual situation, and play a pivotal role in organizing group balances within human communities supporting cognitive niches. This is why they will be crucial for the understanding of the deep reasons why Social Networking Websites seem to afford so perfectly violent behaviors such as cyber-bullying.[2]

In order to do so, I will start by reconnecting my argument to the evolutionary studies (some of which I already mentioned in the previous chapter) carried out in the past two decades, focusing on gossip as a tool for social assortment, thus endowed with a dual function: protect the group from free riders, intruders and bullies...but also violently mob and punish the deviant members. In the second section, I will describe which aspects of gossip, vital for bullying, are co-opted by social network scenarios. A fundamental trait of human social life, that is the subdivision is smaller coalitions, or sub-groups, will be shown as missing in SN dynamics—which therefore constitute themselves as *structurally* violent. The third section will deal with techno-ethical and epistemic concerns regarding how gossip mediated by SN manages to empower cyber-bullying. The fourth section will characterize cyber-bullying as often sparked by self-gossip (often degrading into self-mobbing) in a scenario

[2]Section 8.4 will focus on the relevance of sub-moralities for our argument: the notion is further specified in footnote 9.

where familiar sub-groups, which also mediate defense and mutual understanding, are disrupted. The final section will consist of a philosophical summary, divided in two parts: a *pars destruens* analyzing whether Social Networks, in their actual configurations, are fit for being used by humans-like-us, and a *pars construens* examining the broad potential consequences of highly enforced regulation aimed at contrasting cyber-bullying.

8.3 Mediating Violence Through Gossip

8.3.1 Gossip as Social Communication: The Naturalistic Fallacy

In order to understand most phenomena relating to the perpetration of violence in virtual prosocial environments, I suggest taking a few steps back in the realm of natural philosophy. I suggest that the analysis of cyber-bullying, as it relates to Social Network Websites, must be preceded by the understanding that social networking is rooted in a most ancient social phenomenon: gossip. Traditionally despised as *idle talk*, studies on gossip as a topic endowed with a significant theoretic dignity started in the second half of past century (Foster 2004; Goodman and Ben-Ze'ev 1994).

As I explained in the previous chapter, the revaluation of gossip is chiefly indebted towards two main disciplines: anthropology on the one hand, evolutionary studies on the other hand. Anthropologists focused on gossip as a means of social regulation (see for instance Gluckman 1963; Yerkovich 1977), but the mechanism often maintained a sense of otherness—both a geographical and chronological distance: gossip could be indeed more than mere idle talk, but that was true as far as *other* people, at a different stage of history, or in another land or society, were concerned—keeping gossip as a anthropological topic seemed to shield "decent" Westerners from the (sometimes violent) implications of these studies. Conversely, evolutionary psychology and sociobiology immensely boosted the reputation of gossip (no pun intended) by showing its relevance as far as it concerns the dawn of language and sociality (Dunbar 2004; Wilson et al. 2002): gossip was presented as *natural*. This is where the naturalistic fallacy plays its part: with different formulations along the history of philosophy, the "naturalistic fallacy" refers to the Human *is-ought to be* question—here, it indicates the bias towards accepting as good (that is, as what *ought to be*) what is natural (i.e. what *is*).[3] Partially because of the hard-to-die naturalistic fallacy, which reverberates in the moral justification of whatever is hypothesized to have been part of our evolutionary heritage, these studies managed to obtain a massive

[3]Objections to evolutionary psychology often involve the assumption that proposers of the discipline endorse the (actual) moral implications of their findings, turning everything evolutionarily selected into something morally positive, and thus admit moral no-no's such as rape, adultery, free-riding and so on. In this chapter I do not mean to carry any accusation on evolutionary studies, but to criticize a certain kind of reception.

dissemination in public opinion: the result was an uncritical and widespread acceptance of the common utility of gossip as "social hygiene", which came with a complete obliteration of its violent nature.

One of the most quoted and successful evolutionary approaches on gossip is Dunbar (2004). His main take is that gossip evolved along the lines of grooming to allow hominids the possibility to cope with life in large social groups: in his view, the origin of gossip is strictly connected with the development of language itself. Dunbar's argument, reduced to its core, consists roughly in what follows:

1. *Homo Sapiens* and their close evolutionary relatives—anthropoid primate societies—do indeed display an uncommon degree of sociability, which warrants for a series of benefits concerning hunting, foraging, protection of the individual, control over the local ecology and so on. Such high levels of pro-sociality, though, are extremely costly to maintain. The main issue comes from the fact that hominids are not eusocial species (i.e. ants and bees living in colonies, displaying a kind of "group mind" Wilson et al. 2002), therefore the social integration backfires in the difficulty in coordinating the behavior of the whole group while preserving general welfare. Primates' solution to these problems was the constitution of alliances by structuring a "sense of obligation" between individuals, through pleasurable activities such as grooming (as many animals do as well).
2. Human beings' early social groups would grow in numbers and complexity, therefore one-to-one social bonding activities were not efficacious anymore. This sparked the co-evolution of language and gossip. Grooming would create social obligation through the manipulation of endorphins: gossip—albeit the reception of an unheard news does provoke a small release of endorphins, would create bonding in another way as well. It managed to convey a powerful message of social commitment, akin to "I consider it more important standing here talking to you than being over there with [anyone else]" (Dunbar 2004, p. 102).
3. The effect of using language-to-gossip as a bonding tool had also the epistemic advantage of creating a social group in which everyone is informed about the behavior of all other members, and thus where everyone feels controlled and it is easier to individuate and punish free riders, intruders, etc.[4]

Dunbar's research, together with works that had begun in the middle of the Seventies (such as Yerkovich 1977, who characterized gossip as a "way of speaking", beyond mere *idle talk*), played indeed a pivotal role in the rehabilitation of gossip as something positive and *okay*. The evolutionary element met a rare pro-scientific bias in popular culture so to favor a kind of *naturalistic fallacy*, as I stated at the beginning of this section. Gossip is natural, therefore it is good. Paraphrasing the popular hymn, the feeling would be that since gossip was "good" for our ancestors, it is "good enough" for me. So "gimme that old time gossip!".

---

[4]This aspect of gossip was analyzed in depth, from an epistemological outlook, in the previous chapter.

8.3.2 Characterization and Punishment of Deviants

What popular culture did not retain so much were those researches concerning the pragmatic effects of gossip. In her seminal work, Yerkovich had stressed how gossip relies on, and produces, *judgements*.

> If it turns out that the opinions coincide, *evaluative categorizations* are developed concerning the individual and his/her way of doing things. These evaluative categorizations, developed from the initial, shared recognition of a name, lead to the possibility for gossip (Yerkovich 1977, added emphasis).

The word "evaluative categorizations" about individuals and their "way of doing things" is a hint to the fact that gossip is mostly about moral prescriptions. This view is also shared by mainstream research on gossip: many essays included in Goodman and Ben-Ze'ev (1994) stress the evaluative nature of gossip, which is carried out by comparing gossip against the moral guidelines shared by the gossiping group. The moral nature is dual, because throughout this shared evaluation, what also takes place is the restatement of the group's moral principles in form of approved and disapproved narratives (Linde 2001; Baumeister et al. 2004). The notion of scandal, crucial in Gluckman's investigation of gossip, is also rooted in *moral* disapproval (Gluckman 1963).

As introduced in the previous chapter, the narratives shared by gossipers are not objective and purely descriptive reports of what happened to a certain person or a group of persons. Conversely, they exert, and boost, a given point of view in order to promote certain values rather than others, omitting or stressing particular features of the actors at play. Those narratives make use of stereotypes common to the participants and deal with typifications of individuals, such as "moral characters," based upon traits of the individual perceived as peculiar. This represents a shift from the social level to the moral level: selective categorizations are built on traits of the individual on whose peculiarity the speakers agree—as an example, people can be characterized as *Casanovas* or as *good-hearted yet gullible people*, *slackers* and so on, and bits of gossip will be judged relevant and shared *according* to these representations. The updated representations, ultimately, will inform behavior. After judgement comes action: as remarked by Gabriele Taylor in his essay *Gossip as moral talk*, "the manner in which a person is represented in thought, whether justly or generously, in terms of clichés or caricatures, is bound to affect at least to some degree one's behavior towards her, if not necessarily what one does, at least how one does it" (Taylor 1994, p. 44).

If gossip, as I argued, deals essentially with morally-connoted information, then its first *victim* must have been *perceived* as morally deviant. To individuate who might have been the original deviant, we will need a profitably economic hypothesis about the origins of moral communities.[5] According to anthropologist Christopher Boehm,

---

[5]It is important to bear in mind the highly conjectural and speculative character of the theories I am about to expose: by the use of data derived from social evolutionism, paelo-anthropology and theoretical biology they attempt to give plausible explanations about the development of *Homo Sapiens*' distinctive traits.

it is the practice of meat-sharing that made human being a *social and political animal* (as Aristotle would say, *zoon politikon*) and, most of all, a moral being: humans, as other hominids and apes, display an innate disposition towards flesh and fatty meat from large game. As long as alpha males (who usually prove to be the most successful hunters as well) ruled hunters-foragers groups, meat consumption would undergo a feast-or-famine trend, depending on the willingness of the hunters to share their prey. The need to separate food consumption from political power triggers the constitution of a proto-morality, in order to cope with the ambivalence the community feels towards the hunters.

> In achieving nutritionally significant variance reduction, band-level moral communities would seem to be working with a human nature that they believe to be strongly selfish, but also socially responsive and perhaps innately generous to some significant degree. Accordingly, bands employ a moralistic mix of "sticks" and "carrots". The best hunters are pressured as needed, but they also are praised and given respect. The widely-predictable result, be it contentious or smooth, is reasonably even-handed meat distribution (Boehm 2002).

A proto-morality is a primitive axiology, a set of shared more-or-less tacit beliefs about *how something should be done*, it involves an approval of group-benefiting behaviors and disapproval of the selfish ones. It is not a full blown morality, but rather the shared knowledge within a small group (hence Bohem refers to "band-level moral communities"), for instance, that sharing the meat you hunted is *good* while keeping it all for yourself is *bad*: thus, the sharing of large game would not be at the hunter's discretion anymore. Similar practices constitute what Boehm classifies as *egalitarian syndrome*:

> As a generalized, anti-bullying political program egalitarianism not only spreads out political power quite equitably; it evens out the consumption of large game, and it also makes it unlikely that dominant types could monopolize a band's women (Boehm 2002).

In brief, according to Boehm's theory, we can trace the origins of early moral communities back to the crescendo of hostility between dominants and submitted individuals over the consumption of particular types of food. Consequently, the first moral deviant was probably an outnumbered alpha male clinging to his power, and the first moral community was a defiant sub-coalition of subordinates. It is very important to notice how, according to this approach, moral definitions to not ignite the conflict, but vice versa the conflict produces the characterizations. With this respect, morality has to be considered conflictive because of the simple fact of being born out of a clash. That is to say, most of the time moral rules imply two sub-groups: a coalition underpinned by the rules, and the *other(s)-for-the-coalition*. The original hunters-foragers group broke into those two sub-groups and he who once was the leader of the whole group (namely the dominant) was eventually considered the *other*, the deviant.

This moral revolution at the dawn of society has been allowed by two main factors, both unique to humans: exceptional skills in remote killing[6] and, of course, symbolic and recursive language. Language afforded labeling (i.e. "deviant") and its displacement features made it easy to report on bullying alphas:

> The moral aspect involved not only well-communicated group values and judgments about individual domination behavior, but also some kind of a crude shared template, carried in the heads of group members, about what kind of political society they wished to live in Boehm (2002).

Language allowed a complex juxtaposition between actual members' behaviors and the ideal ones all members carried in their mind: thus, thanks to language and gossip, groups were enabled to act as "judge and jury, as well as policeman" (Boehm 2002). Concluding this section, I suggest state that gossip is not only about asserting rules, but an active instrument to individuate and label deviants and trigger the persecution, that is to say mobbing. As a matter of fact, morality involved more than prescriptions about food sharing, but the constitution of these first moral principles worked as a blueprint for the creation of further ones:

> Once the basics of equalizing political power and meat consumption were taken care of, there was in place a methodology that permitted groups to invent a wide variety of prohibitory bells and whistles such as incest taboos, special exogamy rules, food taboos, menstrual taboos, and many other locally-concocted rules which, as today, were far from being universal and often had little direct impact on the basics of subsistence and social life (Boehm 2002).

An increase in the number of minor and major moral rules and principles caused an increased number of potential deviants: the ambivalence we are dealing with can be explained considering how gossip, through a violent use of moral information about individuals, can label any member of the coalition as *deviant*, thus making her or him eligible for condemnation, sanctioning and eventually expulsion.[7]

---

[6]Bingham (2000) maintains that cooperative behavior was not given since the beginning in human groups, but had to be achieved by discouraging mutual deception. In brief, Bingham suggests that the ability in killing from a distance allowed humans to overcome what he defines "hostile manipulation"—which postulates the likeliness for an individual to cheat his peers whenever the occasion comes. Of course, hostile manipulation is the nemesis of cooperative behaviors, and can be limited only by an effective punishment system. It can be argued that most of animals do not display advanced cooperative behaviors because they are not eager to attack an element of their group who proves to be hostile, as this would result in one or several one-to-one matches, with a 50–50 chance of being killed or severely injured for both the "cheater" and the "punisher". Remote killing, instead, made individuals more committed to undertake the punishment of cheaters since the chances for the attackers to get killed or even injured are reduced exponentially, as the cheater is exposed to a number of contemporary attacks and cannot answer all of them at one time. This theory also claims that only once remote killing constrained human beings' natural inclination to hostile manipulation, language eventually developed into the articulateness we are acquainted with: otherwise it would have proved to be no more than a cognitively expensive cheating tool.

[7]The self-gossip of gossip carried out in this chapters is necessarily not exhaustive, as it is functional to the analysis of cyber-bullying and the authors would not burden the reader with unnecessary notions. Bertolotti (2011) deals specifically with the connection between gossip and social networking—only marginally dealing with cyber-bullying—whereas an epistemological and inferential analysis of gossip as group cognition was provided in the previous chapter.

8.4 Mediating Gossip Through Social Networks

If the background I have just summed up is accurate, social networking websites indeed appeal to an extremely potent and long established human inclination—but often only the eusocial aspect of gossip is pointed out, and not the punitive one. As a matter of fact, when thinking of Social Networks (SN), theorists such as Tufekci (2008) stress in a parallel way how social networks are eminently *social*: my idea is that Social Networks can be studied as Constitutively Violent Networks (CVN), where violence is structural and not to be seen as a by-product of the system, just as in gossip the violence is constitutive (it is a tool to observe and punish deviants) and not a side effect. Structural violence, also called systemic violence by Žižek (2009), is that kind of violence that cannot be attributed to an individual (like a single homicidal act) but is embedded in societal structures, and that social justice deals with. Labeling something as *constitutively* violent is a way of saying that its structures are developed to carry and diffuse a certain kind of violence (not necessarily illegitimate).

Tufekci (2008) points out how most social networking websites are similar in as much as they present a series of profiles, which *correspond* to the website users, and those profiles are linked by relationships of "friendship". The effect of this deflationary use of the word *friendship* is clear: calling virtual acquaintances (who may or may not correspond to real-life acquaintances) "friends" configures the whole exchange in a different way than if—say—they are called "links" or "connections" or simply "contacts". Mimicking the rhetoric of friendship, social networking websites such as Facebook and Myspace can have a particular appeal by simulating some affordances of real-life human relationships, perhaps in an even easier way.[8]

The word "friendship" possesses in real-life an amazingly rich range of hues and variations, while sooner or later every user of social networking website experiences the inadequacy of the word "friend" as a label which is meant to cover roughly anyone from high-school sweethearts to colleagues, including relatives, partners, neighbors, VIPs and so on.

Considering what I exposed so far, we could indeed see how this configuration equals setting a ring, a fight arena. Human beings are fond of divisions, because divisions (no matter how apparently trivial and petty) do mirror the high number of sub-groups and sub-moralities[9] we belong to Magnani (2011), and that often clash

---

[8]This description seems to echo the eco-cognitive epistemological approach I outlined in Chaps. 2 and 3. The aim of this book is indeed to make a case for the cross-specificity of certain inferential patterns: the care for one's surrounding, the avoidance of "traps" and mimicked adversaries and adversities (may they be real or cognitive/intellectual), the construction and individuation of phenomena are processes relevant for ecological survival, for the development of efficacious social cognition and for scientific reasoning as well. Of course, in different settings different aspects become in turn critical.

[9]Magnani (2011) devotes the fifth chapter to the issue of sub-moralities, a concept I already sketched out while introducing this chapter. He contends that not only moral absolutism cannot withstand a reality check, but even old-fashioned moral relativism is short-sighted: not only different human beings partake of different moralities, all claiming to be the right(eous) one without a superior meter to discriminate among them, but even within one single moral group (say catholic

with each other. The concept of friendship is multi-faceted so to acknowledge the existence of regular friends, acquaintances, close friends, best friends, soul mates, old friends, former friends, friends with benefits, office friends, false friends, friends we do not really want to see, and so on. Whereas we put (and love to put) a special emphasis on each of those categories, SN roll a huge dozer over them and flatten them all. This is the first reason why SN can be understood as Constitutively Violent Networks. They are CVN because the first rule a member has to accept is *undifferentiation*. As brilliantly remarked by Girard along his Biblical and anthropological studies, human thought always identified undifferentiation as a condition and trigger for violence, within a game of chaos and mimetic rivalry.[10] When undifferentiation is a forced feature of the system—as in the case of SN, where all hues of friendship are assimilated—then the violence becomes structural and constitutive.

Consequently, the more or less intended confusion triggered by the linguistic abuse of the word *friendship* can spark a wide range of perversions, but in this reflection I am going to focus on those relating to gossip: gossip is so relevant to group-level interactions that Wilson identified it as a signal of the fact that groups can indeed be considered as levels of natural selection (beyond the individual unit) Wilson et al. (2002), but the effects of gossip go beyond that. It could in fact be defined as a "group-projecting" behavior: referring to friends, it is both true to say that gossip is what we do with friends, and that friends are the persons we gossip with.[11] Otherwise said, while gossip was co-opted as a tool serving clashes between sub-groups (and hence sub-moralities) Gluckman (1963), in the undifferentiation of social networks it becomes a tool that anyone can use to potentially harm anyone else.

The gossip-related nature of social networking websites, though, is not immediately clear and therefore calls for a philosophical enquiry: on the one hand, common sense is perfectly aware about the *gossipiness* of social networking websites (to the point that they are renowned to raise severe privacy issues), on the other hand it is rather rare to see on somebody's wall a post such as "Hey! I saw Mary kissing John at the bar yesterday night!! But she hasn't broken up with Peter yet, has she!? What a...", maybe complete with tags to the involved individuals. The gossipiness of

(Footnote 9 continued)
or capitalistic morality) or within a single individual there might be several sub-moralities (sometimes corresponding to parallel sub-groups) that are activated, deactivated and switched several times even during a single day (for instance, catholic morality, self-perceived as a whole, can harbor sub moralities such as a feminist one, a racist one, a money-oriented one and so on). Such sub-moralities, in spite of being extremely opaque to the agent, play a pivotal role in social assortment and feeling of "being part" of a determinate human coalition—and are obviously just as incommensurable as traditional "big" moralities.

[10]Girard (1986) studies the phenomenon known as scapegoat in which a state of crisis, characterized by undifferentiation between different agents and brought over by mimetic desire and rivalry—i.e. wanting something just because it is desired by the other party—is perceivably solved by choosing and sacrificing a scapegoat. The victim is distinguished by signs that make it a perfect victim facing the moralities and sub-moralities of the perpetrators, namely she is different, but she bears no responsibility before the crisis. I already matched the Girardian perspective with gossip studies in Sect. 7.3.3.

[11]The relationship between gossip and groups was analyzed in great detail in the previous chapter.

social networking websites is subtler since it assumes the peculiar configuration of an invitation towards "self-gossip", which often winks at narcissism and voyeurism (Rosen 2007): users are indeed supposed to "post" images, movies, preferences and much more, provided they concern *themselves*.

Following these insights, in the next section I will focus on the precise reasons why SN empower bullying to the extent that several institutions consider cyber-bullying a phenomenon needing actual policies. This question is mirrored by another, more philosophical one: why is cyber-bullying *perceived* as such a threat.

8.5 Why and How Do Social Networks Empower the Distribution of Violence?

The notion of "empowerment" (Amichai-Hamburger et al. 2008) has always been crucial in the technoethical discourse. New technologies *empower* users in a number of ways: they empower their epistemic capacities (Internet, thanks to its pervasiveness, can make us know more information and in a faster way, nearly zeroing the costs of provision), and from this epistemic empowerment follows a civic empowerment (Bertolotti et al. 2011). The most recent IT media also empowered us emotionally and morally, making distant communication easier, more reliable and simulating higher proximity—also moral proximity.

When we think of empowerment, we most often refer to positive or desirable traits improved by new technologies and media. Yet, digital amplification often concerns less than positive characteristics. Among what has been dramatically empowered by IT we must also acknowledge our capacity for fraud, for intruding in other people's personal data to harm them and so on. Stapleton (2012) most interestingly points out how, over the past few decades, the emergence of advanced technology (mostly relying on data processing and retrieval) empowered a progressive forsaking of our values via an externalization of our functions to technological processing: referring to an anticipatory analysis offered by Sproull and Kielser (1991), he states that:

> Prophetically, [they] understood that what was shaping our social selves was the combination of medium and message. […] A picture emerged of some now familiar issues. These authors hinted at what would become known later as cyber-bullying and noted the fragmenting power of both email and online conferencing as social media. Whilst it seemed to many that these technologies would bring us together in new ways through massively integrated systems architectures – Sproull and Kiesler warned us about the flying apart. They also saw hope in the democratising affects of the online processes we would struggle with. But they also forecast difficulties arising from, for example, the lack of social cues and the amplification of extant organisational fault lines, which could accompany these new online spaces (Stapleton 2012, p. 5).

Such negative empowerment, empowerment that is not for good (at least considering the consequences), is the same that lead, according to Stapleton, to the tremendous economic crises which has been plaguing the west for the past 6 years (at the moment this book is being written). Human beings were fooled by the same intelligent systems they had created and imbued with what they thought were their values, but ended up creating an environment (this time financial and economic) that is not livable by its very creators.[12] As I am going to explain, I believe that the problem of internet-mediated violence should be seen the same way, that is at least partially created by a misperception (and mis-reception) of the values mediated by that peculiar part of IT technology that are SN.

What is it so tremendous about cyber-bullying? As often suggested, its perceived ugliness rests in its pervasiveness. It is a form of empowered gossip, and it can follow the target so deep inside her emotional safety to be felt as a bunker-busting bombshell. This is caused by a complex intertwinement of reasons. The first can be drawn back to what I explained in Sect. 8.4, that is the *undifferentiation* between levels of friendship. Sub-moralities, as described by Magnani (2011), are not only factors triggering violence, but also (and consequently) defense mediators. Sub-moralities do not depend only on debates about moral issues, but on random elements that are turned into moral issues, and hence become *chances* to trigger violence (Barnett and Littlejohn 1997).[13] Therefore, one must not think that only adults can partake of clashing sub-moralities, but children and teens as well. A dress code, a taste in music, one's appearance are all elements that can inform a moral decision (hence an aggressive decision) against somebody. Of course, also traditional bullying factors such as ethnic belonging, social status, religious orientation and so on inform sub-moralities (and clashes between them).

Those sub-moralities are disrupted when transposed in the digital *arena* of SN (which we are now able to see as Constitutively Violent Networks), and we witness the activation of a free-for-all situation. Everybody is perceived as a potential foe, and the same connections regulate both in-group and between-group dynamics, as far as the RL sub-groups are concerned.

The further implication is that this new undifferentiated situation makes it hard for users to rely on the same sub-group defense mechanisms: as suggested by many studies (for instance the comprehensive Sontag et al. 2011) teenage cyber-bullying comprehends many cases of reactive bullying (that is violence which is a reaction to a previous violence), and curiously the targeted victims of cyber-bullying are statistically likely to be bullies themselves in their "real lives" (Rigby and Smith 2011).

It is worth remembering here Boehm's theory about the evolution of hominid egalitarian societies explained in Sect. 8.3.2: basically, he claims that egalitarian societies can develop when individuals with a lower status overthrow the "bullying" rule of the alpha male. Biologist Paul Bingham interestingly enriches this

[12] Whereas Stapleton's argument is slightly apocalyptic, I do not suggest that highly technological environments, also decision-making ones, are necessarily bound for ruin: one way to avert this is to contemplate knowledge as a *duty* and foster a sufficient number of human beings that are knowledge-committed and warrant for a safe and proper functioning of the technology, without being blinded by expectations, greed and laziness (Magnani 2007). I will discuss Stapleton's perspective on the economic crisis in the next chapter, especially in Sect. 9.4.2.

[13] As stated in the introduction, this process is mutually implicating with over-moralization: a sub-morality could take a random, morally indifferent element, such as being skinny, fat, foreign, wearing odd clothes, or being different from most people as an implicit sign of punishable moral deviancy, and hence trigger violence.

kind of speculation, suggesting a correlation between egalitarianism and aggressive technology (Bingham 2000)[14]: only remote killing could allow subs to efficaciously challenge an overwhelmingly powerful alpha male. Now it should be clear why some evolutionary understanding is necessary to fully grasp phenomena such as cyber-bullying: it is easy to see an analogy between a group of hominids stoning a boorish alpha male, and a group of teens virtually harassing the bully who harasses them at school—just to make his cyber-life miserable the way he makes their real-lives a pain. In case of reactive violence, the computer screen does not only shut off the moral proximity, but the possibility to be harmed in return.

Provocatively, it could be said that the continuous appeal to the separateness of the virtual dynamics as something that encourages violence by diminishing empathy and so on is rather *wishful* if not misleading. Most of the time children and teen really mean the horrible things they say each other. They mean them even when they say it to each other in real-life, but on SN it is much less likely to get slapped back.[15]

The efficacy of cyber-bullying does not only rely on the *remoteness* of attacks, but also on their (actual or perceived) epistemic quality. Once again, the phenomenon of bullying has to be traced back to its trigger, that is gossip. Gossip, indeed, is about sharing information—most often about people who would much rather not have that same information divulged and spread—but, as I showed in the previous chapter, what ordinary gossip is not about is copy and paste mechanisms. When gossiping, information is always subject to ending up in the wrong hands—*willing* or *unwilling troublemakers* in Yerkovich's words—and thus cause potentially unwanted violence. Still, one of the ambivalences of gossip—responsible of making it such a fundamental device—is that, as if aware of its poor epistemic value, gossip can always be downplayed as being "mere" gossip, probably not even transmitted in its correct form. In real life, gossip turns into mobbing or bullying only after *words* have reached a critical state and *trigger* some kind of violent punitive behavior.[16]

---

[14]Bingham's proposal is sketched out in footnote 6.

[15]Concerning this issue, a good research question would be to investigate whether teens belonging to weak sub-groups in their real lives, and are thus exposed to bullying and violence, adopt more quickly to the new digital environment and are thus more likely to make use of the *undifferentiation* for their own benefit. With this respect, it is also important to *separate* those cases where cyberbullying (proactive or reactive) newly arises on the internet, and those where the cyber-bullying is *just* an extension of precedent real-life violence, but also to focus on the *interactions* between the two categories.

[16]Several essays in Goodman and Ben-Ze'ev (1994), often quoted in the previous chapter, stress that gossip is *idle*, hence purposeless, in its nature. Gossips are often not interested with the truth-value of their utterances: it is therefore easy to share something that, albeit armful, is just being produced and shared together in the gossiping interaction (Ayim 1994). Conversely, one might have second thoughts before spreading some information that is known to be a *fact*, and that it will necessarily harm somebody's reputation. Furthermore, gossip tends to mature into rumor, where the individual who originated it progressively dissolves: whereas a fact, in order to detain its higher epistemological value, often requires a contact with its original source for validation. The matter comes down to the "epistemic responsibility" one is willing to take on.

Conversely, in technology-mediated gossip, gossipers can rely on copy and paste functions to make sure that the information is transmitted with nearly total copy fidelity, whereas information in regular gossip has usually been only verbal—at least as ordinary citizens were concerned,[17] in social networking websites gossip is enriched with images and movies which further increase its epistemic value: text and pictures posted on Facebook, for instance, can circulate and become source of nuisances or harassment in a way that cannot compare to traditional forms of gossip. A picture cannot be downplayed just as a word can be, in this sense gossip on SN-CVN ends up being perceived as a swarm of potentially threatening facts looming above the victims head. In a way, we could say that, in SN, gossip can automatically turn into bullying, because publicly gossiping about somebody resorting to actual facts (proved by pictures, screen captures and so on) can be charitably called "defamation" and is a fully aggressive behavior even if no direct effort is moved against the target.

8.6 Self-Gossip and Self-Mobbing in the Light of the Disruption of Sub-Moralities

The last point I mean to analyze in this chapter is still related to the disruption of sub-groups performed by SN, and concerns on the one hand the "authoritativeness" of the first-person posting, and on the other hand the moral distance presumably afforded by the computer screen.

The rupture of moral proximity brought about by computer screens and avatars is often advocated as one of the causes of the lack of empathy which results in particular verbal violence, threats and so on that would not be carried out so openly in real life. My claim is that this presumed moral gap is a biased artifact of the analysis, informed by the honest (but untrue) answers of the subjects: what is perceived as a distance between the virtual world and real life is actually caused by the loss of references as far as sub-moralities and sub-groups are concerned. But let me analyze the two issues separately, for better clarity.

8.6.1 From Enhancing One's Cyber-Reputation to Self-Mobbing

Reputation, that is the common image generally held about somebody, can be defined as the product of a distribution of knowledge constructed *by negotiation* around that very person. It is not the same as a moral characterization inasmuch as it displays a further shared nature, but it is surely influenced by the different characterizations at play in gossiping scenarios: gossip is in fact one of the main forgers of reputation

[17] Tabloids have long since accustomed us with visual gossip, but this concerned only VIPs, celebrities, politicians and so on, and informs another conception of the word "gossip".

(Gluckman 1963; Yerkovich 1977; Hunter 1990; Goodman and Ben-Ze'ev 1994), and the accurateness of reputation does also depend on the amount and diversification of circulating gossip (Sommerfeld et al. 2008; Ayim 1994). One's reputation is not univocal, and depends on the sub-morality we use as a reference. After some particular event (say the behavior held at a party), a kid can be considered "cool" by his friends and "lame" by another party (for instance those a practical joke was played on). Gossip, within the coalition the kid belongs to, will reinforce his reputation as a local hero, while gossip in the opposing coalition will stress the negative image and—if reaching a critical point—it will trigger a reaction of mobbing and possibly bullying against the kid.

Traditionally, the distinction between reputation and identity has always been tacitly acknowledged: reputation is an *artifact* superimposed on an agent by others, with a chaotic dependance on the agent's own actions. There are no monotonic correlations between one's deeds and the effects they have on the person's reputation: the very same action, perceived from different outlooks or performed at different moments—even if the intention is the same—can affect positively or negatively one's reputation. The proof of this complexity is that, reputation-wise, we are often trapped in *catch-22* situations, knowing that one course of action might hinder our reputation with certain respect, but abstaining from it would damage it from another perspective. Furthermore, we know we have no *single* reputation but several, and when we speak of "our reputation" we intuitively consider the reputation of the sub-group we belong to.

With respect to social networking websites, a different phenomenon might arise. First of all, we must keep in mind that, as I suggested, the form of gossip afforded by social network is in the first place "self-gossip:" users think they are publishing only what they want and they might be unconsciously led into believing that their self-gossip might improve their reputation.

Indeed, self-gossip does not enter the gossiping networks unmediated, but becomes the object of gossip itself: gossip, in fact, is about the information being offered together with a sense of characterization, which is also what makes it relevant and interesting, as I suggested in the previous sections: therefore, if social networking websites delude the user into believing that one's profile can be a simulacrum of her identity, and if this simulacrum is thought to be able to project one's reputation as the subject wants it to be, this process might very likely see failure as its ultimate end.

Mere rumors have always been able to seriously damage people's reputation, and hence their life, sometimes beyond repair: still, it is important to stress how there is a fundamental difference, if only at a psychological level, between the rumor being introduced by somebody else rather than by the object of the rumor herself.

This phenomenon can be individuated as emerging right now, inasmuch as many of the violent consequences of technology-mediated gossip originated with the subject sharing sensible information about herself: furthermore, the subject usually shares high-epistemic quality information about herself, such as images and movies, which corroborate the *authoritativeness* of the first person. If I say something about myself, it is quite likely to be true. If I give some proof of it, it is even likelier to be true. The further problem with SN is that the disruption of sub-moralities has not always been a feature of internet-mediated communication. If you consider the phenomenon of forums, message boards, thematic chat rooms, it is clear that they more-or-less mirror the real life distinction between different subgroups and sub-moralities. Reputation within these environments is more similar to its real-life connotation: because of some constraints such as the theme of the board (you do not want to go off-topic) the information you post, also about yourself,[18] is generally *appropriate*.

Reputation and *appropriateness* are strictly connected, and both have an eminently local nature, and can often be judged only over the consequences: while, for instance, for an anonymous avatar (the alias identity of the internet user) it could be appropriate to forge her virtual identity by publishing sexually explicit images of herself in a dedicated board, an "innocent" bikini set shared by a teenager on Facebook might encourage comments that the user would not foresee. This could heavily damage her real-life reputation, when all she meant was to display herself as a solar, energetic person among her peers.

Indeed, the prerequisite for something to be perceived as "appropriate" is that the partners in conversation (but also the onlookers) apply the correct frame, and this frame is evoked by the sub-morality the subject and their peers belong to. So, a new frame does not conflict or erase all the pre-existing meanings that a certain activity—for instance, sharing sexually explicit images of oneself over the internet—may have, but it somehow becomes a new layer that re-frames it adding new interpretations (for instance in a BDSM virtual community). At the same time, it also favors the emergence of dissonance, in case the participants are not able to successfully re-frame the activity and hence do not perceive them as *appropriate*. Consider the following example: Jane is an erotic photographer in her free-time, but nevertheless she is very skilled, knows a lot about cameras, computer editing and so on. In brief, erotic photography is something serious for her—requiring skills and knowledge. One day, in order to understand some more about it, she agrees to model for a friend with similar interests. After the shooting and the editing her friend decides to post the picture on his wall and tag Jane (as a pragmatic way to say "Go and see that picture of you I took the other day"). Jane and her friend know that erotic photography is a moral issues to most, but they apply a shared frame because they see it as something artistic and beautiful. Jane's other friends and relatives might see the picture, and not be able to apply the same frame and just consider the image as gross and vulgar: the girl might start being bullied as a promiscuous and dissolute girl, with all the imaginable consequences. In real life, would she be walking down High Street handing around pictures of her erotic modeling to friends, relatives and acquaintances?

---

[18] Yet another element of this phenomenon is anonymity. As a matter of fact, internet communication has always been characterized by *anonymity* obtained through the use of avatars. The concept of avatar, etymologically linked to that of incarnation, represented the essence of virtual social environments (Meadows 2008): the avatar is a partial embodiment of the operating user, and their identities can be told one from the other. Furthermore, the avatar has usually been a warrant of anonymity for the user, allowing the Internet to become a reenactment of Victorian morality, where everyone could pursue more or less licit perversions protected by the mask of their avatar. Of course, avatars have reputations as well, but unless data leaks are the case, they do not affect those of their *puppeteers*.

There is already a discrete amount of literature dealing with the reason why users feel like sharing that much information about themselves on social networks, whereas they would not do it in real life. A particular bias towards Facebook and other social networking websites embed users in a kind of "It's just Facebook" state of mind: to paraphrase a common motto, the common feeling is that *what happens in Facebook stays in Facebook*. Research shows that, even after facing (mild) consequences caused by excessive divulgation of personal information, users tend to go back to being active on Facebook. The main idea is well exemplified by Debatin et al. (2009): "The technical strategy is to tighten the privacy settings; the psychological strategy is to integrate and transform the incidents into a meaningful and ultimately unthreatening context. Knowing or at least believing to know who the perpetrator is creates a feeling of control and reassurance. The perpetrators are branded immature individuals or creeps, and the incidents tend to be minimized as pranks, exceptions, or events from a different world: 'In the end', said Brian, reflecting on the hacking incident, 'it's just Facebook... There was a time before Facebook. You can do without it. It's ok'." (p. 100).

This kind of reaction against cyber-bullying basically resorts to storing Facebook as a separate cognitive niche mediated by the screen, but it is rather wishful because it hides from the view of the subject (and, perhaps, from many researchers as well) the core of the matter and the kind of disruption brought about by SN.

8.6.2 Separate (Moral) Words Trigger Bullying

Many accounts of cyberbullying seem to corroborate the *(perceived) separate worlds* thesis. It is interesting to consider a case that had made headlines in the US in January 2012. A 15-year-old girl committed suicide, reportedly after experiencing a history of school bullying, both in her real and virtual life. In this case, cyberbullying seems to be more like something that *adds up* to the traditional phenomenon, rather than a case of its own. Nevertheless, a statement on the ABC News report is quite relevant: the uncle of the victim said that "girls at Cummings' high school had been tormenting Amanda, and continued to leave inappropriate comments on her Facebook profile even as she lay in a coma at Staten Island University Hospital" (Curry 2012). Additional details include that the cyber-bullying went on *for six days* as she was hospitalized before her death, it was carried out by both male and female teens, and not all of them knew the victim personally. The first reaction of (self-perceived) moral and appropriate human beings (as we are) can be of two kinds: either these girls and boys are totally disconnected from the real life, and perceive the digital one as utterly separated with no influence upon the real world, or they are a shocking example of pure evil.

In this case, what strikes the *bien-pensant* as unacceptable is the fact that horrible things are said *when the girl was in a coma*. In my analysis, conversely, the striking part is the public statement of those feelings. If we consider political matters, during the legislations of President Bush in the US and PM Berlusconi in Italy, political polarization reached unprecedented levels and in small talk among sustainers of opposing parties it was not unfrequent to hear wishes of death for those political leaders, even in the apparently meekest people. I do not mean to foster yet another natural fallacy claiming that we have the right to be violent people, but as contended by Magnani (2011), understanding our violent nature helps us becoming at least consciously violent beings, which is a way of limiting the negative effects. Much of the research on cyberbullying seems to be blind to the inherent violent traits that are to be found in teenagers as well.

Even the author and any reader of this book are aware that not everybody likes us, we might have a bad reputation among some people and so on. We would be fools to think otherwise. What we do not expect is that such negative opinions and feelings be expressed publicly and (in our perception) gratuitously. As long as the offending opinion is held private, circulated among similar-minded people and does not trigger offending behavior, nobody thinks she is being bullied.[19] Conversely, as soon as the offenses are made public and publicly enforced, the rightful feeling of being bullied arises. This is to say that, in the case mentioned before, the "sociopathy" does not rest in meaning ill thoughts towards a dying, helpless person (otherwise we would be all sociopaths), but the sociopathy is to be found in the public sharing of such ill thoughts.

Conclusively, what I meant to stress is that the kind of sociopathy displayed by cases of cyber-bullying is an artifactual-driven one: cyber-bullies are not so because it is easier to be bullies online, but because the digital environment *de facto* invites such sociopathy by flattening (or making transparent) all of those fences—that are also communicative ones—between subgroups informed by sub-moralities. I would say that bullying, just as violence, has two characteristics: (1) bullying is distributed[20]; (2) bullying is what is perceived as such. The second characteristic is particularly relevant as far as cyber-bullying is concerned: even more than in real-life, the semantic divide between a practical joke and bullying is extremely blurred. When the sharing of offensive information and thoughts about someone is an activity that—because of the construction of the vehicle (that is the SN)—becomes public (and *consequently* humiliating), then this informational exchange (that we could call gossip) becomes immediately bullying even if it was not explicitly meant to be so at the beginning.

8.7 Sharing Thoughts: Can the Philosophical Reflection Still Impact Society and Policies?

The violence of SN seems to be originated by the intersection of biological, technological and moral constraints, some rooted in our evolutionary history and other more recent. The first actor (evolutionary) is gossip, a tool for social cognition that

[19] With this respect, Seligman et al. (2008) make an interesting plea for external rituality, characterizing it as much more vital for a peaceful sociality than sincerity, which—in their opinion—ultimately elicits extreme conflictuality.

[20] That is to say it is not about one's intentions, but it has to be enacted in the real or virtual world.

was selected for social bonding but for social distancing as well, within a complex interplay of subgroups and sub-moralities—the second actor. The third actor is social networks themselves, entailing a massively egalitarian structure that makes no distinction between the different ways of relating and communicating displayed by human beings within the different subgroups. This is a potentially lethal mix that exposes everybody's feelings (positive but especially negative ones) to everyone, as the power of gossip for social assortment powerfully emerges again. The main issue is that such group-assortment activity is a paroxysm of social cognition because the mega-groups (sometimes including thousands of individuals) created by SN-dynamics have no correspondent in real life.

Cyber-bullying came to the attention of media (and hence policy makers) after a number of cases in the past few years, especially in the US and UK, where a number of teens and young adults took their lives *after* being cyber-bullied. It would be cynical and wrong to dismiss these saddest cases as *post hoc propter hoc* fallacies (meaning that since they happened *after* the episode of cyber-bullying, then we *fallaciously* deemed they were *caused by cyber-bullying*). Yet, it seems reasonable to agree with Agatston (Kowalski et al. 2012) and other practitioners who contend that in many of these cases the cyber-bullying was the proverbial *straw that broke the donkey's back.* In most cases victims of cyber-bullying do not feel the need to end their suffering by taking their lives: not only statistics suggest that stand-alone cyber-bullying is less widespread than traditional bullying—and the latter is decreasing, as suggested by Rigby and Smith (2011)—but it is obviously less harmful from the physical point of view. I do not mean to philosophically support the idea that bullying toughens people up and therefore it is good, nevertheless recent philosophical and multi-disciplinal research, such as Magnani (2011), Žižek (2009), Mucchielli (2010), suggest that human experience is indeed a constant transition from one kind of violence to the other, some structural (like the disruption of sub-moralities and sub-groups by SN), some traceable to peculiar agents (like bullies).

If the analysis I provided—by integrating eco-cognitive elements with themes belonging to naturalized morality and evolutionary studies—is correct, and the violence (including cyber-bullying) afforded by Social Networks can indeed by reduced to the disruption, operated by the same SN structure, of the isolated sub-moralities and sub-groups that are essential to real life, then even an issue such as cyber-bullying is much deeper than originally thought.

8.7.1 Pars Destruens

As stated in the introduction of this Part, human beings can be defined as cognitive engineers, inasmuch as their main evolutionary task is the production and fixation of *cognitive niches.* The concept of *cognitive niche* (Odling-Smee et al. 2003; Magnani 2009), indicates a series of externalizations of knowledge into the surrounding environment, through for instance material culture, resulting in a modification of the selective pressure that an organism has to face. Cognitive niches should make our survival in a determinate ecology easier, inasmuch as the selective pressure should be modified so to favor the life of their developers. There can be many example of cognitive niches, some more vague and some more precise: language and social living can be seen as level-zero cognitive niches, while other environments such a classroom, a highway or a laboratory are extremely "objectified" cognitive environments. In the middle we can consider niches such as architecture, agriculture etc. Over the past decades, our environment was enriched by digital cognitive niches where the externalization of knowledge, and chances for change in welfare, are brought about via computer-mediated interactions (these will be analyzed in Part III, Chap. 11), and Social Networks can be rightfully considered as a cognitive niche.

Yet, as I explained in this chapter, the results of this peculiar externalization of knowledge can be counterproductive inasmuch as they clash with some evolved dispositions of human beings towards social cognition, even if they were engineered to improve social dynamics. Admitting a bit of apocalyptic rhetorics, social network could be labeled as a good example of *Terminator niche* (the next chapter, concluding Part, will be entirely dedicated to the analysis of this concept). Social networks, while in fact created to improve connections between human beings, could ultimately denature and break down the very connections they mean to foster because their architecture is actual-human-blind and moral-blind. One reason for this is that they might have been developed referring to an idealistic conception of human beings and not to an actual biological and psychological one.

Social networking are extremely handy for their way of integrating social agenda, messenger, entertainment in a unique app-based approach, but it should be questioned whether the benefits outweigh the social and moral downsides, part of which were explained in this chapter. In other words, I do not exclude that future debates might come to the conclusion that social networks are not *fit* for human beings and their use should therefore be strictly regulated as smoking, alcohol consumption, pornography and so on.

8.7.2 Pars Construens

I recognize that accepting this framework could transform policy-making into an even harder task than it is, but the right aim should be to develop policies that are not human-blind just as what they mean to police. In this sense, on the one hand policies might go towards a progressive limitation of use of SN to children and pre-teens whose sense of moral dynamics, understanding of gossip and so on has not fully formed yet.

In order to replicate the decently healthy functioning of real life sub-groups, social networking exchanges should be modeled around the famous "Dunbar's number", that is 150 individuals: Hill and Dunbar (2003) researched 150 as the highest number of individuals that a social group can count in order to function healthily. If every user had a top of roughly 150 contacts, she could have a fair mental image of who receives the information she shares, and share consequently. Implementations of the social networking websites could comprehend a progressive raising of the number of friends as the age of the user increases: this would correlate the experience with the age, and if a kid has, say, 20 available friends slots she will be less likely to add strangers (or peers she does not really like) and consequently her "friends of friends" are likely to be peers as well.

The final question could be rightly more philosophical and pedagogical: a debate is needed about how we want to calibrate our response to cyber-bullying. Namely, about what importance we assign to freedom in the education of teenagers, and whether the response should favor safety over freedom or the opposite. Even though I stated at the beginning of this reflection that I do not believe in any positive pedagogical role of "bullying" as something that toughens people up, I think it is important to issue a warning against *excesses* of safety (such as the British wave of "bump proof" playgrounds): a recent sociological article by Mucchielli (2010) about the development of violent behaviors in interpersonal relationships in France (and, generally, in Europe) since the 1970s, usefully stresses how the pacifying process, implied by the related civilizing process starting back at the end of the Middle Ages (postulated by Elias), is still at work and has not been reversed. Only verbal violence (insults, threats) has increased in police and judicial statistics: "if a pacifying process tends to reduce recourse to violence, it is because its primary consequence is the stigmatization and delegitimization of that violence" (p. 213). Yet, as crime news show us everyday, we face an increasing *inability* to engage the correct reaction to discomforting events: a widespread phenomenon such as road rage shows that even if we are less violent in general, we easily become more violent for *futile reasons* (Harding et al. 1998).

In our opinion, the risk is to produce something similarly unbalanced in the digital world: the emergence of more and more agents that cannot cope in a proportionate way with unpleasing situations. An excessive policing of cyber-bullying following top-down approaches[21] might produce cyber-citizens that are unable to face any kind of situation without the help of institutions, and when the institution cannot (or they perceive it does not want to) help them they recur to *go-for-broke* extreme solutions.[22]

Considering our biology and the violent traits we inherited from apes and hominids, the eradication of violence (also in the form of cyber-bullying) seems to be potentially more costly than benefiting, and the costs could be paid by those very traits upon which violence and bullying seem to prey, such as sociability, fun, creativity, cooperation and so on. The first step towards the eradication of violence might have to go through the removal of those sub-moralities (of any kind, not only those embedding for instance rhetorics of violence, such as racist sub-moralities) that are in themselves responsible for a number of violent conflicts and bullying (I am thinking,

---

[21] An example of bottom-up approach would be the progressive implementation of Dunbar's number in SN structure.

[22] Totally shielding human beings, especially teens, from proactive and reactive online violence could also prevent future citizens, in a more and more digitalized state organization, from being able to protect themselves from institutional and structural bullying, or to recur to strategies of rightful political activism that could itself be labeled *cyber-bullying* by opponents. I specifically addressed the issues relating to the evolution of activism into cyberactivism in Bertolotti et al. (2011).

mass scaled, about policies akin to school uniforms in certain schools, whose original aim was to prevent the clothes of pupils to signal their socio-economical situation). At the same time, as suggested along this chapter, those sub-moralities are indissolubly connected to the differentiation of human beings in groups and subgroups that on the one hand are the cause of conflicts, but on the other hand originate the unique variety of lifestyles, cultures and uniquenesses that characterize and give worth to human life. That is to say, we might come to understand that, within the disruptive structures individuated in SN and other digital environments, some of the *uniqueness* (seen as a positive trait) of real life is the ultimate cause of digital unruliness.

On an extremely provocative note, it could be said that a world where there can be an efficacious and systematic control of (cyber-)bullying already exists and it is the world constructed by SN websites. It is a world where the violence is produced because we cannot fit our evolutionary and moral heritage into that world. Since we are still at a stage of transition (that is we are not close to giving up our evolutionary and real-life moral heritage to turn into fully virtual beings), cyber-bullying is an ugly sign that we are not *just* fitting in Social Networks. Policy-makers might have to decide whether they want to continue improving the real life of people at the cost of this misfitting, or achieve a violence-free cyber-world, not knowing the cost of this—and how it might fire back on our real lifestyles.

As far as we—21st century adults—are concerned, freedom seems to be more valuable over safety. Totalitarian regimes of the past century taught us this lesson the hard way, but we should not blindly transpose this assumption on developing individuals—because of their different needs. The very idea of *education* implies that younger generations are in need of precautional constraints in order to protect them in their development, so that they can become full-grown individuals. As stated at the beginning of this chapter, minors do not perceive legal tools as adults do, in order to use them for their protection—or are just not that prone to use them[23]: consequently, minors can hurt each other much more freely and mindlessly. Of course, such dichotomy between safety and freedom of information has been a legal and political issue over the past few decades, especially since the unprecedented shift brought about by the internet. In a famous ruling of the US Supreme Court in 1997, *Reno v. American Civil Liberties Union*, justice John Paul Steven concluded that:

> We are persuaded that the CDA [Communication Decenty Act] lacks the precision that the First Amendment requires when a statute regulates the content of speech. In order to deny minors access to potentially harmful speech, the CDA effectively suppresses a large amount of speech that adults have a constitutional right to receive and to address to one another. That burden on adult speech is unacceptable if less restrictive alternatives would be at least as effective in achieving the legitimate purpose that the statute was enacted to serve. […] It is true that we have repeatedly recognized the governmental interest in protecting children from harmful materials. But that interest does not justify an unnecessarily broad suppression of speech addressed to adults. As we have explained, the Government may not "reduc[e] the adult population … to … only what is fit for children".

[23]As stated in the introductory section, I do not claim that teenagers cannot be protected by the Law as adults are: more subtly, I mean to stress that teenagers might lack the awareness and the knowledge to comprehend that their actual cases could deserve treatment in a justice court.

This view, favoring freedom over safety, could be acknowledged as the mainstream *institutional* position adopted in the United States. This position emerged in answer to a kind of *all or nothing* question, namely whether the Institutions should block the entire diffusion of a message to prevent harm to part of the population. Efforts in Europe are following a different direction. Policy-makers have been struggling to avoid all or nothing, one-size-fits-all solutions, aiming at the development of dynamic, tailor-fitted solutions: famously, in 2011 the German state of Schleswig-Holstein had ordered state institutions to shut down their Facebook Pages and remove the "Like" button from their websites, or face fines.

On another account, policy makers could now prescribe the use of technologies such as profiling and biometrics (Cantore 2011) to make sure that only appropriate audience can access determinate content (it could be a way to enforce age-requirements for joining social networking websites, similar to fingerprint-sensible safe locks on firearms): these technologies were not widespread and available when the *Reno v. American Civil Liberties Union* was ruled. Is this a viable solution? In my opinion it falls short of being ideal because, whereas it could indeed provide a significantly lie-proof system to regulate the access to certain features, what would happen is just a shift of the burden: instead of having a dichotomy between safety and freedom, we would have another one between safety and freedom on one side, and privacy on the other one.[24] Indeed, this would mean to entrust a third party with an incredibly vast and relevant amount of personal data, and furthermore, our activity online would become integrally trackable.

In conclusion, it is worth signaling that the first draft of the forthcoming EU Report on *Social Networks and Cyber-bullying in the Teenager Population* (Rizza and Pereira 2013) (comprehending results from the workshop where preparatory work for this chapter was originally presented) stressed the positive role—confirmed by literature—that families could play in preventing and assessing the emergence of cyber-bullying in teenagers. Mass media have accustomed us, at least as far as the case of Italy is concerned, to overprotective families whose children are always the victims and never the perpetrators of violence. Nevertheless, once this idiosyncratic detail—perhaps an artifact of public opinion—is overcome, the role that families could play is clearly pivotal. Yet, because of generational digital-divide, however well-intentioned the intervention of the family risks to become inevitably *clumsy* and perceived as violent and irrationally intrusive by the children. This could lead to a further *submersion* of their online activity which would make it even harder to monitor and could be exploited by ill-meaning individuals. In this sense, the familiar educational effort should be canalized onto an informed—but not necessarily expert—curation of the teenagers' *moral character* (Kohlberg 1964). Therefore, notwithstanding the much-needed and appropriate effort to develop polices on

[24]It is easy to imagine a database containing data of citizens, including fingerprints: instead of entering one's age, or stating she is older than 14, 18, 21 etc., the system might ask for a fingerprint scan. In Italy, automated cigarette resellers require that customers enter their National Health Card—which is shaped like a credit card—before buying tobacco products, to prove they are older than 18: the same process, with the National Health Card, is used for automated sex shops (sic!). The customers must enter their National Health Card, and then the device displays its available content.

cyber-bullying (and to enforce those policies), a significative improvement could be also individuated in a less specific but more powerful statement of *traditional* values such as the old-fashioned "Golden Rule:" a teenager that has incorporated among her moral guidelines that one should not do onto others what she would not like others do onto her *in real world* will probably adopt the same principle in her cyber-life. Conversely, focusing on cyber-bullying as an intrinsic phenomenon, whose causes and solutions reside uniquely in the medium it pervades—and in how this medium is approached—will probably lead to an ultimate failure.

8.8 Conclusion

As I suggested in this chapter, many issues concerning the change of the role played by gossip in the construction and maintenance of a cognitive niche seem triggered by the shift in the structure gossip is enacted upon (in this case, the techno-cognitive niche actualized by prosocial internet services and websites). What should be focused, then, is the radical change brought about by technology in cognitive niche construction and maintenance.

The next chapter will focus directly on the construction and maintenance of techno-cognitive niches, and will explore how some particular aspects of technological innovation can be a revolutionary actor in the way human beings order their own cognitive niches.

References

Amichai-Hamburger Y, McKenna K, Tal S (2008) E-empowerment: empowerment by the internet. Comput Hum Behav 24(5):1776–1789

Ayim M (1994) Knowledge through the grapevine: gossip as inquiry. In: Goodman RF, Ben-Ze'ev A (eds) Good gossip. University Press of Kansas, Lawrence, pp 85–99

Barnett PW, Littlejohn SW (1997) Moral conflict: when social worlds collide. Sage, London

Baumeister RF, Zhang L, Vohs KD (2004) Gossip as cultural learning. Rev Gen Psychol 8(2):111–121

Bertolotti T (2011) Facebook has it: the irresistible violence of social cognition in the age of social networking. Int J Technoethics 2(4):71–83

Bertolotti T, Bardone E, Magnani L (2011) Perverting activism. Cyberactivism and its potential failures in enhancing democratic institutions. Int J Technoethics 2(2):14–29

Bingham PM (2000) Human evolution and human history: a complete theory. Evol Anthropol 9:248–257

Boehm C (2002) Variance reduction and the evolution of social control. Paper presented at annual workshop on the co-evolution of behaviors and institutions, Santa Fe

Cantore D (2011) On biometrics and profiling: a challenge for privacy and democracy? Int J Technoethics 2(4):84–93

Curry C (2012) Staten island teen bullied before taking her life. http://abcnews.go.com/US/staten-island-teen-bullied-suicide-family/story?id=15287910#.UKIUpqU2VU9

Debatin B, Lovejoy JP, Horn AK, Hughes BN (2009) Facebook and online privacy: attitudes, behaviors, and unintended consequences. J Comput-Mediat Commun 15(1):83–108

Dunbar R (2004) Gossip in an evolutionary perspective. Rev Gen Psychol 8(2):100–110
Foster EK (2004) Research on gossip: taxonomy, methods and future directions. Rev Gen Psychol 8(2):78–99
Girard R (1977) Violence and the sacred [1972]. Johns Hopkins University Press, Baltimore
Girard R (1986) The scapegoat [1982]. Johns Hopkins University Press, Baltimore
Girard R (1987) Job: the victim of his people [1985]. Stanford University Press, Stanford
Gluckman M (1963) Papers in honor of Melville J. Herskovits: gossip and scandal. Am Econ Rev 4(3):307–316
Goodman RF, Ben-Ze'ev A (eds) (1994) Good gossip. University Press of Kansas, Lawrence
Harding RW, Morgan FH, Indermaur D, Ferrante AM, Blagg H (1998) Road rage and the epidemiology of violence: something old, something new. Stud Crime Crime Prev 7(2):221–238
Hill R, Dunbar R (2003) Social network size in humans. Hum Nat 14(1):53–72
Hunter V (1990) Gossip and the politics of reputation in classical Athens. Phoenix 44(4):299–325
Kohlberg L (1964) Development of moral character and moral ideology. In: Hoffman ML, Hoffman LW (eds) Review of child development research, vol 1. Russel Sage Foundation, New York, pp 383–432
Kowalski RM, Limber SP, Agatston PW (2012) Cyberbullying: bullying in the digital age, 2nd edn. Wiley-Blackwell, Malden
Linde C (2001) Narrative and social tacit knowledge. J Knowl Manag 5(2):160–171
Magnani L (2007) Morality in a technological world. Knowledge as duty. Cambridge University Press, Cambridge
Magnani L (2009) Abductive cognition: the epistemological and eco-cognitive dimensions of hypothetical reasoning. Springer, Berlin
Magnani L (2011) Understanding violence. Morality, religion, and violence intertwined: a philosophical stance. Springer, Berlin
Meadows MS (2008) I, Avatar: the culture and consequences of having a second life. New Riders, San Francisco
Mucchielli L (2010) Are we living in a more violent society? Br J Criminol 50:808–829
Odling-Smee FJ, Laland KN, Feldman MW (2003) Niche construction. The neglected process in evolution. Princeton University Press, Princeton
Rigby K, Smith PK (2011) Is school bullying really on the rise? Soc Psychol Educ 14(4):441–455
Rizza C, Pereira AG (eds) (2013) Social networks and cyber-bullying among teenagers: EU scientific and policy report. Publications Office of the European Union (Forthcoming)
Rosen C (2007) Virtual friendship and the new narcissism. New Atlantis J Technol Soc Summer 15–31
Seligman A, Weller R, Puet M, Bennett S (2008) Ritual and its consequences: an essay on the limits of sincerity. Cambridge University Press, Cambridge
Sommerfeld R, Krambeck H, Milinski M (2008) Multiple gossip statements and their effect on reputation and trustworthiness. Proc R Soc B Biol Sci 275(1650):2529–2536
Sontag LM, Clemans KH, Graber JA, Lyndon ST (2011) Traditional and cyber aggressors and victims: a comparison of psychosocial characteristics. J Youth Adolesc 40(4):392–404
Sproull L, Kielser S (1991) Connections: new ways of working in the networked organisation. MIT Press, Cambridge
Stapleton T (2012) Zarathustra and beyond: exploring culture and values online. AI Soc 28(1):95–105
Taylor G (1994) Gossip as moral talk. In: Goodman RF, Ben-Ze'ev A (eds) Good gossip. University Press of Kansas, Lawrence, pp 34–46
Tufekci Z (2008) Grooming, gossip, Facebook and Myspace. Inf Commun Soc 11(4):544–564
Wilson D, Wilczynski C, Wells A, Weiser L (2002) Gossip and other aspects of language as group-level adaptations. In: Hayes C, Huber L (eds) The evolution of cognition. The MIT Press, Cambridge, pp 347–365
Yerkovich S (1977) Gossiping as a way of speaking. J Commun 27:192–196
Žižek S (2009) Violence [2008]. Profile Books, London

Chapter 9
The Crises of Techo-Cognitive Niches: From Maladaptive to Terminator Niches

Abstract The aim of this chapter is to analyze in full detail a concept I merely introduced at the end of the previous Chap. 8, that is the possibility of a *Terminator Niche*. The concept of terminator niche should intuitively represent a cognitive niche that instead of benefitting its users with a decreased local selective pressure, has eventually a negative impact on the population's welfare. I will specify this concept along the chapter, first characterizing the specificity of techo-cognitive niche construction, then showing how some maladaptive niches can be individuated in human beings' evolutionary past. Finally I will properly consider the techno-cognitive terminator niche and elaborate upon two relevant cases.

9.1 Introduction

The aim of this chapter is to analyze in full detail a concept I merely introduced at the end of the previous Chap. 8, that is the possibility of a *Terminator Niche*. The concept of terminator niche should intuitively represent a cognitive niche that instead of benefitting its users with a decreased local selective pressure, has eventually a negative impact on the population's welfare. I will specify this concept along the chapter, first characterizing the specificity of techo-cognitive niche construction, then showing how some maladaptive niches can be individuated in human beings' evolutionary past. Finally I will properly consider the techno-cognitive terminator niche and elaborate upon two relevant cases.

The original version of this paper, here revised and expanded, was presented at the *Laval Virtual* Conference in Laval, France (April 2013). I am grateful to prof. Colin Schmidt, to the organizers and the participants for the useful comments that were later incorporated in this revision.

© Springer International Publishing Switzerland 2015

T. Bertolotti, *Patterns of Rationality*, Studies in Applied Philosophy, Epistemology and Rational Ethics 19, DOI 10.1007/978-3-319-17786-1_9

9.2 Virtualized and High-Tech Cognitive Niches

As we now well acquainted with, the theory of cognitive niches is extremely valuable because it affords us not only to understand human cultural development in its traditional meaning but, as I began to show in the previous chapter, its frame can be extended to comprehend hyper-technological cultures as well: the "standard cognitive niche" model I sketched out in the beginning of this chapter could be say to work at least until the Fifties of past century, then something changed. Before, a cognitive niche could be described as a relationship between *biota*, *abiota* and dead organic matter. Either you are alive, and then you can be a constructor, either you are not, and then you are a constructed. What is constructible is the *object* of cognitive niche construction: it is the target and the materiel on which the externalization of knowledge was built. And that was it. Since the computational revolution, though, cognitive niche construction was enhanced by something that was neither a biota, nor an abiota or DOM (Dead Organic Matter): it was the category of *constructed constructors*.

The main effects of this enhancement belong to two categories: first, the so-called "virtualization" produced a multiplication of cognitive niches that goes beyond traditional ecologies, their ontologies and what they can afford. Computers and internet allowed the creation of a "sub-environment" (as I will contend in the next part, Chap. 11) that is able to interact with our natural environment (or to simulate it), but that can be manipulated in a much easier way, also because it is ontologically consistent with the way knowledge is expressed: this is to say, the environment and its modifications are expressed in the same coding.[1]

Secondly, but perhaps more importantly, the activity of high-tech niche construction involves the production of more or less complicated "artificial minds" (Magnani 2007a). Understood within the notion of eco-cognitive engineering, the notion of artificial mind is interesting: it can be seen as an help, or as a "maid-mind," but the aim is the same, that is to obtain a new kind of eco-cognitive engineer that contributes to the activity of niche-construction: this is an extremely relevant addition to the eco-cognitive paradigm.

Virtual niches, and high-technological niches, are populated by a number of *constructed constructors*, that is by agencies that were constructed (or programmed) externalizing knowledge on abiota materiel, but can actively engage a more or less extended range of active behavior within the niche. These new actors can either chiefly serve either as assessors, maintainers and mediators of existing externalizations, or as engineers of new externalizing solutions in the niche, or as full-right agents in the cognitive niche.

These actors need not be "material:" those interacting within traditional cognitive niches (such as driving supporting systems) tend to be material, but they can also reside in a bit of coding, such as a data mining software, and yet be able of causing

[1] Think of virtual realities such as Second Life: avatars can easily "create" things because there is no gap between information and matter. Matter is reduced to coding, and the only space requirement is available memory to host the coding.

significative modification to the global structure of the niche. In all of these cases, the crucial feature is the presence of non-human cognitive agents, usually embedded within a broad cognitive niche, that are able to:

- Assess a situation.
- Make an appraisal.
- Take a decision based on that appraisal.

The final decision, which can be seen as the artificial agent's contribution to the cognitive nice (for instance in the shape of an affordance) is meant to be for the good of the human user—or at least of some human users, as in the case of "intelligent" weaponry (Krishnan 2009). As I already stated several times in this section, the revolutionary steps consisted in the assumption of non-biological material to the status of actor in a cognitive niche: it is not the same as stating that, for the first time, the new status was given to something different than a human being, since animals have traditionally been actors of cognitive niches, also as assessors and decision makers (a trivial example: watchdogs are expected to be able to tell a friend from a foe and act consequently), but animals are part of the biota, they are trained and not constructed, and do sometimes actively resist niche construction activity. Conversely, in high-tech cognitive niches new actors are introduced, and they are shaped precisely as their creators want them to be.

One interesting feature of high-tech cognitive niches is the presence of *cyborgs* (Clark 2003; Magnani 2007b). This is not the place for a discussion of cyborgs, but they are worth mentioning because not only we witness the delegation of cognitive niche construction to artificial agency, but also biological agents, the traditional constructors, are further and further hybridized with the technological artifacts, so that the limit situation could be described as a combination of robotic niche construction activity and cyborg niche construction activity. In other words, the high-tech cognitive niche could be seen as supporting artificial decision maker and hybridized (part biota and part abiota) decision maker, with the exclusion of the biological decision maker.

Of course, much more could be said about high and hyper-technological cognitive niches, first of all about the abundance of available information stored in the environment, its accessibility and ubiquity and so on (again, some of these issues will be addressed further on in the third part, when I will deal with a small phenomenology of religious virtual behavior in Chap. 11), but the point I am making here is philosophically more subtle. Since the dawn of material culture onwards, human beings built cognitive niches by externalizing knowledge onto their surroundings (Mithen 1996). Technological developments are still part of material culture, and supported this externalization just as stone and paper did: it is a matter of a quantitative enhancement, and not qualitative. The qualitative jump comes with the delegation of niche-construction activities, that is to say of *decision making*, to artificial (abiota) within the niche. This, I suggest, is the boost afforded to niche construction by the computational era, and furthermore by artificial intelligence. Selective pressure is dramatically affected if the agents that counteract it to make it more favorable to human life are not only human but also technologically crafted "minions."

9.3 Terminator Niches

What happens, though, when artificial-minds *as* eco-cognitive engineers cease to collaborate with human beings? Actually, the question is not accurate, since it would bean to imbue them not only with passive moral rights, but also of an intentional moral will: more properly, could it happen that such agents *keep pursuing the tasks they were programmed for by their human programmers in a way that is not beneficial to human beings anymore?* In order to answer this question, we must not forget that the essence of niche construction is in fact to lessen selective pressure, not to increase it making life more difficult or simply unsustainable: terminator niches need not be necessarily high-technology niches. In fact, the conditions for the emergence of a terminator niche are simple: the niche must turn maladaptive because of some of the structures that chiefly achieved (or were thought to achieve) the ease in selective pressure; and, the more the conditions cause by the cognitive niches grow sever, the harder it gets to revert and dismantle the cognitive niche. Within an hyper-technological niche, as we will see, the terminator phase can acquire some peculiar characteristics—that depend on the discussion I just sketched out about high and hyper-technological niches—but cognitive niches have already happened to turn the change of selective pressure *against* the human beings who had engineered them.

9.3.1 In History

One should not be lead astray by the label of "terminator niche." We are not (necessarily) thinking of androids chasing human beings, of machines rebelling and such things. In fact, terminator niches have been developed since the dawn of humankind. Consider this very fitting example provided by paleoanthropologist Steven Mithen. It is about the Natufian culture, which existed in Eastern Mediterranean from 13,000 to 9,800 years ago, and their way of managing hunting.

> When the Kebaran people had used the Hayonim Cave, five thousand years before the Natufian became established, they killed male and female gazelles in equal proportion. *By preferentially selecting the males, the Natufians were probably attempting to conserve the gazelle populations.* Although both sexes were born in equal proportions, only a few male animals were actually needed to maintain the herds. Carol Cope thinks that *the Natufian people decided that the mass were expendable* while recognizing the need to ensure that as many females as possible gave birth to young.
>
> If this was their *aim*, it went *horribly wrong*. The Natufians made the mistake of not just hunting the males, but selecting the biggest that they could find to kill. So the female gazelles were left to breed with the smaller males – unlikely to have been their natural choice. As small fathers give rise to small offspring, and as the Natufians killed the largest offspring, the gazelles reduced in size with each generation. [...] Smaller gazelles meant that there was lees meat available to feed an ever-growing population. This shortage was compounded by over-exploitation of the 'wild gardens': too many stalks of the wild cereals had been cute and excessive quantities of acorn and almonds had been collected for natural replenishment to occur. The health of the Natufian people began to suffer, especially that of the children.

> [...] Food shortages can also lead to poor physical growth (Mithen 2004, pp. 47–48, added emphasis).

The final result of this process was emigration and the eventual abandonment of Natufian settlements. The point of his historical example is quite clear: not everything that *goes horribly wrong* can be the sign of the development of a *terminator niche*. For instance, natural mishaps can take place. Climate changes before they were linked to human activity. Or else, human beings can develop strategies that are ultimately harmful for themselves. Yet, I do not think that *smoking*, for instance, can be labeled as a terminator niche. Why? Because smoking, in a strict sense, was not developed as a mean to reduce selective pressure. It is an intrinsic activity, carried out for its own sake because it is found pleasurable.[2] Conversely, the Natufians' situation can be resumed as follows:

1. *Hunting* is a cognitive niche because it can be described as a set of heuristics, techniques and affordances aimed at providing better food income than if preys were chased randomly.
2. Hunting heuristics favor the intentional killing of larger males as this seems to provide two advantages:
   - More food is purveyed to the group.
   - Herds are maintained since the male to female ratio can be very small and yet permit the numerical prosperity of the herd.
3. The hunting heuristics, on the long term, induces a diminution in the size of the prey, hence in the quantity of food purveyed to the group: this is the opposite of the original strategy.
4. The survival of the group is ultimately jeopardized by the cognitive niche it structured

A characteristic of cognitive niches is that they have to be maintained in order to function, hence there is a *conservative* drive towards making externalizations persistent and resilient. This accounts for the fact that many human achievements are hard to eradicate even when they prove less than ideal. Cognitive niches resist being easily wiped away because their original scope is to contrast selective pressure. Externalizations must be therefore persistent to oppose exogenous changes. This, of course, makes things harder when the increased weight of selective pressure is happening *because of* the niche that should shield us from it.

In the case of the Natufians, the niche goes *terminator* because of the interplay between stages (2) and (3). This spells out a very interesting character of terminator niches. It is possible to individuate a *point of no-return*. The Natufian crisis did not happen overnight, so some individuals could—at a certain point—individuate the

---

[2]It could be argued that smoking is an adaptive trait inasmuch as it allows smokers to better cope with stresses and hardships of modern life, but this seems rather as a post hoc justification, or better as a by-product. One does not start smoking, and develop an addiction, *in order to* be able to better cope with life events. I, so far, have not found any paper discussing the evolutionary weight of smoking.

negative trend, relate it to the hunting practice, and call for it to stop. We do not know if this happened among the Natufians. Historically, even when such enlightened individuals made their appearance, they often went unheard.[3] Yet, after the stage where the negative trend is apparent to some, the *terminator niche* goes in full bloom at the moment when any solution is not, or seems not, viable anymore. Imagine that at time t^1 there were no "big" male gazelles anymore, but there were medium-sized one, small and smaller ones left. At this time, it could be sensible to refrain from preying on the medium sized ones, and concentrate on smaller ones: if the niche had developed and incorporated such heuristic, it would not have turned terminator. The situation at t^2 is different: you do not have any medium sized gazelles left, you only have small, and smaller, so you have to harvest empty most of the spontaneous natural cereal cultures. At this point, there is no viable solution left: when small gazelles are not enough to sustain the group, it makes little sense to shift to hunting even smaller ones. The situation could only degenerate, and apart from intellectual honestly, human beings tend to dislike accelerating their doom, however unavoidable this seems to be. As history tells us, the Natufians' terminator niche culminated in the abandonment of the niche itself, with the emigration of the population.

9.3.2 Hyper-Technological Terminator Niches in Popular Culture

Science fiction did often anticipate many topics that were later fully comprised by ethics of advanced technology. In this case, terminator niches were envisaged in popular fiction in ways that stressed two key elements. The irreversibility, on the one hand, and the progressive nature of the phenomenon on the other hand: it is very hard to acknowledge one has entered a terminator niche before one feels the effects on her own ability to live well, or survive.

The notion of *Terminator Niche* is obviously modeled upon 1984 sci-fi action blockbuster *The Terminator*, in which Skynet, an AI system originally engineered to protect human beings from nuclear warfare, perceives humans as a threat and attacks first, causing nuclear holocaust and then building androids ("Terminators") to hunt down remaining humans.

The idea of conflicting goals between programmers and programmed was not invented by James Cameron, though. We could say that, in a far less gruesome way, Isaac Asimov had had the same insight in the 1940s when he wrote the short stories that would later become *I, Robot* (first published in 1951), where the buried topic is the progressive loss of self-determination of humans, in favor of the wise rule of robots, for the wellbeing of humankind itself.

---

[3]This aspect will be further examined in Sect. 9.4.2 about the emergence of contemporary finance as a terminator niche. As cognitive niches acquire also a moral value as orthodoxies, we can expect a violent reaction against those who suggest safer alternatives just because they are alternatives (Magnani 2011).

Of course these are two different ideas of *terminator niche*, Asimov's vision comes after the brutalities displayed by the rule of men during the two recent World Wars, while The Terminator was conceived at the end of the Cold War and thus stresses liberal-democratic values such as the importance of self-determination, but in both major examples the core could be seen as similar: artificial minds apply the "algorithmic values" programmed within the artificial mind (protection, safety, and so on) that were thought for the good of mankind, but which ultimately prevent human beings from *being human* (not necessarily by killing them, but for instance by making them less free to decide their own destinies (Magnani 2007b)).

9.4 The Actual Emergence of Hyper-Technological Terminator Niches

Although we are clearly far from the apocalyptic scenarios described by science fiction, it could be useful to compare the high-contrast models offered by sci-fi with what is already taking place nowadays.

9.4.1 Smart Traffic as a Potential Terminator Niche

Many examples could be made, but the automobile industry seems one of fittest to understand the phenomenon. Cars benefit human beings in their logistic needs, but are a source of potential harm as far as accidents are concerned, many of which as we know are caused by driving under the influence of alcohol and other substances. Therefore, engineers have been implementing cars with more and more "intelligent" systems to prevent losses of human lives. For our concerns, two classes of systems are particularly interesting:

- Autonomous braking systems.
- Breath alcohol analyzers able to prevent ignition.

How can these incredibly precious systems sometimes impair our possibility of "being human," to the extent of potentially impairing our survival? Let us consider autonomous braking systems first: it is about technologies that are able to make an active appraisal of what is in front of the car and, in case of imminent collision (1) warn the driver, (2) power charge the brakes so that the driver's pressure will be empowered, (3) break autonomously should the driver fail to respond (Kusano and Gabler 2010). It is a widely popular topic in car safety, and is making headlines even among laypeople, as proved by a recent article on the Telegraph entitled *Auto braking: a quantum leap for road safety*: studies and tests show that implementing such technologies can reduce the occurrence of crashes by 25–30 %. The issues with obstacle avoidance systems are trivial, but it is not impossible to imagine a situation in which the driver's survival might depend on the possibility that she breaks down an obstacle with her vehicle. A car that, for safety reason, prevents it from happening (because crashing into something is assessed as bad and therefore inhibited) could reduce the driver's chance of survival (imagine she is caught in a fire in a parking, and her only chance of survival comes from crashing into the gate).

The alcohol starter seems to be less trivial. Alcohol starters are devices (such as those marketed by Smart Start, Inc.) that connect the ignition to a breath alcohol analyzer: if the breath alcohol level is too high, the car will not start. Just as in the case of autonomous braking system, this device can save a number of lives (and a significative number of fines and charges for DUI), but can produce some negative effects as well. Consider the following situation: a girl drives to a party, she has a couple of drinks so that her alcohol levels are slightly above the limit. At the party she meets a stranger and he starts harassing her. She leaves earlier than expected, and the man follows her outside and his ill intentions are quite clear by now. She could take the car and drive off, but the alcohol starter assesses the alcohol in her breath and deems her unfit for driving, and therefore will not start the car, leaving her as a potential prey of the harasser.

It is true that in this case the discussion becomes easily philosophical, and one has to debate *whether we should prefer a single girl likely raped or injured or a potential car crash, but what matters is that the decision is being taken by an artificial mind.* A computational intelligence is taking a decision which might result in a girl being raped rather than incurring in a potential accident.

Especially the latter case gives the flank to an easy objection: usually, devices such as the breath alcohol starter have a bypass system (as suggested by a 2010 patent released in the US, about an *Alcohol immobilizer having an emergency drive option, US Patent App. 12/678,447, 2008*). *Easy Start* offers a one-time unlock via the insertion of a code, which is not clear whether it is provided with the device or only upon request—for immediate use. Yet, the objection is easily answered: the bypass might not be achievable by a person in an altered emotional state, as in case of fear, or peculiar environmental settings.

A number of other fitting examples could be make concerning the car industry. Think for instance of another common feature: doors automatically unlock shortly after a collision. Of course, the system is devised to avoid passengers being blocked inside a vehicle, but if the collision is not casual, but provoked in order to allow a robbery or a kidnapping, the unlocked doors are one less obstacle for the criminal. Most of those examples are arguably limit-cases. The development of safety devices in automobile industry is usually utilitarian from a moral point of view, pursuing the greatest good of the largest number of drivers. So, even if the cases I mentioned were actually widespread, they would not question the goodness of car safety measures. I am not contending that car-safety *is an actual terminator niche*. Nonetheless, these examples are significative to show how an hyper-technological cognitive niche (as in the case of computer aided safety features such as autonomous braking, which sets off after the computational assessment of a situation) *could* turn in a terminator niche: in the cases I briefly analyzed, the striking part was a direct correlation between what is designed *to protect* human beings and what *ends up harming* them, or making them, for instance, less free.

9.4.2 Neo-Liberal Finance as a Terminator Niche

> In the early 2000s, financiers believed that, through our integrated financial systems, we could fragment and disperse loan risk so much as to make that risk completely negligible. Risk itself would magically disappear in the ecstasy of post-structuralist communications [...]. Like a starship, financial risk would at last achieve escape velocity and reach the financial galactic beyond. And so we created the giant Ponzi scheme known as the international financial system based on almost infinite hedging and fund fragmentation and dispersal, all made possible by our integrated global financial technologies. Like pre-enlightenment financial alchemists, we could turn base sub-prime loans into gold. Instead, we found that we turned it into a global bank debt crisis and eventually a sovereign catastrophe. (Stapleton 2012, p. 5)

Stapleton's analysis is unforgiving, as it was in the case of virtual pro-sociality (cf. Sect. 8.5). Or even better, it is "curiously" forgiving inasmuch as he does not approach the crisis from the financial point of view, but from that of hyper-technological cognitive niches. Focusing on the crack of the Anglo-Irish Bank, he claims that the fault is not to be found in masterminds of crime or "slackerism," but rather in the decision-making system that was cyborg-like, shared between humans and the computers they had—themselves—programmed.

> What was the role of management information systems in all this? It was surely these systems that facilitated financial imprudence and light tough regulation, simultaneously providing a sense of a controlled and well-monitored business. Rather than deliver solid management information to support wise decision-making processes, the systems not only failed, but created an illusion that all was well. [...] Thus, management does not gain a real-time, true and integrated picture of their firm. Instead, technology and culture operating together in this Faustian tryst produce the very opposite effect: an illusion of prudence and effective risk management. A technoculture of deceit, of hiding and cover-ups, is therefore potentially enabled by our technology-cultural system. (p. 6)

Can finance be defined as a terminator hyper-technological niche? I suggest that the categorization is fitting.

1. With the benefit of a powerful charity principle, we can say that finance is a cognitive niche constructed in order to increase total welfare, albeit in a capitalistic conception of markets.
2. Finance is a high-tech cognitive niche, as it is greatly virtualized and its actors are not only human beings but software and algorithms (such as those for risk-assessment) and other forms of robotic intelligence. This, in turn, means that the human agents involved are significantly cyborgized.

3. Karl Marx had already theorized that crises are endemic to the structure of capitalism, but crises following speculative bubbles such as the 1929 one that spurred the Great Depression, and the 2008 subprime loans one that turned into the ongoing global crisis, seem to be shifting from something structural to something that is jeopardizing the welfare (and potentially the survival) of those who populate and maintain the niche.

Albeit they did not use the term "terminator niche," since they are not adopting the niche theory at all, many economists (for instance in the Post-Keynesian school) have argued about the intrinsic unstable nature of financial markets (seminal work of Minsky) and have stressed the pervasiveness and the disruptive nature of an excessively *financialized* economic system. Neoclassical finance considers economical agents as entirely rational (and this trust was extended to the hybrid and artificial agents of the past few decades), and—basing on this philosophically uncertain assumption—developed models that too often mix up "risk," as something that can be *measured*, and Keynes' concept of fundamental "immeasurable uncertainty."[4]

Minsky, in 1963, claimed that financial markets are intrinsically unstable because of debt structured built by economical agents (namely Ponzi schemes), that will sooner or later cause the collapse of the whole system (Minsky 1963). Recently, when the latest economical crisis was far from exploding, Structural Keynesian economist James Crotty showed that:

> NFCs [US large Nonfinancial Corporations] were eventually placed in a *neoliberal paradox*: intense product market competition made it impossible for most NFCs to achieve high earnings most of the time, but financial markets demanded that NFCs generate ever-increasing earnings and ever-increasing payout ratios to financial agents or face falling stock prices and the threat of hostile takeover (Crotty 2003, p. 1).

For the sake of brevity, I have to make very short a story that would be much longer. We have a cognitive niche (finance) which impose itself over market competition, but which cannot make the necessary gains from market competition (which conversely it impairs), therefore it creates some proper schemes for increasing its welfare by assuming *counterfactual* principles such as the rationality of economical agents and the illusion of control by calculating risk through unrealistic mathematical models. It is not necessary to postulate any *evil* in the game, this is how cognitive niches work: furthermore, as argued by Stapleton, the reliance on an artifactual hyper-technological niche blissfully blinded (and still blinds) many operators: finance in many cases is not a mere self-fulfilling prophecy, but a prophecy that aims at being self-fulfilling, but falls short of it because prophets are not even human but cyborgs or artificial intelligences.[5] This whole mechanism rings a bell, but where did we see

[4]For specific events (for instance a roulette table) we can calculate the probability of the outcome. Conversely for others—such as catastrophes and other events, which have been often used as the underlying of many derivatives instruments—we just cannot measure the probability of the outcome.

[5]One could make a pun and claim that these AI's are not that intelligent after all, but the tricky part of hyper-technological niches (so far) is that they cannot outsmart human beings at what they were programmed for: if a financial algorithm is programmed to compute risk as something calculable, and inform consequent market strategies, it cannot be blamed for doing so.

it...Ah! Of course, the good old Natufians![6] Actually, the description of finance I just sketched out, which is quite an approximation but consistent with serious economical analyses, is not that different from adopting the hunting decision to kill the biggest gazelle males, so to get more food *and* let the population of preys thrive by not subtracting females to the herds. History proved the Natufians wrong: human beings have an innate desire to have their cake *and* eat it. A certain kind of terminator niche can be seen as the externalization of this desire. In particular, hyper-technological cognitive niches can make the actualization of this desire as something more possible, and at least at the beginning they make it happen: human beings have a conscience, most of them anyway, but computational intelligences, if "properly" programmed, can create whole systems of meaning and whole possibilities of action which, albeit in traditional cognitive niches they would be quickly debunked as unfeasible, do appear as viable.

Another echo between the financial crises and the case study of Natufians can be found in the (feeling of) ineluctability. As I suggested, terminator niches are not simply maladaptive, but are characterized by a point of no-return (or of very-difficult-return). It can be hypothesized that the Natufians, when the food emergency was in full bloom, could chose between preying on small or smaller gazelles. Preying on smaller gazelles meant being very hungry today and hungry or very hungry on the morrow. Preying the small ones (but not the smallest) meant to be very hungry on the morrow, but not so hungry today: it is a matter of hot/cold collective decisions.

Finance as a terminator niche plays the same game: as everybody witnessed, current politics, aimed at *regulating* markets, coupled with generous insertions of liquidity from the Federal Reserve and European Central Bank, have not achieved a stable recovery yet. According to economists such as Palley, the only way out would be to revert the financialization of the entire world economy (Palley 2007).[7] But asking this seems like telling a hungry Natufian to eat an even smaller gazelle because the situation *might* improve.

As we saw in the introductory chapter on cognitive niches (Sect. 5.2) cognitive niches are plagued by "doxastic irresistibility," that is the desperate need to believe what is commonly said. This, summed to a tendency towards resilience and persistence that is vital for the maintenance of cognitive niches, triggers a sclerotization of terminator niches (Magnani 2011, Chaps. 4 and 5): the more they fail in offering a positive trade-off in selective pressure, the harder human beings cling to them. Each time the current financial crisis seems to be touching an all-time low,neoliberal think-tanks (such as the Tea Party movement in the US) call for

[6]The sad story of Natufians has been described in Sect. 9.3.1.

[7]I am talking about dismantling a cognitive niche. History shows that, in order to break the resilience of a cognitive niche, significant impetus is required: for instance, massive invasions, cataclysms and similar things.

harsher neoliberal politics. Telling people "It has worked till now, it will recover and work again"—notwithstanding the epistemic scarcity of inductive reasoning—is more welcome than alternatives such as "This is not working anymore, we have to look somewhere else for a solution."[8]

It is interesting to note that, whereas terminator niches are a concept inspired by science fiction, it is actually quite possible to make some cross readings. In popular culture, terminator niches are characterized by a progressive loss of rational hope, as the everyday scenario turns grimmer and more distopian. This general depression comes hand in hand with religious and mystical hope, prophecies and *deus ex machina* solutions,[9] but also with human beings getting violently carried away by a sacrificial state of mind: as spelled out by Girard in *The Scapegoat* (Girard 1986), crises spark victimary mechanisms in which random victims, fittingly individuated, are blamed for the situation and punished, expelled from the community or killed, hoping that this will bring relief. This theme, ever-present in sci-fi terminator niches, could be traceable also in finance-as-a-terminator-niche: after all, if the cause of the actual crisis is the neoliberal finance-dominated economy, the perseverance in blaming as culprits whole nations (such as Greece, or Spain, or the next in line) or the occasional crook or rogue trader is nothing but the spectral appearance of a sacrificial mentality. Monetary bloodshed—fostering poverty, and degrading the potential of whole nations—may dope the terminator financial niche into believing that *hope* and *optimism* (ironic keywords in a system believing in the rationality of all agents and in the–unreal—calculability of any kind of risk) are finally justified. I can hope to have my cake and get to eat it at the same time, and this can make me merry for the whole day, and during the day I can celebrate with a shopping spree: this is the kind of illusion that finance, as a terminator niche, fosters. But come evening I will discover the bitter truth: if I eat my cake, I do not have it anymore and on the morrow I will be cake-less and miserable.

9.5 Terminator Niches and Supernormal Stimuli

One of the aims of this chapter was to frame the issue of technological dilemmas within the more comprehensive theory of cognitive niches (which in turn connects with evolutionary studies, referring to the modification of local selective pressure and heritage systems). Another concept from biological studies, then, could be of help: *supernormal stimuli*. Naturalist and nobel laureate Niko Tinbergen had coined the expression in the 1930s to indicate the imitations of natural signals that would

[8]Such a reflection can help us further understand the dramatic intervention performed by US President Franklin Delano Roosevelt with his *New Deal*: the New Deal is indeed a formidable cognitive niche reconfiguration, but it could be enacted as a gambit, only once the precedent (terminator) niche, *Hoover's Blanket*, was proved inevitably doomed. The New Deal was enacted because Americans felt that there was nothing left too loose in a gambit involving counterintuitive politics such as deficit spending and so on.

[9]Movies such as The Terminator, Matrix and so on are perfect examples.

induce in animals a stronger reaction than the natural thing would, even in presence of the natural signal itself: famous examples include fishes attacking unnatural-looking dummy fishes with brighter colors than their natural counterparts, and trying to mate with unrealistic female dummies with exaggerated fertility attributes, even in presence of a "normal" living female. Recently, Barrett (2009) comprehensively suggested how this very concept could be applied to human beings and explain the exponential emergence of phenomena such as the pursuit of "cuteness" in artifacts and people, the diffusion of pornography and sexual esthetic enhancements, and even human beings' fondness for shopping malls and wars that are always more destructive: in all of these cases, according to Barrett, artifactual or artificial stimuli engage our primeval dispositions in a far more convincing way than their natural counterparts, even if the latter is at hand. Barrett's account, however illuminating, sets the debate on supernormal stimuli at the individual level (with an occasional penchant towards moralism): the next step could be to relate supernormal stimuli to an ecological, niche-wide level.

It is interesting to investigate whether supernormal stimuli could be related—or not—to the development of terminator niches. Let us bear in mind that in order to have a terminator niche, the cognitive niche must be: (1) hardly reversible (it might even defend itself preventing users from defusing it), (2) impairing the survival and the welfare of most of its users instead of enhancing them.

Some of the examples provided by Barrett (2009), such as the diffusion of *fast food* restaurants in the US (serving an offer of meals that corresponds perfectly to the definition of supernormal stimulus, inasmuch they are more fatty and tasty than the "natural" counterpart would be, and hence more attractive) seem to be perfectly good instances of terminator niches.[10] If we consider SN websites, they could be seen as well as a peculiar terminator niche, providing a new kind of "doped" sociality (that appeals to evolutionarily-selected drives) but tends to replace traditional social institutions (such as friends) without displaying similar affordances (Bertolotti 2011). In spite of many of their well-known and much-debated issues (e.g. privacy concerns, lack of quality relationships, diffusion of slackerism and desultory behaviors, commodification of personal preferences, and so on), there is common agreement between scholars and experts that "there's no going back." Public and private institutions and think-tanks work to develop policies for better coping with the issues produced by SN, but the whole thing cannot be eradicated—albeit no visible improvement was brought about in our lives.

Supernormal stimuli, therefore, are probably not the explanation behind the development of a terminator niche, but can play a pivotal role in the crystallization of preferences (ranging from food habits, to the pursuit of personal safety to the management of one's sexuality) that, on the long run, may become harmful to the groups that select them.

---

[10]Notwithstanding the legitimate criticism, one must acknowledge that the *fast food culture* is indeed a product of a high level knowledge society, that can support studies on chemical and mechanic engineering of food—so to reduce costs and augment taste, logistic planning, merchandising and so on. Therefore it can be rightfully be defined a *cognitive niche*.

The issue with bringing supernormal stimuli in the discussion is that, albeit they contribute with an impressive theoretical tool, they can easily become a moralistic peg, or anyway one of unclear application. Barrett's argument on supernormal stimuli is, in fact, less convincing when it comes to deal with sexuality and pornography than when it focuses on food or on the economy. It becomes therefore problematic to define what counts as supernormal stimulus and what not: consider the increasing presence of robots in factories since the Seventies. We now know that this kind of technology, which made it possible that a man alone did (with far less effort) the work of several, massively contributed to reducing the number of people employed in manufactories. It is arguable whether the introduction of robots in factories constitutes a terminator niche (surely, we are far from the Asimovian utopia of a plentiful world where machines toil for men and men benefit from their labor), and one could also argue that they provide a supernormal stimulus, responding excessively to human beings' dislike for labor, but anyway—in this case as well—there is no going back: even if workers acknowledge that the introduction of robots is one of the causes of industrial unemployment, no one would like to go back to toil on a Ford 1913—like assembly line. Could it then be possible to categorize, and understand, human beings ever-increasing desire for safety, control, gain (as in financial speculation) as a morally-neuter but biologically existent drive towards available stimuli that are more powerful than the natural ones?

9.6 Terminator Niches and "The Singularity"

The concept of terminator niche is somewhat similar to that of technological singularity. The singularity can be assimilated to a critical point in human development determined by technological development, usually concerning advancements in the fields of biomedical and artificial intelligence. The Technological Singularity should represent an unmistakable moment in space-time disrupting most expectable regularities as for the human imprint on the biosphere and semiosphere. "It is a point where our old models must be discarded and a new reality rules. As we move closer to this point, it will loom vaster and vaster over human affairs till the notion becomes a common-place. Yet when it finally happens it may still be a great surprise and a greater unknown" (Vinge 1993, p. 90). Terminator niches are the exact opposite: one cannot identify one crucial moment because they are diffused both in space and time, but they disperse a kind of diffused structural toxicity (Magnani 2011). A cognitive niche (such as car safety, to make an example, or the kind of finance that has been characterizing the crisis since 2009) could have become *terminator* long before users realize it. And even once we realize it, we might continue making use of some niches even once they are "gone terminator:" finance is a self-describing example of a technological niche, mediated and augmented constantly by artificial-minded engineers, which has clearly turned out to be a terminator niche and yet is always being used. Terminator niches could contribute to reaching such metaphysical singularity, but the singularity could also wipe out the very notion of terminator niches by projecting a general state of wellbeing. Therefore, albeit they can be easily juxtaposed, and perhaps even fruitfully, it seems better—least as the concept of terminator niche is being defined—to keep it separate from that of singularity.

9.7 Conclusion

Upon concluding this chapter, it could be interesting to explicit a few remarks connecting what I just analyzed to the broad eco-cognitive outlook I have adopted in my research and consequently reverberating in this book. In Part I, I devoted the first two chapters to understand the eco-cognitive role of modeling: that is to say, I stressed the extent to which models serve a pragmatic role in fostering understanding of one's ecology. As I showed in the second chapter, one could easily follow a red line connecting natural mental modeling to the heights of scientific modeling, without facing any qualitative shift, and it is fundamental to retain the complex relationship that the model entertains with the modeled: models, be it perceptual or scientific modeling, poietically imprints what is modeled, thus *being able to create* phenomena as the agent knows them.[11]

Nevertheless, from this does not follow that in the process of modeling *anything goes*. Paraphrasing the Peircean motto on abduction (which is of course strictly linked to modeling), it could be argued that "good modeling is akin to truth." Clearly, truth has to be understood in a pragmatic, utilitarian conception. In biological/natural modeling (think of perception and the discourse in Chap. 2 about camouflage), good modeling is what frames and structures phenomena so that the organism may survive; similarly in science good modeling affords better understandable phenomena and accurate predictions. Connecting with the debate I illustrated and joined in Chap. 3 about epistemological *fictionalism*, it can be argued that the kind of models that, when enacted, bring their users to a condition of *terminator niche*, these are indeed fictional. Modeling the DNA as a double helix punctuated with four kinds of bases, and modeling economic agents as perfectly rational and able to rely on accurate calculations of any kinds of risk cannot be both *equally* labeled as fictions: the impossibility of this equation is apparent if one considers the practical outcomes of those two different ways of making and using models.

Magnani, in his counter-manifesto against fictionalism entitled *Scientific Models are not Fictions: Model-Based Science as Epistemic Warfare* (Magnani 2012), interestingly contends that the fictionality of models is something that can only be assessed *a posteriori*, when a model is outdone by a better one, or anyway proved wrong.

> [For] example in a scientific discovery process, the scientific model is simply eliminated and labeled as "false", because *new material has come to light* to provide a better model which in turn will lead to a new knowledge that supersedes or refines the previous one, and so the old model is buried in the necropolis of the unfruitful/dead models. Still, similarly, in the whole

[11] As shown in Chap. 3, in the previous part, not all scientific models should be understood as creative and able to poietically inform new phenomena.

scientific enterprise, also a successful scientific model is sometimes simply eliminated (for example the ether model) together with the theory to which that model belonged, and so the old model is buried in yet another necropolis, that of the abandoned "historical" models, and yes, in this case, it can be plausibly relabeled as a fiction (p. 16).

The appreciation of Terminator Niches may allow the emergence of another way of attributing *fictionality* to a model, and might be connected to the obstinacy displayed in the use of counterfactual, however intuitive, models that are continuously proven wrong, witnessing outcomes that do not justify holding onto those models if not for a kind of *doxastic inertia*. In Sect. 3.2.4 I introduced the logical concept of "negation as failure," and extended it to describe how certain models could be *softly* negated because of the ecological failure their caused (possibly including the death of the modeler/user): in the case of the terminator niche, one could almost say that the cognitive niche serves to avoid a negating failure that should be a matter of course. To use a metaphor, a certain reliance on counterfactual models could be likened to encouraging the assumption of a lethal poison while providing at the same time the antidote (and not the assurance that the antidote will be available forever): the toxicity of the whole endeavor is rather apparent!

While many of this topics will be dealt with in the next part, as I will discuss irrationality and "epistemic immunizations" connecting them with debated issues in religious cognition, one last point can be relevantly made now. In Chap. 3 I had contended that the difference between natural and scientific use of models should not be individuated at the level of models themselves, but in their use and in the problematization of the use itself. The analyses I am carrying out about cognitive niches should enforce the eco-cognitive and epistemological awareness that attributes such as "fictional," and other more positive ones, have little meaning if conferred absolutely to a single object, or class of objects. Models, indeed, have little sense in themselves if not properly understood in a triadic system including the modelers, the model and the modeled (this latter can naïvely be indicated as "realtity," or "environment" or "surroundings"): with certain respects, a model can be said to be fictional with respect to what is modeled, but with some other respect, the *use* of the model is fictional, and not the model itself.

References

Barrett D (2009) Supernormal stimuli: how primal urges overran their evolutionary purpose. W.W. Norton, New York

Bertolotti T (2011) Facebook has it: the irresistible violence of social cognition in the age of social networking. Int J Technoethics 2(4):71–83

Clark A (2003) Natural-born cyborgs. minds, technologies, and the future of human intelligence. Oxford University Press, Oxford

Crotty J (2003) The neoliberal paradox: the impact of destructive product market competition and impatient finance on nonfinancial corporations in the neoliberal era. Research Brief 2003–5 of the PERI—Political Economy Research Institute, University of Massachusetts-Amherst

Girard R (1986) The scapegoat [1982]. Johns Hopkins University Press, Baltimore

Krishnan A (2009) Killer robots: legality and ethicality of autonomous weapons. Ashgate, Burlington

Kusano K, Gabler C (2010) Target population for injury reduction from pre-crash systems. SAE Technical paper 2010-01-0463

Magnani L (2007a) Semiotic brains and artificial minds. How brains make up material cognitive systems. In: Gudwin R, Queiroz J (eds) Semiotics and intelligent systems development, Idea Group Inc, Hershey, pp 1–41

Magnani L (2007b) Morality in a technological world. Knowledge as duty. Cambridge University Press, Cambridge

Magnani L (2011) Understanding violence. Morality, religion, and violence intertwined: a philosophical stance. Springer, Berlin

Magnani L (2012) Scientific models are not fictions: model-based science as epistemic warfare. In: Magnani L, Li P (eds) Philosophy and cognitive science: western and eastern studies. Springer, Berlin, pp 1–38

Minsky HP (1963) Can 'it' happen again? In: Carson D (ed) Banking and monetary studies. Homewood, Illinois

Mithen S (1996) The prehistory of the mind. A search for the origins of art, religion, and science. Thames and Hudson, London

Mithen S (2004) After the ice: a global human history, 20,000–5000 BC. Harvard University Press, Cambridge

Palley TI (2007) Financialization: what it is and why it matters. Working paper no. 25, The Levy Institute and Economics for Democratic and Open Societies. Washington

Stapleton T (2012) Zarathustra and beyond: exploring culture and values online. AI Soc 28(1):95–105

Vinge V (1993) The coming technological singularity. Whole Earth Rev, 88–95

Part III
The Eco-Cognitive Epistemology of Counterfactual Beliefs

This third part, connecting to the first two in a kind of dialectic opposition, focuses on what is commonly labeled as "irrational" by *contrasting* the standard ecological and social inferences analyzed so far (that is, those usually employed to satisfy our need to understand the environment and other human beings) with another kind of inference characterized as epistemic, or cognitive, "immunization." The epistemological and cognitive interest will lie in the analysis of *religious and magical thinking,* in order to uncover other regimes of rationality that allow agents relying on them to cope *in a different way* with the demands of *ecological-cognitive problem solving,* both in the past and in contemporary cognitive niches.

Chapter 10
Natural Religion, Models, and the Invention of Supernatural Beings

Abstract My research was not aimed at drawing a history of cognitive "progress," which emerges through a juxtaposition of different "stages" of thought (Barnes, Stages of thought: the co-evolution of religious thought and science 2000); if it were, then it would have made sense to begin with religion and the origins of culture, and track the "evolution" leading to science also through the consolidation of social cognition. Conversely, at this point of my research it was possible to frame religion through the conceptual tools developed so far, such as the modeling of external agencies and the construction of cognitive niches. Thus, religion can appear as a model of a class of inferences, traditionally perceived as irrational (or having to do with counterfactual beliefs), but which can be very interestingly studied by an epistemological analysis. Such an outlook can be regarded as analogous to the psycho-anthropo-cognitive effort to frame religion "as a natural phenomenon" (Boyer, Religion explained 2001; Atran, In gods we trust: the evolutionary landscape of religion 2002; Dennett, Breaking the spell 2006): similarly, my aim was to investigate religion as a "philosophical phenomenon" without adhering excessively to what is traditionally understood as *philosophy of religion*. The local interest was in fact to apply an epistemological toolbox in order to study and model the religious inferential regime, how religion could be defined as an ecological-cognitive activity, and how religious pragmatics can regulate behaviors that are typically connected with religion such as sacrifice and forgiveness.

10.1 Introduction

The most interesting approach to studying "irrationality" in an eco-cognitive perspective consists, in my opinion, in an analysis of religion: in particular, I mean to found my claims on the cognitive studies of religion chiefly because of two reasons. As I will try to demonstrate, while religion is intuitively connected with utter abstraction and immateriality, recent studies have shown how at its base one can find a plethora of cognitive needs rooted in the ecological reality of what will eventually become the religious agent. Secondarily, religious belief has always been characterized by a certain level of ecological *counterfactuality*: the complex relationship

© Springer International Publishing Switzerland 2015
T. Bertolotti, *Patterns of Rationality*, Studies in Applied Philosophy, Epistemology and Rational Ethics 19, DOI 10.1007/978-3-319-17786-1_10

between counterfactual and irrational can therefore be unraveled by means of an epistemological study of the matter—akin to the one employed in the first Part of this book—which will provide a clearer image of the *inferential* ground supporting religious cognition and other kinds of reasoning broadly labeled as *irrational*.

Indeed, each and every day, we deal with a number of entities that do not belong to our everyday ecologies: our minds are full of items that we could not see nor perceive in the outside world. This description does not only match entities such as daemons, fairies, spirits, aliens, gods and deities, but also square roots, quarks and gluons, gravity fields, sinusoids: in brief, not only typically religious entities, but also many scientific objects cannot be detected by our standard human sensory apparatus and hence be grasped "naturally" by our minds, and therefore belong to the broad category of *counterintuitive* (see for instance Atran and Medin's book for a wide approach to the relationship between nature, folk-sciences and culture (Atran and Medin 2008)).

A meaningful difference, though, keeps scientific concepts separate from religious ones: science may deal with what is counterintuitive *insofar as its understanding requires a consistent intellectual effort*, while religion's objects are counterintuitive *because ontologically counterfactual*.

Both of those categories regard things which are not immediate for us, i.e. objects that do not belong to our normal ecologies. Still, counterintuitive objects like an imaginary number (i.e. the square root of -2) or the process of replication of DNA or a chemical reduction-oxidation reaction, in spite of their being extremely unappealing to "ordinary" human minds, can be integrated in a more complex system, analyzed within the proper artifactual framework, they can be the object of controlled and reproducible experiments and their *existence* can be eventually confirmed: I dedicated the two first chapters of Part I to this issue, debating the ecological relevance of models (see Chaps. 2 and 3). Conversely, human minds seem particularly eager to absorb as many counterfactual objects as they can: such objects can be defined as those entities that violate (more or less systematically and at different levels of magnitude) common expectations about the world around us.

As mentioned above, counterfactual entities are amazingly appealing to human minds notwithstanding the factual impossibility to investigate them on a "scientific" basis and the lack of plainly empirical evidence on their account: systematic skepticism about counterfactual belief is in fact a distinctive trait of a scientific stage of rationality, which has never been exceedingly common among average human beings. The unattainability of an empirical proof for counterfactual elements is intrinsic to their definition, and it is not due to the rarity of episodes in which they are involved: over the last two decades, the popular distribution of portable camcorders first, and then the advent of digital media devices would have been expected to cause a massive rise in sightings and consequent documentation of UFOs, ghost and paranormal activities, Big Foot's, Loch Ness monsters and so on, but this was clearly not the case.[1]

[1]This argument is adapted from Carl Sagan's *The Demon-Haunted World* (Sagan 1997): "Many homes in America now have moderately sophisticated burglar alarm systems, including infrared sensors and cameras triggered by motion. An authentic videotape, with time and date denoted,

The seemingly unstoppable diffusion of religion among human beings can be regarded as the best example of the success that counterfactual beliefs met within our minds. Scott Atran stresses how this commitment to counterfactual entities should have proved particularly maladaptive for mankind: "[...] to take what is materially false to be true (e.g. people think and laugh and cry and hurt and have sex after they die and their body disintegrates) and to take what is materially true to be false (e.g., people just die and disintegrate and that's that) does not appear to be a reasonable evolutionary strategy" (Atran 2002, p. 5). The key to solving this issue, as I already suggested, seems to be that humans usually know when to believe (and behave) "as if" counterfactual entities really belonged to their ecologies and when to suspend this belief. Let us consider an example from Christian faith: in the Gospels, Jesus is reported to have said "I tell you the truth, if you have faith as small as a mustard seed, you can say to this mountain, 'Move from here to there' and it will move. Nothing will be impossible for you." (Matthew 17:20, NIV) and "And these signs will accompany those who believe: in my name they will drive out demons; they will speak in new tongues; they will pick up snakes with their hands; and when they drink deadly poison, it will not hurt them at all" (Mark 16:17–18), but the great majority of Christians, even if they might sometimes believe that to be true, do not behave in their everyday life as if they could move a mountain by the power of faith or try to handle poisonous snakes fearless of the likely injures (or death) they might incur into. There are some exceptions to this *moderate* and widespread approach to Christian belief, but they all fall down within the laws of Nature: we can consider for instance the practice of snake-handling, typical of a Pentecostal minority in the eastern United States (Scott 1994). Those believers take literally the verses of the Gospels I just quoted, and bare-handedly manage venomous snakes, resorting to prayer in case of bites: laws of physiology cannot be changed by mere conviction and, as a matter of fact, such practice has resulted in 70–80 reported deaths over the past century: a clear case of the lack of cognitive firewalls I will describe in a few pages, at the end of the next chapter!

While in Chap. 12 I will attempt a "functional" reply to Atran's question by introducing the cognitive architecture definable as the *religious bubble*, I am going to deal along this chapter with a possible causal explanation about the origin of the supernatural entities that populate our beliefs: our explanation will hint at a solution of the paradox opposing religion's persistence and its likely negative effects with respect to individuals that sincerely believe in things that cannot be proven to exist, and that cannot be proven to affect our lives.

(Footnote 1 continued)
showing an alien incursion—especially as they slip through the walls—might be very good evidence. If millions of Americans have been abducted, isn't it strange that not one lives in such a home?" (pp. 174–175).

10.2 Hypotheses About the Naturalness of Religion

The cognitive origins of religion have become a hotly debated topic over the last decade: the attempt to *naturalize* religion and investigate it as any other *natural* phenomenon overflowed the boundaries of strictly academic debate. Scholars (coming from various fields, i.e. anthropology, psychology, biology, etc.) were prompted to write widely successful books that would convey to a broader public the latest findings in the domain: the most famous example of this trend is probably Daniel Dennett's *Breaking the Spell* (Dennett 2006).

This literature can provide us with useful introductory hints for this research, as I move to a level of greater details. Cognitive anthropologist Pascal Boyer, whose book *Religion Explained* (Boyer 2001) can be regarded as one of the forerunners of Dennett's best seller, pointed out how religion was *natural* to human beings, and conversely science itself appears to be one of the most *unnatural* activities they ever undertook.

Seemingly in conformity with this claim about the *unnatural* character of science, Dixon—in his appreciable *Introduction* to the confrontation between science and religion—stated that many of the issues starting with Galileo's clash with the Church can be brought back to the problem of scientific *realism*, that is believing (or not) that we are able, through science, to provide an account of phenomena that cannot be observed, and therefore counterintuitive insomuch as *not* intuitable (Dixon 2008).[2]

Boyer highlighted how religion, as a complex phenomenon, originates from a wide range of stages of the mind, starting from the most automatic and unconscious levels. According to Boyer's imagery, religion is natural because most of the cognitive activity crucial for religion takes place in the mind's *basement* (Boyer 2001, Chap. 9), that is to say, hidden from the believer's awareness. Basements are gloomy, damp, poorly lit: the basement is a topical setting for horror and mystery fiction, because what happens in the basement is not usually clear to the upper floors of the house (otherwise there would be no mystery to solve at all!). The aim of the basement metaphor is to discharge the naïf conception about how our perceptions and our cognitive performances inform our beliefs: oddly enough, it can be said that the perceptions we take more for granted (such as the visual ones) come *from the basement* and not simply from the outside world, as if from a window.

There is no *homunculus*, no self-in-the-head who perceives the outer world as if senses were openings in the body shell. If we truly want to investigate the formation of our beliefs (any belief, and hence of religious ones, too), it is important to accept the neuroscientific findings suggesting that there is no such thing as a central controller that makes sense of the different perceptual stimuli. Rather, it is just the different perceptual stimuli that make sense into the global picture that we are aware of, thanks to a sort of *self-assessment* operated by each different system Driver and Spence (1998): in Part I, Sect. 3.3.1, I have explained the creative modeling enacted

---

[2]As stated in the Introduction, this Part does indeed establish a kind of dialectical opposition with many issues faced in the first two Parts. I dedicated the first chapters books to issues such as scientific realism and the nature of scientific models.

by perception as it mediates the cognizant access to her surroundings. This is the kind of work going on in the less conscious layers of the mind, so crucial for the very possibility of religion. On the other hand, science is called unnatural because most of the work seems to take place in the fully conscious levels of the mind: science is about which method should be used to obtain the best knowledge, and even perception and attention are strictly controlled and directed to a target: this positive conception of science can be the result of a favorable bias, but the higher standards of science (and in a lesser measure, philosophy until it was outclassed by modern science as the prime provider of knowledge) are actually aimed towards a critical analysis of the immediate self-assessment of perceptual channels.[3]

It might be argued that few things are less intentional and consciously ordered than religious hierarchies, but it is important to bear in mind that I do not mean to deal with one particular religion, nor with religion *as a whole*. To make use of Boyer's useful distinction, I am not interested in theological religion but in practical religion, that is to say that our investigations focus on those beliefs that are "[...] not concerned with the general question of God's existence or powers, but more understandably with practical questions: what to expect now and what do do next" (Boyer 2001, p. 313). Nearly any contemporary religion rests on a theology: a theology is a corpus of ordered beliefs and (recommended) inferential rules and the very etymology of theology is the same as for biology, geology, neurology. A *logos* can constitute only in the upper, well illuminated floors on the mind. Conversely, I am interested in what actual human agents enact in their consciousness when their behavior can be described as "religious". Of course, theology is accepted also because of these *naturally originated* beliefs, and vice versa those beliefs are partly informed and shaped by theology, but it is important to stress how theology and practical religion are not the same thing and, hence, the former cannot be used to invalidate explanations concerning the latter.[4]

So far, so good: still, it is clear that, by now, one might wonder what kind of "work" is exactly ingoing in the recesses of the basement of the mind, that Boyer describes chiefly in an operative fashion. Such description is perfectly suitable to his purpose, but I shall attempt to work my way out of the analogy and shed further light on those underground mental activities in order to acquire a better understanding of the cognitive traits that make religions one of the constitutive characteristics of *homo sapiens sapiens*.

---

[3]Remember Galileo's development of a new kind of scientific model, able to "bend" reality inasmuch it proved our senses to be misguiding and thus created a new array of phenomena (cf. Sect. 3.3.3).

[4]Recent debate about the cognitive science of religion focused on this specific matter. Much of the critique against works such as Boyer's was that they concerned only magical thinking, and religion could be considered as such only if devoid of theological commitments. Latest research argued that the Cognitive Science of Religion may warrant for theological beliefs as well (De Cruz 2013).

10.3 Abduction as a Multilevel Model for Perception

I will significantly elaborate further on one of the main takes of this book, i.e. that human cognition chiefly displays an *abductive* nature: I shall describe how, within an abductive framework, the human brain operates in a *similar* way over a wide range of stimuli, thus drawing a continuous line from the most basic levels of perception to the origins of belief in supernatural agents and, *a fortiori*, of the complex series of phenomena we label as religion. As I already pointed out, this discussion complements the one I carried on in Chaps. 2 and 3, as it faces a different, yet closely related facet of abductive cognition because of its ecological relevance.

It can be useful to remember that abduction, as understood within the Peircean framework, can be accounted for as the process of inferring certain facts and/or laws and hypotheses that render some sentences plausible, that explain (and also sometimes discover) a certain (eventually new) phenomenon or observation: it is the process of reasoning in which hypotheses are formed and evaluated (Magnani 2009). It must not be regarded as a merely sentential inferential process, concerned with the discovery of physical laws or resolving mysteries (Doctor Watson makes a constant mistake in Conan Doyle's novels by complimenting Holmes for his brilliant *deduction*: his hypotheses are actually abductions from clues to the situation that originated them). Abduction opens a much wider field of investigation: survival, for any animate organism, is a matter of coping with the environment and the relationship with the environment is mediated by a series of cues the organism must make sense of in order to generate, even if tacitly, some knowledge it did not possess before. This *making sense of signs* is an abductive activity that human beings share with any organism endowed with a nervous system or, on an even bigger perspective, any organism capable of reacting actively to modifications of its environment (from bacteria to *homo sapiens sapiens*), as shown in Magnani's paper about animal abduction (Magnani 2007a).

As hinted in the first chapter, our brain cannot operate directly on the outside world, and instead needs *inner representations* of the outside world on which to operate. As for all cognitive processing, both data and hypotheses can have a full range of verbal and sensory representations, involving words, sights, images, smells... but also kinesthetic experiences and other feelings such as pain, and so for all sensory modalities: all representations are brain structures and abduction certainly is a neural process in terms of transformations of neural representations. We can also re-conceptualize abduction neurologically as a process in which one neural structure representing the explanatory target generates another neural structure that constitutes a hypothesis.

It is important to focus on how this *reduction* to a neural representation necessarily applies both to lower and higher cognitive processes: as a digital memory storage unit contemplates only one's and zero's in a bistable activation state, so a human brain can only process input *encoded* as neural structures. A neural structure can be seen as a set of neurons, connections, and spiking behaviors, and their interplay, and the behavior of neurons as patterns of activation (as maintained by the connectionist tradition) also endowed with an important exchange of chemical information. In this perspective it is clear we can say all representations are brain structures and abduction is a neural process in terms of transformations of neural representations. Such a description means to transcend a mere computational analogy, it is useful insofar as to negate that everything in our brain exists as and *only as* a structure of neural activation would be like breaking a computer open and insisting that we can actually *see* images, physically *write* on documents and *hear* the sounds that are just stored in its memory. Now, let me consider a consequence of this, crucial for our next steps: different kinds of cognition, be it sentential, visual, emotional, and so on are encoded within the same framework.

Neural processing of externally originated input is clearly not a random activity: I am not contending that the brain could be seen as a unique and undifferentiated abductive inferential machine. Conversely, I am about to provide a brief recapitulation of scientific hypotheses that suggest the exact opposite: our brain seems to display an extreme specialization in the accomplishment of basic and more complex tasks. The unicity of the neural and abductive ground, though, might encourage both the hypotheses that those neural systems are able to influence one another and that each specialized pattern within the brain displays an abductive endowment.

At this point, it seems proper to briefly refer to a theory, that of the *modular mind*, that became very popular in the 1990s and the first half of the years 2000s, and was often advocated when explaining matters pertaining to the cognitive science of religion. Indeed, the patterns of neural activation I have just mentioned could be comprehended by the definition of *mental modulmental moduleses*. Mental modul*mental modules*es were introduced in literature by Fodor (1983, 2000) who argued that mental phenomena result the operation of multiple distinct processes. Basically, according to Fodor modul*mental modules*es are reflex-like, hardwired devices that process information in a specific, domain dependent, and stereotyped ways. Such multiple specialized processes have two basic features: the first is *domain specificity*, whereas the second is *encapsulation*. Domain specificity refers to the fact that a domain is specialized to accept as input pre-determined and pre-defined classes of information. Encapsulation refers to the fact that any activity of information processing is influenced by nothing external to the process, but only by what the system accepts as "legal inputs" (Barrett and Kurzban 2006). To have a better idea of how a modul*mental modules*e works, it is been advocated the idea of a pipe: as a piece of information is accepted, then it is processed according to procedures that cannot be externally re-engineered.

Later on, Tooby and Cosmides (1992) have developed the idea of modul*mental modules*es claiming that they evolved because of evolution in a Darwinian way, namely, descent with modification. Accordingly, modul*mental modules*es would be described in terms of specialized functions designed to adapt to the relevant aspects of the ancestral environment humans lived in. The theory of mental modul*mental modules*es (or massive modul*mental modules*arity) has been hotly questioned during the last two decades, and it still results highly controversial (Carruthers 2007 provides a quick yet complete review). The main objection raised against modul*mental modules*arity is concerning the presence of a "central" system and its ability to be flexible in handling information from multiple source and for which there is no specific mechanism designed.

More recently, Barrett (2005) put forward an alternative perspective on modul*mental modules*arity that is potentially able to avoid the objections usually raised by those who do not support any modul*mental modules*ar view of the mind: he suggested that specificity and encapsulation should be re-defined respectively in terms of access and processing. Both access (the information accepted as input) and process (the procedures devoted to handling the information) can be specific or general, so that some mechanisms, for instance, might "have access to large amounts of information in the mind but only process information that meets its input criteria" (Barrett and Kurzban 2006, p. 631), and vice versa. This alternative perspective on the role of modul*mental modules*arity is less narrow than the original one allowing to escape from the pipe-like trap. For instance, it has not to assume complete isolation from other systems regarding the kind of information a single modul*mental modules*e can accept and then process. In fact, a modul*mental modules*e can guarantee flexibility and specificity at the same time.

From an epistemic point of view, our abductive perspective on human cognition can explain many characteristics of the modul*mental modules*ar mind theory, but with the advantage of providing a coherent account of low-level brain functions as well. Let us consider a very simple cognitive performance: vision. Vision is not an immediate perceptual performance: as I stated before, vision is simple only if we accept the naïve conception of a inner controller peeking through our eyes, mere openings in the head. Vision is a complex series of physical, chemical and neural reactions that turn a raw mixture of signs into something that is readable by other systems in the brain.[5] As Jacob and Jannerod illustrate in their monograph about visual cognition (Jacob and Jeannerod 2003):

> [T]here are many stages on the way from the retina through the optic nerve to the higher levels of information-processing in the visual cortex. Each such stage carries some information about the distal stimulus and about everything the distal stimulus stands in some non-accidental correlation with. However, neither the retina nor the optic nerve represent everything they carry information about (Jacob and Jeannerod 2003, p. 5).

Being mediated, sensorial perceptions are the result of an inferential activity: according to Peirce, perceptions are abductions, and thus they are hypothetical and it is always possible to withdraw them. Moreover, given the fact that judgments in perception are fallible but indubitable abductions, we are not in any psychological condition to conceive that they are false, as they are unconscious habits of inference: this is precisely what informs our naïve conception of an immediate sensorial perception. Peirce considers perception a fast and uncontrolled knowledge-production process. Perception is a kind of vehicle for the instantaneous retrieval of knowledge that was previously assembled in our mind through inferential processes:

[5] "[D]ifferent visual attributes of objects are processed in separate cortical areas in the visual brain of primates: neurons in area V3 respond to moving shapes; neurons in area V4 respond to colors; neurons in areas MT and V5 are specialized for the processing of motion" (Jacob and Jeannerod 2003, p. 8).

"[…] a fully accepted, simple, and interesting inference tends to obliterate all recognition of the uninteresting and complex premises from which it was derived" (Peirce 1931–1958, 7.37).

Perception is abductive in itself: as Peirce would say, "[a]bductive inference shades into perceptual judgment without any sharp line of demarcation between them" (Peirce 1992–1998, p. 224). As I maintain that perception is an inferential activity, it is easy to understand why the distinction between retrieving and producing knowledge becomes fuzzy: working on a set of signs, the result of perception is always something new and different from the initial set of signs. Therefore, many types of perception display the characteristics of *semi-encapsulation*, insofar as they work as both a *bottom-up* mechanism but are endowed with some *top-down* cognitive affections as well (Raftopoulos' contribution to this research is fundamental, for instance in Raftopoulos 2001a, b). This is not only true for human vision, we can for instance consider animal sensorial perception such as a dog's ability to make sense of olfactive cues, or a cat's perception of vibrations through its whiskers, or a bat noticing the presence of a moth through its ultrasonic echolocation.

In all of these examples, perceptions are the result of an "inferential" (in the Peircean sense) activity, more or less complex, always mediated: what they all have in common is the striving to make sense of *per se* unreadable, raw signs, and make the output available to other cognitive performances. In fact, as Peirce himself had stated, "[a] mass of facts is before us. We go through them. We examine them. We find them a confused snarl, an impenetrable jungle. We are unable to hold them in our minds. […] But suddenly, while we are poring over our digest of the facts and are endeavoring to set them into order, it occurs to us that if we were to assume something to be true that we do not know to be true, these facts would arrange themselves luminously. That is abduction […]".[6] When contending that all perceptual activity is inferential, and hence a low-level yet mediate form of new-knowledge generation, I suggest that the meaning of the word inference is not exhausted by its *logical* aspects but is referred to the effect of various sensorial activities.

If awareness, whether propositional or perceptual, is semiotic, then all awareness involves the interpretation of signs, and all such interpretation is inferential: semiosis not only involves the interpretation of linguistic signs, but also the interpretation of non-linguistic signs, and this will be of crucial importance for our account of the origins of belief in supernatural agents.

Going beyond a conception of abduction which is merely logical, say, related only to its sentential and computational dimension, we can move towards a broader semiotic dimension worth investigating (Magnani 2007b). Peirce himself fairly noted that the all thinking is in signs, and signs can be icons, indices, or symbols. In this sense, all inference is a form of sign activity, where the word sign includes "feeling, image, conception, and other representation" (Peirce 1931–1958, 5.283).

Thesen and his colleagues provide an interesting review of neurological data showing how neural inputs coming from different sensorial system are merged at

[6]cf. Peirce's "Pragmatism as the logic of abduction", in Peirce (1992–1998, pp. 227–241), the quotation is from footnote 12, pp. 531–532.

very early stages of cognition: "[t]raditionally, multisensory processing in the cortex has been assumed to occur in specialized cortical modul*mental modules*es relatively late in the processing hierarchy and only after unimodal sensory processing in the so-called 'sensory-specific' areas", while recent imaging studies hint that "senses influence each other even at the earliest levels of cortical processing, that is, at the level of the primary sensory cortices" (Thesen et al. 2004, p. 85). Pettypiece's team explored the interaction of visual and haptic sensorial input and highlighted how the interplay is not a fixed one but varies with each task (Pettypiece et al. 2010). Emotions seem to play a crucial role in those processes as well: Phelps and her colleagues reported that their behavioral experiments were the first able to demonstrate how emotions influence not only superior faculties but also processes as low as early vision are influence by them (Phelps et al. 2006).

The claim that abductive inferences display a multi-modal tendency does not contradict the presence of pre-wired patterns for neural activation—each dealing with its proper input and output—within the shared internal semiotic environment (Magnani 2009). Even beliefs and desires, that philosophy traditionally interpreted as merely propositional attitudes, can be usefully seen as brain structures (stages of neural activation), and moreover, in this extended framework the concept of inference can be reinterpreted to include non-verbal representations from all sensory modalities and their hybrid combinations, going beyond its merely logical meaning in terms of arguments on sentences. Such a perspective can really help us understand how the nature of abduction is indeed multimodal on the basis of its very neurological ground: on that account it is possible to understand how abductive inferences are not exclusively sentential, or iconic, or model-based, and so on, but all of these aspects can meddle in the neural circuits that make up the basement of our mind, so that emotions (such as fear of predators), desires (i.e. sexual desire) are processed together with a neural structure denoted by a different origin (visual, for instance), in order to achieve an output that can be emotional, visual or neither of the two, and represented in still another formulation.

Human beings, much like any organism, do not *contemplate* the environment uninterestedly, but as *survival machines* they are actively involved in continuous problem-solving activities and the surroundings are hardly ever uniformly relevant to us while, on the contrary, our scope (on the short or on the long term, if it is the case of a complex plan) leads us to select which traits of the environment are relevant.[7] This is extremely important as far as a process like the detection of other agents is concerned: as it will emerge in the next section, actual inferences concerning the presence of other agents in the surroundings are extremely dependent upon emotions, contextual features and previously acquired knowledge.

[7]A quite famous experiment was conducted on *inattentional blindness*: subjects were asked to observe a certain performance (a ball being passed) in a movie and, as they were focused on it, they would miss the appearance of a gorilla beating his chest among the actors (Simons and Chabris 1999).

10.4 From the Detection of Natural Agency to the "Invention" of Supernatural Agents

Nearly any cognitive account of the origins of religion takes as a fundamental assumption the predominance of agent concepts: as a matter of fact, such predominance is displayed not only by religion but by nearly all fictional and folkloric templates as well. We ourselves are agents, we consider our fellow humans (and hominids) as agents and we would file under "agency" our preys and predators as well. Atran highlights three fundamental characteristics connected with the identification of others as agents (Atran 2002, Chap. 3): "observable, short-term productions of complex design", evidence of "internal motivations" directing behavior, and finally the display of teleological acting. A fundamental characteristic of an agent's behavior are, in fact, "telic" actions: not only they imply the possibility of reaching a discrete goal but also the eventuality of stopping the particular action.[8]

As I extensively explained in Part I (Chap. 2), all these aspects are picked up by human brains as *signs*. Our semiotic brains (Magnani 2006) perform a series of abductive operations upon the vast amount of signs surrounding us: what is sometimes referred to as agency-detection module (as said by Atran 2002) or device (in Barrett's formulation Barrett 2009) could in fact be only a pre-disposed abductive pattern. Picking up certain signs and cues, our brain would infer the most likely subject that originated them. What I am referring to is the capacity to detect the presence of another agent from a series of clues in the surrounding environment, i.e. another human (or hominid), a friend or a foe, or animals that could be predators or preys: such an ability is clearly of extreme importance, especially when humans had not reached their ecological dominance yet (Flinn et al. 2005). Yet, even once the ecological dominance was achieved, that is to say, when other human beings became the main major threat for the survival of an individual, the detection of agents remained of crucial importance: we just became more sensitive and thus more capable of recognizing cues signaling complex conscious volition, moral behavior, for instance something bad happening to someone who just committed a mischief, *as if* she was being punished.[9]

Of course, the cues hinting to the presence of another agent are not picked up randomly: any sign is not any sign. They must respond to certain patterns in order to be received and acknowledged by the cognitive systems dedicated to the detection of agents: I admit the existence of a proper domain on which such inferences are operated, so that they can provide us with a sensible output (Sperber and Hirschfeld 2004). Such proper domains comprehend the signs that are effectively produced by agents. When inferring from signs that are fairly symptomatic of the presence of an agent, the quality of our abduction is likely to be elevated. The

---

[8] *The tiger is attacking the goat* describes a telic action, because it is oriented toward the discrete goal of killing the goat but the action (attacking) could be interrupted by the tiger, if needed.

[9] For further reference about anthropomorphism see Gebhard's (Gebhard et al. 2003) and other essays in Clayton and Opotow's edited book about psychological relevance of nature (Clayton and Opotow 2003).

call of a blackbird or the howling of a coyote are fairly symptomatic of the presence of such creatures. Similarly, spearheads, artistic artifacts, evident manipulations of the environment and similar signs displaying complex design are symptomatic of the presence of other human beings. This inferential activity displays an instinctual and subconscious, hence self-assessing, nature: it is deeply integrated in our neural wiring and we usually assume its output to be correct and reliable.

As far as a non-human animal or a human being is concerned, if the quality of the output inference is low and defective, e.g. when we fail to detect a predator, or we think that a tiger is actually a goat, or we fail in distinguishing a friend for a foe, death can be a most likely result.

Consequently with what I have described about animal camouflages techniques in Chap. 2, it is important to add that our systems had not only to detect ordinary signs of other agents but also to infer their presence when they were *actively* trying to hide it by concealing its signs or producing incoherent ones. This explains why our mental systems for the detection and inference of other agents in our surroundings could develop to be so "touchy and hypersensitive", as Barrett claims:

> In our evolutionary past our best opportunities for survival and reproduction and our biggest threats were other agents, so we had to be able to detect them. Better to guess that the sound in the bushes is an agent (such as a person or tiger) than assume it isn't and become lunch. If you reacted unnecessarily (e.g., because of the wind blowing in the brush), little is lost (Barrett 2009, p. 85).

Still, the distinction between what our mind-brain *should* process and what is *actually* processed is rather fuzzy and we often commit errors and we take one kind of sign for another, different one (i.e. we see some dust moving and we think we saw an insect). Shifts in *magnitude* can be a common cause for such—so to say—"abductive error": that is, a similar *kind* of sign but with different intensity such as a very loud noise or the movement of large bodies, such as stones and clouds. Such misperceptions are likely to happen, especially if we consider our ancestors' impossibility to rely on the scientific knowledge we are accustomed to: if we know nothing about what a geyser is, its shrieking and hissing is likely to be cognitively *filed* as the hissing of a snake or the shrieks of a bird. The only problem is magnitude: dimensions divert from what we are used to processing. It is clearly the case with natural conformations of stone that resemble human artifacts, if not for the size. At any rate, once these signs are picked up by agent-dedicated abductive systems, then they *necessarily* become the signs of an agent which originated them.

The common scientific approach to religion, inherited from the 19th century, claimed that "myths [and therefore supernatural beings] prevailed in an early, usually the earliest, stage in an evolutionary scheme, or that myths were the result of primitive mythopoeic man's attempt to explain such natural phenomena as the rising and setting of the sun" (Dundes 1984, p. 3). I contend, though, that such an explanation does not quite grasp the essence of the very beginning of belief in supernatural and can only apply to a later, more structured evolution of "religion:" most of all, it makes sense only as far as we benefit of an alternative, that is scientific knowledge, to match with religious belief.

My point is that supernatural beings were *invented* indeed, but according both to the actual meaning of the word, i.e. to *create or design (something that had not existed before)* and the original latin etymology, that is *in venire*: to *come upon*, to stumble on. The generation of belief in something *super*natural, inferred from certain signs, is just as creative and non *theoretical* as the generation of the belief in an antelope hiding in the bushes: it is the same kind of inferential pattern, just operating on different kinds of signs. The first time our ancestors felt the cognitive need to invent-and-discover the existence of supernatural agents, they were not behaving as theologians. They were not engaging in highly speculative reasoning about the essence of what goes beyond our reason. The first glimpses of belief in supernatural agents might not have involved words, let alone a specific *logos* like the ones we are accustomed to when dealing about religious and mystical matters. Historically, supernatural agents proved indeed to be ideal components of theories explaining different puzzling aspects of our world, but I maintain that their very origin did not display such theoretical and consciously fictional character. That is to say, at the very beginning of belief in supernatural, those super-agents were held to be as real as everyday agents, animal and human.

> Many events lead to agency detection without a known agent as a possible candidate. Suppose a woman walking alone through a deep gorge rounds a bend in the trail and rocks tumble down the steep wall and nearly hit her. HADD [Hypersensitive Agency Detection Device] might reflexively search for the responsible agent. A man hiking through an unfamiliar forest hears something behind a nearby shrub. HADD screams, 'Agent!' If, after detecting agency in these sorts of cases, a candidate superhuman agent concept is offered and seems consistent with the event, belief could be encouraged. Similarly, when a god concept is already available as a good candidate, events that HADD might have overlooked become significant (Barrett 2009, p. 86).

The reason why this kind of inference, trying to make sense of the surroundings, is not critically questioned once it gets to higher levels of consciousness (i.e. not fully unconscious anymore) is that it rests on the same pre-assumptions that inform higher conscience. As Barrett puts it, we can stop and consider the evidence,

> But the evidence (if available) is always filtered and distorted by the operation of mental tools. We never have direct access to evidence but only processed evidence – memories. When asked 'how many colors are in a rainbow?' I might recall the last time I saw a rainbow and what it looked like. But this 'evidence' has already been tainted by non-reflective beliefs (Barrett 2009, p. 81).

Non-reflective beliefs are the base components of folk-physics, folk-psychology, folk-biology and so on. They embody many kinds of regularities that human beings witness during the development of their mind-brain system, and they become the arguments of low-level abductive inferences, similar to:

1. Situation X causes Effect Y
2. *Hence, Effect Y is likely to be symptomatic of Situation X*
3. I notice Effect Y
4. *Therefore, I must be in presence of Situation X.*

If we transpose this simple model into our exemplar narrative about the woman witnessing some falling rocks, we can obtain something like this:

1. An animal climbing on a cliff causes some gravel and rocks to move and fall when he treads over them
2. *Hence, falling rocks are likely to be symptomatic of an animal stepping up hill*
3. I notice rocks falling down
4. *Therefore, I must be in presence of an animal stepping uphill.*

As said, shifts in magnitude allow the same interpretation: if an animal-agent can cause gravel to fall, a big and heavy stone must have been displaced by a mighty powerful agent, but if those stones were seen falling, then the existence of such powerful agent is not minimally questioned. It requires a powerful institutional agent[10] such as a scientific endeavor to transcend the dimension of the singular observer and break our intuitive pre-assumptions, that is our non-reflective beliefs about the world. Consider the movement of the Sun: we know and we believe, because we were taught so, that the Earth revolves around the Sun and still in our everyday life we believe and we behave according to our intuitive assumption that the Sun revolves around the Earth.

Non-reflective beliefs cannot be erased or overwritten no matter how strong and convincing—from a conscious point of view—the evidence against them can be.[11] This can explain why, in spite of the amount of positive scientific knowledge we can rely on, beliefs concerning supernatural agents are extremely resilient: as a matter of fact, many of us still fear presences in the dark, or pray and put their trust in almighty supernatural beings whose might cannot be—at least *scientifically*—tested.

As for this analysis, I am not concerned with cultural refinements of god-concepts and other religious objects: what needs to be stressed at this moment is how beliefs in the existence *supernatural* agents are automatically produced because of the very way our mind-brain system deals with simply *natural* objects. Hence, the first invention of supernatural agents is not a matter of being credulous, or *irrational*, and far less an epistemic struggle to provide an explanation to the mysteries of the universe: this is not to say that religion never deals with such aspects, but they are not the heart of the matter. Conversely, one may say that if our ancestors had not been able to respond to external stimuli with the generation of supernatural concepts, that would have been a sign of poor capacity to cope with their natural and physical environment, and this could have had a terrible evolutionary impact.

[10]By introducing the term "institutional agent" I mean to stress the difference between the approach on reality displayed by a first person agent and an agency (such as *Logic*) that by method, distribution of knowledge and commitment manages to transcend the first person dimension. For a further discussion of John Woods' concept of *institutional agent* see for instance Magnani's monograph on abduction (Magnani 2009, Chap. 7 and Woods 2013).

[11]In Sect. 3.3.3, in Part I, I argued that certain scientific modeling are able to impose a new understanding of phenomena constituting our access to reality, as in the case of Galileo's experiments. Non-reflective beliefs concern our intuitive expectations about the functioning of our surroundings and indeed cannot be erased: also if we *know* that a feather and a bowling ball in a vacuum bell fall at the same speed, we nevertheless *expect* the ball to fall faster: this expectation is systematically disappointed.

10.4.1 The Thunderbird Example

Let us make a clear example of how a super-agent can be extracted out of natural sign configurations. North and South American native populations would draw a supernatural agent from the phenomena usually connected with thunderstorms: our unreflective beliefs about the world (that is, those informing folk-physics) suggest us that an effect must always have a cause. As stated before, to our mind-brain system, "cause" means "agent doing something." If we experience an effect in our ecology, then an agent must have caused it. The bigger the effect, the more powerful the agent *must* be. It is important to stress once again how these universalizing inferences from the cause-effect situation do have an abductive nature: if any cause has a discrete effect, then anything perceived as a discrete effect hints to a cause.

Thus, agent-detecting abductive processes would pick up some of the signs originated by the storms as if originated by an agent, and the resulting *super*agent would be elaborated as a Thunderbird, whose enormous wings stirred the wind and whose powerful cry was thunder itself, as the South American tradition held:

> According to the Ashluslay Indians of the Paraguayan Chaco, thunder and lightning are produced by birds who have long, sharp beaks and who carry fire under their wings. The thunder is their cry and lightning the fire which they drop over the earth. They were also the owners of fire and their enmity against mankind began after they had been deprived of that element (Métraux 1944, p. 132).

North American Indian myth provide a very similar account of the Thunderbird. The cultural relationship between the two supernatural agents is more of an anthropologist's matter, and hence I will not deal with it; conversely, it is important to notice how in both cultures the same kind of signs—related to the weather and the phenomena of sky and air—are processed *as if* symptomatic of an avian super-agent.

> When it is stormy weather the Thunderbird flies through the skies. He is of monstrous size. When he opens and shuts his eyes, he makes the lightning. The flapping of his wings makes the thunder and the great winds. Thunderbird keeps his meat in a dark hole under the glacier at the foot of the Olympic glacial field. That is his home. When he moves about in there, he makes the noise of thunder there under the ice (Reagan and Walters 1933, p. 320).

Adding to this example, once the Thunderbird-concept is ready, it can be inserted in another explanation, which makes sense to those who are already comfortable with the original Thunderbird concept. In this case, also the avalanche is considered as an effect of the Thunderbird's actions:

> Some men were hunting on Hoh mountains. They found a hole in the side of the mountain. They said, "This is Thunderbird's home. This is a supernatural place". Whenever they walked close to the hole they were very afraid. Thunderbird smelled the hunters whenever they approached his place. He did not want any person to come near his house. He caused ice to come out of the door of his house. Whenever people came near there, he rolled ice down the mountain side while he made the thunder noise. The ice would roll until it came to the level place where the rocks are. There it broke into a million pieces, and rattled as it rolled farther down the valley. Everyone was afraid of Thunderbird and of the thunder noise. No one would sleep near that place over night (Reagan and Walters 1933, p. 320).

The inferential process leading to the *invention* of the Thunderbird can be reassumed as follows, within a clearly abductive structure:

1. A bird, as it is taking flight or landing, raises dust whirlwinds and emits a distinctive sound flapping its wings
2. *Hence, whirlwinds and flapping sounds are symptomatic of a the presence of a bird*
3. I notice impressive whirlwinds accompanied by what sounds like a deafening flapping sound
4. *Therefore, I must be in presence of some huge and mighty bird, that is the Thunderbird.*

In brief, I share with the received view on religion the idea that the origins of supernatural rest in its ability to *explain*, but I contend that its very genesis has a much less intentional, conscious and theoretic nature than commonly thought. It is an *explanation* indeed, but one that belongs to the cognitive urge of human beings to constantly make sense of the surrounding environment. Of course it quickly merged and was structured within human beings' *constitutive* curiosity, and within the struggle to knowing causes and origins of natural phenomena: still, in a diachronic perspective, the very first pulse of belief in supernatural agents might have rather sprouted from an essentially wired[12] (and mostly unconscious), neuronal-cognitive processing of environmental signs.

10.5 Embodying Supernatural Agents by Disembodying Cognition

If so far I have attempted to provide a mind-based explanation about the origin of everything loosely "supernatural" within our brains, the question "How did supernatural entities get in our mind *as we know them*, if they do not exist in our ecologies?" is still unanswered. How did humans come up with the precise graphic representation of angels, and the same with Egyptian gods, dragons, fairies and so on? This section will deal with the processes of distribution of "raw" beliefs about supernatural agents into material culture, which allowed a verbal and/or iconic blending of characteristics from different cognitive domains, and eventually lead to the recapitulation of the supernatural concept *in the final form* shared by human minds.

[12]The term *wired* can be easily misunderstood. Generally speaking, I accept the distinction between cognitive aspects that are *hardwired* and those which are simply *pre-wired*. By the former term I refer to those aspects of cognition which are fixed in advance and not modifiable. Conversely, the latter term refers to those abilities that are built-in prior the experience, but that are modifiable in later individual development and through the process of attunement to relevant environmental cues: the importance of development, and its relation with plasticity, is clearly captured thanks to the above distinction. Not all aspects of cognition are pre-determined by genes and hardwired components. For further considerations about this issue refer to Barrett and Kurzban (2006).

In order to proceed in this investigation, I have to elaborate on a concept I already introduced earlier in this chapter and along this book:[13] *semiotic brains*. That is, brains that can construct a series of signs and that are engaged in making or manifesting or reacting to a series of signs: through this semiotic activity they are at the same time engaged in "being minds" and so in thinking intelligently.

Several studies (Mithen 1996, 1999; Humphrey 2002; Lewis-Williams 2002) in cognitive paleoanthropology—even if rather speculative—contend that reflective and high-level consciousness (in terms of thoughts about our own thoughts and about our feelings, that is consciousness not merely considered as sensation) is intertwined with the development of language (speech) and material culture. 250,000 years BP several hominid species had brains as large as those of modern day humans, but their archaeological remnants do not provide any major evidence of art or symbolic behavior. If we consider high-level consciousness as related to a high-level organization of human cortex, its origins can be related to the active role of environmental, social, linguistic, and cultural aspects.

As a matter of fact, the production of new artifacts (such as *hand axes* as in Mithen's account (Mithen 1999)) had to rely on two crucial factors:

1. a good degree of *fleeting consciousness* (thoughts about thoughts).
2. the exploitation of *private speech* (i.e. speaking to oneself) to allow an overall supervision and appraisal of the various activities involved in the development of an artifact (as for hand axes, private speech served to trail between planning, fracture dynamic, motor control and symmetry). In children as well we may witness a kind of private muttering which makes explicit what is implicit in the various abilities.

It is extremely important to stress that material culture is not just the product of this massive cognitive chance but also its cause. "The clever trick that humans learnt was to disembody their minds into the material world around them: a linguistic utterance might be considered as a disembodied thought. But such utterances last just for a few seconds. Material culture endures", as stressed by Mithen (1999, p. 291). Fleeting consciousness and rudimental private speech provided a kind of inner semiotic blackboard where previously distinct cognitive resources can be exploited all together and in their dynamic interaction. The result of this synthesis can be similar, from the phenomenological point of view, to the isolated application of the single components, but from the psychological and semiotic perspective it sparks a revolution, because it allows the blending of several cognitive domains, thanks to a distribution of cognitive tasks into the external environment.

From this perspective the semiotic expansion of the minds is in the meantime a continuous process of disembodiment of the minds themselves into the material world around them. In this regard the evolution of the mind is inextricably linked with the evolution of large, integrated, material cognitive semiotic systems. It may take a little effort to find this argument compelling (and especially the following

[13] Sects. 2.2.2 and 3.2.1

one strictly concerning supernatural beings), because of our modern-humans' linguistic bias. Indeed, it could be said that the highly-organized mind of modern-day human beings has already been pre-wired *according to* the effects of such cognitive distributions. Most of all, our semiotic brains make use of a powerful symbolic language and an advanced private speech: we are thus allowed to rehearse—in our symbolic imagination—cognitive distributions and blending we would otherwise externalize on material supports. Nevertheless, when we have to infer a meaning from a set of data we cannot understand, we often rely on *model-based reasoning* in form of recurring schemas, diagrams or other kinds of visual manipulations: similarly, the need for conceptualizing agents that went beyond mere biological ones sparked the need to manipulate and hybridize already known features by distribution over external supports.

A wonderful example of meaning creation through disembodiment of mind is the carving of what is probably a mythical being from the last ice age, 32,000 years ago: a half human/half lion figure, carved from mammoth ivory, found at Hohlenstein Stadel, Germany, often displayed as an example of the Aurignacian culture (Fig. 10.1).[14]

> An evolved mind is unlikely to have a natural home for this being, as such entities do not exist in the natural world, the mind needs new chances: so whereas evolved minds could think about humans by exploiting modul*mental modules*es shaped by natural selection, and about lions by deploying content rich mental modul*mental modules*es moulded by natural selection and about other lions by using other content rich modul*mental modules*es from the natural history cognitive domain, how could one think about entities that were part human and part animal? Such entities had no home in the mind (Mithen 1999, p. 291).

A mind consisting of different separated intelligences cannot come up with such entity. The only way is to extend the mind into the material word, giving the environment a primitive organization and exploiting in a semiotic way external materials (such as stone, ivory, *etc.*) and various techniques to impress a modification on them: "[...] artifacts such as this figure play the role of anchors for ideas and have no natural home within the mind; for ideas that take us beyond those that natural selection could enable us to possess" (Mithen 1999, p. 291).

In the case of our figure we deal with an anthropomorphic thinking created by the material representation, serving as a semiotic anchor for the cognitive representation of a supernatural being. In this case the material culture disembodies thoughts, that otherwise would soon disappear—without being transmitted to other human beings—and realizes a systematic semiotic delegation to the external environment. The early human mind might have possessed two separate intelligences for thinking about animals and people. Through the mediation of the material culture, the modern human mind can manage to think *internally* about the new concept of animal and human at the same time. But the new meaning occurred *out there*, in the external material world from where the mind picked it up.

[14]To appreciate the importance of this finding, consider that the first subsequent evidences of human-animal hybrid go back to the third millennium BC: that is, more than 25,000 years after the Aurignacian "Lion-Man".

Fig. 10.1 The Hohlenstein Stadel *Lowenmensch*, via Wikimedia Commons

A little digression might be interesting at this point: in Part I (Sect. 3.2), relying on Holland's (1995) intuitive definition, I had described a model as "allow[ing] us to infer something about the thing modeled." Hence, without a model it would not be possible to infer certain things about what is modeled. My current argument about knowledge distributions permitting the creation and recapitulations of supernatural beings is very similar, to the extent that it seems plausible to describe proto-religious artifacts such as the Lion Man as *models*. I will further elaborate on this in the Concluding Remarks of this book, but it can be useful to illustrate the insight now as the discourse about the generation of beliefs in supernatural beings is still fresh: my broad claim, already sketched along this book, is that many activities have an eco-cognitive dimension and display similar inferential patterns albeit in different domains. In this case, the generation of supernatural concepts can be seen as a way of creating *models* of external (and internal) phenomena, relying on construals that allow the externalizations and distributions of thoughts, as I am arguing right now: from the cognitive point of view, the "Lion Man" is not that different from many scientific models that came millennia after it was carved, and what they share is their affording certain inferences about the modeled objects that would not be possible without them. Furthermore, this kind of modeling is indeed creative and does imprint the phenomenon, because supernatural explanation do indeed reconfigure, at several levels, the agent's cognition.

Indeed, in this perspective one should acknowledge that material artifacts are tools for thoughts as language is: tools (and their related new "signs") for exploring, expanding, and manipulating our own minds: the evolution of material culture is inextricably linked with the evolution of consciousness and thought.

Through the mediation of the material culture the modern human mind could arrive to *internally* think the new meaning of animals and people at the same time. This process involves two fundamental representational activities:

- *external representations* are formed by external materials that express (through reification) concepts and problems either already stored in the brain and extracted for manipulation, or concepts that do not have a natural home in the brain;
- *internalized representations* are internal re-projections, a kind of recapitulations (learning), of external representations in terms of neural patterns of activation in the brain. They can sometimes be "internally" manipulated like external objects and can originate new internal reconstructed representations through the neural activity of transformation and integration.

As for the Hohlenstein Stadel "Lion-Man", the *external representation* stage concerns the single parts of the figurine, the human body and the lion head, while the *internalized representations* comprehends the Lion-Man as a meaningful whole.

It is plausible to imagine that most anthropomorphic and zoomorphic hybrid deities—and supernatural creatures in a wider sense—were *models*, generated in a similar way. This is not to say that a Hindu should not be scandalized if I labelled her pantheon a series of primitive iconic hybrids: as human beings progressed in the use of their mind as a "semiotic sketchpad," they could use it as a *virtual* support reproducing part of the tasks they would face in their environment. Imagination supported by advanced language and full consciousness is affected by a much lesser number of constraints than material culture.

As shown in the previous section, the new hybrid super-agent—once internalized in the mind of human beings—could become the *explanation* of the original signs. It is not a *best explanation* (from an epistemological point of view) but still a satisfactory one: further occurrences of the same signs will mechanically lead to the identification of the super-agent as causing the phenomenon. Once we have the result, the externalization-blending-and-recapitulation process is obliterated and the new hybrid concept is accepted within the human brain as if it had been there *originally*. Cryptozoology and akin *borderline sciences* seem to rely on the naïve assumption that if we have the concept of some creature in our minds and our cultures, and furthermore we have depictions of it, then it probably exists (or at least it existed sometimes in the past). Human beings exhibit a similar behavior very often when, for instance, after watching an horror or sci-fi movie they expect the monster, alien, undead or vampire to sneak up in a dark hallway at night in their own house, even though they consciously know they just witnessed a production of human fiction.

With this respect I could suggest one last, but not least, consideration for this particular study. Once we externalize and distribute our thoughts about supernatural beings into the material culture, a dramatic increase in *manipulability* resulted. By associating a supernatural agent with its representation the agent became immediately present *there and then*. It could be seen, contemplated, touched, transported from one place to another, shown to other people who in turn possessed different ones: we can easily imagine a never ending continuity stretching from the *Lion Man* to present day Crucifixes.

The abductive theoretical framework on which I have been relying from the beginning of this book is useful once again, with the concept of *manipulative* abduction (Magnani 2009, p. 1.6), already deployed when dealing with experimentation (Part I, Chap. 4): the externalization and embodiment of agent-concepts in artifacts was the result of an hypothetical eco-cognitive distribution in the environment, so that thinking and discovery were achieved *through doing*. The possibilities brought about by the externalization and materialization of thoughts exceed those of the same thought when it was *just* in the mind.

The manipulation of (what would become) the religious artifact generates *new* knowledge concerning it, which is promptly re-absorbed in the manipulator's mind: the process that produced the new knowledge is obliterated and this information gathers with what was already known about the supernatural agent before its materialization: thus, in a self-reinforcing dimension, what was as a matter of fact the consequence of the externalization process is considered to be the very knowledge that permitted it.

10.6 Conclusion

The aim of this chapter was to show how a philosophical analysis could engage cognitive science in fruitful dialogue, and lead us to a partial revision of mainstream theories about the development of religion in human societies. If belief in supernatural agents has a constitutive origin within our mind-brain system, we can imagine religion to have a much less generative role as far as the extramundane is concerned. That is, from a eco-cognitive perspective, it might be argued that, at least at the very beginning, religion did not create beliefs in supernatural beings, but conversely religion might be seen as a cultural tool aimed at answering the emergence of such entities in our minds: thus, the issue at stake here is not about religion being maladaptive[15]

---

[15]My use of adaptive and maladaptive can be said to be *loosely darwinian*. I do not wish to enter the debate on wether religion is an evolutionary adaptation: its mal-*adaptiveness* would just signify a negative impact on the fitness and the welfare of the concerned individuals. Arguments about the contended evolutionary role of religion tend to focus rather on the social side than on

per se but rather religion being a way to regulate and control potentially dangerous beliefs about *super*natural entities conceived by our minds.

In the next chapter, I will try to describe the formation and the evolution of counterfactual/religious beliefs in connection with the discourse about cognitive niches I undertook in the second part. In particular, my aim will be to investigate a peculiar kind of counterfactual belief, pervading or even generated in techno-cognitive niches: the interests resides in the fact that, whereas techno-cognitive niches are often regarded as the epitome of scientific rationality, our cognitive systems could be so biased towards agency detection—and subsequently religious beliefs—to the point of spontaneously producing a number of "ghosts in the machine".

References

Atran S (2002) In gods we trust: the evolutionary landscape of religion. Oxford University Press, Cambridge

Atran S, Medin D (2008) The native mind and the cultural construction of nature. MIT Press, Cambridge

Barnes MH (2000) Stages of thought: the co-evolution of religious thought and science. Oxford University Press, Oxford

Barrett HC (2005) Enzymatic computation and cognitive modularity. Mind Lang 20(3):259–287

Barrett HC, Kurzban R (2006) Modularity in cognition: framing the debate. Psychol Rev 113(3):628–647

Barrett J (2009) Cognitive science, religion and theology. In: Schloss J, Murray MJ (eds) The believing primate. Oxford University Press, Oxford, pp 76–99

Boyer P (2001) Religion explained. Vintage U.K. Random House, London

Carruthers P (2007) Simple heuristics meet massive modularity. In: Carruthers P, Laurence S, Stich S (eds) The innate mind: volume 2: culture and cognition. Oxford University Press, Oxford

Clayton S, Opotow S (eds) (2003) Identity and the natural environment: the psychological significance of nature. The MIT Press, Cambridge

De Cruz H (2013) Cognitive science of religion and the study of theological concepts. Topoi 33(2):487–497

Dennett D (2006) Breaking the spell. Viking, New York

Dixon T (2008) Science and religion: a very short introduction. Oxford University Press, Oxford

Driver J, Spence C (1998) Crossmodal attention. Curr Opin Neurobiol 8(2):245–253

Dundes A (1984) Introduction. In: Dundes A (ed) Sacred narrative: readings in the theory of Myth. University of California Press, Berkeley, pp 1–3

Flinn MV, Geary DC, Ward CV (2005) Ecological dominance, social competition, and coalitionary arms races: why humans evolved extraordinary intelligence. Evol Hum Behav 26:10–46

Fodor J (1983) The modularity of the mind. The MIT Press, Cambridge

Fodor J (2000) The mind doesn't work that way. The MIT Press, Cambridge

Gebhard U, Nevers P, Bilmann-Mahecha E (2003) Moralizing trees: anthropomorphism and identity in children's relationships to nature. In: Clayton S, Opotow S (eds) Identity and the natural environment: the psychological significance of nature. The MIT Press, Cambridge, pp 91–111

Holland J (1995) Hidden order. Addison-Wesley, Reading

(Footnote 15 continued)
the cognitive one: for further reference on this subject see Wilson's *Darwin's Cathedral* (Wilson 2002); reflections on the evolutionary weight of religion can also be found in the rich edited book *The Believing Primate* (Schloss and Murray 2009) and of course in Atran's monograph (Atran 2002).

Humphrey N (2002) The mind made flesh. Oxford University Press, Oxford
Jacob P, Jeannerod M (2003) Ways of seeing: the scope and limits of visual cognition. Oxford University Press, Oxford
Lewis-Williams D (2002) The mind in the cave. Thames and Hudson, London
Magnani L (2006) Mimetic minds. Meaning formation through epistemic mediators and external representations. In: Loula A, Gudwin R, Queiroz J (eds) Artificial cognition systems. Idea Group Publishers, Hershey, pp 327–357
Magnani L (2007a) Animal abduction. From mindless organisms to artifactual mediators. In: Magnani L, Li P (eds) Model-based reasoning in science, technology, and medicine. Springer, Berlin, pp 3–37
Magnani L (2007b) Semiotic brains and artificial minds. How brains make up material cognitive systems. In: Gudwin R, Queiroz J (eds) Semiotics and intelligent systems development. Idea Group Inc, Hershey, pp 1–41
Magnani L (2009) Abductive cognition: the epistemological and eco-cognitive dimensions of hypothetical reasoning. Springer, Berlin
Métraux A (1944) South American thunderbirds. J Am Folk 57(224):132–135
Mithen S (1996) The prehistory of the mind. A search for the origins of art, religion, and science. Thames and Hudson, London
Mithen S (1999) Handaxes and ice age carvings: hard evidence for the evolution of consciousness. In: Hameroff A, Kaszniak A, Chalmers D (eds) Toward a science of consciousness III. The third tucson discussions and debates. MIT Press, Cambridge, pp 281–296
Peirce CS (1931–1958) *Collected papers of charles sanders peirce*. Harvard University Press, Cambridge, vols 1–6, Hartshorne C, Weiss P (eds); vols 7–8, Burks AW (ed)
Peirce CS (1992–1998) *The Essential Peirce. Selected Philosophical Writings*. Indiana University Press, Bloomington, vol 1 (1867–1893), ed by Houser N, Kloesel C; vol 2 (1893–1913) ed by the Peirce Edition Project
Pettypiece CE, Goodale MA, Culham JC (2010) Integration of haptic and visual size cues in perception and action revealed through cross-modal conflict. Exp Brain Res 201(4):863–873
Phelps EA, Ling S, Carrasco M (2006) Emotion facilitates perception and potentiates the perceptual benefits of attention. Psychol Sci 17(4):292–299
Raftopoulos A (2001a) Is perception informationally encapsulated? The issue of theory-ladenness of perception. Cogn Sci 25:423–451
Raftopoulos A (2001b) Reentrant pathways and the theory-ladenness of perception. Philos Sci 68:S187–S189 (Proceedings of PSA 2000 biennal meeting)
Reagan AB, Walters LVW (1933) Tales from the Hoh and Quileute. J Am Folk 46(182):133–297 465
Sagan C (1997) The demon-haunted world: science as a candle in the dark. Headline, London
Schloss J, Murray MJ (eds) (2009) The believing primate. Oxford University Press, Oxford
Scott SL (1994) They don't have to live by the old traditions: saintly men, sinner women, and an appalachian pentecostal revival. Am Ethnol 21(2):224–227
Simons DJ, Chabris CF (1999) Gorillas in our midst: sustained inattentional blindness for dynamic events. Perception 28:1059–1074
Sperber D, Hirschfeld LA (2004) The cognitive foundations of cultural stability and diversity. Trends Cogn Sci 8(1):40–46
Thesen T, Vibell JF, Calvert GA, Österbauer RA (2004) Neuroimaging of multisensory processing in vision, audition, touch, and olfaction. Cogn Process 5(2):84–93
Tooby J, Cosmides L (1992) The psychological foundations of culture. In: Barkow J, Cosmides L, Tooby J (eds) The adapted mind. Oxford University Press, Oxford
Wilson DS (2002) Darwin's Cathedral. Chicago University Press, Chicago
Woods J (2013) Errors of reasoning. Naturalizing the logic of inference. College Publications, London (Forthcoming)

Chapter 11
Digitalizing the Religious Niche (and Vice Versa)

Abstract This brief chapter deals with the developments of online religiosity and its possible perversion. The first section examines whether the impact of Internet on traditional religions is fundamentally helpful in adding anything new to the world of spirituality and devotion. The second section deals with some instances of religious and spiritual behaviors that are being produced by digitalized lifestyle, even though they are not concerned with traditional religious beliefs, but emerging through the same inferential patterns as the one we witnessed in the previous chapter. The question underlying this research is whether we are looking *the right way* when we mean to study the link between computers, Internet and religiosity.

11.1 Introduction

Religion, understood as a set of beliefs and practices, has such a clear counterpart within material culture that it can be easily seen to individuate a full "religious niche," a notion modeled on the concept of *ecological* and *cognitive niche*, introduced and analyzed in Part II.[1] In everyday life, this religious niche is understood as overlapping with other niches (cultural, moral etc): most readers could at this very moment stand up, leave their chair or desk for the nearest window and see at least one religious building (be it a church, a mosque, a synagogue or another kind of temple: some might even be able to see different ones), and some readers could have a cross hung onto the wall, or some other religious symbol adorning the neck of a colleague or classmate. The religious niche is also made up of *external gestures*, and *words*,

I must credit Carolina Cinerari for significantly helping me in focusing the concept of *religious niche*, and in general for her contribution to the research for this chapter.

[1]cf. Chapter 5 in Part II. To put it very briefly, an ecological niche could be defined, following Gibson, as a "set of environmental features that are suitable for an animal" (Gibson 1979), and differs from the notion of habitat in the sense that the niche describes *how* an organism lives its environment, whereas habitat simply describes where an organism lives. The concept of *cognitive niche* (Clark 2005; Magnani 2009) indicates a series of externalizations of knowledge into the surrounding environment, through for instance material culture, resulting in a modification of the selective pressure that an organism has to face.

© Springer International Publishing Switzerland 2015

T. Bertolotti, *Patterns of Rationality*, Studies in Applied Philosophy, Epistemology and Rational Ethics 19, DOI 10.1007/978-3-319-17786-1_11

that have a distributed nature—being accessible to other cognizant agents in a local ecology.

In many cases, the significance brought about by the religious niche can be seen as another layer juxtaposed over a pre-existent one. Consider churches and temples, for instance: first of all, religious buildings are ecological niches inasmuch as they offer shelter from the weather. Then, they belong to a cognitive niche since they are externalizations of skills concerning architecture, arts, material studies etc., but they differ from all other buildings since they embed the further signification of being *holy* places (Magnani 2011, Sect. 6.4). People with different levels of religiosity do not react the same way when, moving about in their local ecologies, come to face parts of the religious niche, but these encounters hardly ever fail to spark peculiar glimpses of thought in the subject: it could be in fact suggested that most elements of the religious niche often act as "moral mediators."

The concept of moral mediator refers to the intentional or unwilling use of external artifacts (including spoken or written words or body gestures and positions), and particular natural phenomena and landscapes, as a prop for unexpected or enhanced moral deliberation and reflection. Western literature is teeming with examples of sudden conversions and abrupt changes in one's course of action following the sight of a religious artifact or building: those are clear testimonies of how moral mediators work. This reflection on moral mediators was meant to help the understanding of how everyday, in our more or less usual itineraries in anthropomorphic environments, we are faced with a number of elements from religious niche, and these *external* encounters are often the prop either for a moment of religious considerations, or for a new moral appraising of peculiar situations in our lives.

11.2 The Enhanced Religious Niche and Its Perversion

For a better understanding of how the various facets of the religious niche are strictly intertwined with our everyday life, and how the enrichment of the former can deeply affect the way we cope with spirituality, it is useful to consider some characteristics of enhanced cognitive niches: this allows to sketch out some analogies in order to appreciate the religious niche as an enhanced niche, and to foresee some potential consequences of its digitalization and its becoming more and more autonomous.

More specifically, the widespread use of new technological artifacts in our everyday environments (already considered in Chaps. 8 and 9) brought about two peculiar phenomena: the omnipresence of technology within the niche and a dramatic incrementation in the niche's predictive capacities. As a matter of fact, on the one hand we are surrounded by digital artifacts, accumulating an astonishing quantity of data and information of every kind, on the other hand those same artifacts are able to interfere with our preferences and habits, as they continually monitor our behaviors. Both these phenomena can be witnessed to take place in digitalized religious niches and could be responsible for interesting changes in our relationship with spirituality especially after the massive advent of the Internet, and the so-called Web 2.0 displaying its hunger for *user generated content*.

The World Wide Web and other Internet-related technologies are also part of our cognitive niches, and they are mirroring at an ever growing rate many activities that were once carried out chiefly in the physical reality.[2] Such a displacement must certainly concern also a number of activities concerning religion. As a matter of fact,

> [...] the online presence of religion is growing daily. One Pew Internet and American Life study reports that, among Americans, "25% of Internet users have gotten religious or spiritual information at one time." Once again, though, this impressive figure must be tempered with other Pew-generated data showing that the number of Internet users who seek online religious information on a daily basis is considerably smaller, at just under 5% (Dawson and Cowan 2004, p. 6).

Data like these are extremely meaningful. First of all, a noteworthy hiatus jumps to one's attention: in our ordinary human niches, as suggested before, it is nearly impossible to go through a whole day without encountering bits of what I defined as the *religious niche*. In physical reality, we have far less power to make the environment depend on our intentions than in virtual reality. Statistics concern when and why people *seek* for religion on the Internet: most of the time, it is unlikely that a user will stumble on a religious website why browsing for a cheap flight for her summer holidays, whereas it could happen to her to find a church while walking to the travel agency. In the physical world, bits of the religious niche come to us even if we do not *seek* it. Then, the first intuitive difference between religion online and in the virtual world is the extent to which the religious niche is juxtaposed as a sometimes different layer of meaning over other contexts: to find about religion on the Internet, we usually must want it. It rarely happens *serendipitously*: if a user is not in the mood for anything more religious than anything likely to appear in the headlines of a news website, she must be well intentioned to look for it, maybe already know where to find what she is looking for.

It is already possible to draw a reflection from what was laid out so far: the role of moral mediation[3] supported by the religious niche in physical reality was twofold, and while the intentional side is preserved (I can look up a religious website to ponder some moral issue in my life), the unintentional one is weakened.[4]

---

[2]I am going to refer to physical reality without any ontological commitment, but just to distinguish it from the virtual, cybernetic reality.

[3]It must be acknowledged that the effect of a moral mediator is not only local, but the consequences of a single moral deliberation can extend far beyond the agent's immediate surroundings, both geographically and chronologically.

[4]With the exception of occasional references as this one, I preferred not to tackle the relationship between religion and morality as relatively eccentric to the "inferential" analysis I would undertake. A wide amount of literature already exists concerning the relationship between moral behavior and the evolution of religion (cf. for instance Wilson 2002; Johnson and Krüger 2004; Johnson and Bering 2006), especially as far as sociobiology is concerned. Conversely, I refer to the relationship between religion and morality (always considered in a pragmatic and naturalistic fashion, exemplified in Part II), when the latter strictly concerns the *inferential* dimension, as in the case of moral mediators, or religious pragmatics such as sacrifices and forgiveness (cf. Chap. 13).

The first consequence that is immediately perceived is that the digitalization of the religious niche involves a separation of the niche itself from other dimensions of everyday (digital) life: this is coherent with the view expressed by a recent article in *Wired* magazine (Anderson and Wolff 2010), suggesting that the idea of the Web as a free, open-range environment is about to see its end, supplanted by an "App-based" conception of Internet (also pushed by the diffusion of smartphones and tablet pc's). This way of structuring the Internet seems to echo the original subdivision in thematic *portals*, that were popular in the late 1990s and early 2000s, in a way that on the one hand facilitates users' retrieval of needed information, contacts etc. but on the other hand jeopardizes the level of serendipity that would characterize a session of "surfing." As I already contended, such specialization of virtual environments could further *encapsulate* occurrences of online religiosity, further alienating them from the diffused dimension characterizing its non-virtual counterpart.

The ever-growing digitalization of the religious niche radically changed people's way to perceive and share their spirituality: the world wide web has become, since the Nineties, an ideal setting to share one's doubts and beliefs in forums and, later, blogs.

> Using text to communicate in forums like ORIGINS, individuals began using the Internet to express their religious beliefs and concerns, as well as simply to talk about religion. In a sense, these Internet bulletin boards became a computer-generated, unofficial, religious environment (Helland 2004, p. 24)

> This illustration of the shift from the offline world to the online indicates two very important social consequences of the Internet: a crisis of authority and a crisis of authenticity. [...] The obviously constructed and pluralistic character of religious expressions online tends to have a relativizing effect on the truth claims of any one religion or its authorities. Rather than appearing unreal, with enough exposure to the Internet religious people may come "to doubt the absolute claims of sacredness and permanence that a religious site can make in the 'real world'" in light of the obvious "ephemerality and heightened access [to] religion in cyberspace" (Dawson and Cowan 2004, pp. 2–3).

Keen (2007) has provocatively denounced the *mass amateurization* processes triggered by the diffusion of user-created content on the Internet. His claim is coherent with Dawson's, as the latter stresses how the communicational capacities offered by new technologies are coupled with the increasing presence of statements whose authenticity and authoritativeness cannot be guaranteed by anyone. This is partly an effect of the major delegation of responsibilities onto the artifactual dimension of the religious niche, and—instead of fostering its enrichment—it could ultimately pervert the relationship between the individual and its environment, in this case consisting chiefly of knowledge, so that the environment could finally become utterly unusable by those who were originally meant to use it: what is at stake is therefore the possibility that the enriched digital niche supporting religion online eventually became *maladaptive* for believers and religion itself.

11.3 The Feasibility of Online Religion and the Problem of Disembodiment

My aim in this section is to contend that the limit of online religion, understood as the virtual transposition of devotional practices originated in the physical world, arises inasmuch religion cannot be body-blind: recent studies about religion as a natural phenomenon do stress indeed the importance of the embodied dimension of religious behavior and beliefs [Atran for instance pays extreme attention to this aspect in his exhaustive analysis of religion (Atran 2002)]. Religion is indeed an *eco-cognitive* activity. For instance, religious rituals are something that take place in our ordinary ecologies, and we take part to rituals using our minds and our bodies: it should be enough to think of an ordinary Christian mass to realize to what extent *body* activities are crucial: we express our inner cognitive states by standing up, sitting down, moving our hands, using our vocal cords and by touching other believers as well: the work of the body is part of what turns rituals into "acts that work" (Søresen 2007).

Even though a virtual dimension of ritual seems to exist, it is usually either informing the user about what should be done in the world of the body, or it is supposed to be mentally and symbolically rehearsed by the user, in a way which completely excludes the bodily dimension of ritual. Let us take as an example a part of a ritual where one must light a candle: in the first case, the Internet will mediate the non-virtual action of lighting the candle, for instance telling the believer when and why to do so; in the second case, the believer-user will light a "virtual candle," usually in a text-based or graphic environment. In the latter case, rather than acting the user could be said to witness their own actions and interpret them symbolically, according to something non-virtual that has already been lived: in this sense, interactivity turns indeed into *interpassivity* (Wilson 2003) since believers use themselves as passive screens for projecting parts of their own psyches (e.g. past experience) instead of living the real experience interactively. Campbell's contention is that this kind of virtualized religion is rather similar to the idea of broadcasting services on the radio or on TV, where the only spiritual inspiration comes from the fact of having, *upon another time*, *been there* with one's body: in this sense, the virtual service has not a full dignity in itself but works as a prop to recall previous fully embodied experiences (Campbell 2004). Löveim also stresses the extent to which online religion lacks *necessarily* the persuasiveness that comes from the involvement of the body:

> The ritual is seen as primary; belief in the conventional sense of that term is almost beside the point. By participation in the ritual, the actors invoke a goddess who may well be seen as a collective fiction but who nevertheless provides some spiritual sustenance and comfort to her followers. [...] What the online ritual lacks, in and of itself, is precisely the quality of physical presence that enables ritual actors to become so deeply embedded in the belief system that they will end up in an underground chamber, clutched with each other in a death embrace (Löveim 2004, p. 55).

Campbell's main objection to the possibility of online religion refers to its possibility of being nothing but a prop to consider something else, and not the real thing *in se*: it is nearly a Platonic argumentation, sustaining that because of its radical disregard of the bodily dimension, online religion is a weaker, stranger and sometimes just more comfortable alternative to real-life religious commitment:

> Consider Brasher's description of the implications of the Internet for the evolution of religion: "A fantasy universe that stimulates the imagination but ignores the rest of the body, cyberspace is a nonenvironment that sucks attention away from the immediate surroundings in which most traditional religious life occurs". Such a statement is stark, perhaps even foreboding; but it must be asked whether instances of online Christianity really bear out such a characterization of Internet religion as being so utterly disembodied [...]. These suggestions make explicit reference to an offline, embodied aspect of the online prayer experience. For example, in the stage of prayer that involves cultivating an awareness of the presence of God, one is instructed to accompany the verbal and/or mental prayer with body and breathing exercises. [...] Again, the experience of the body, this time through the sense of sight, is used to assist the online prayer. Hence despite its presentation of an opportunity for online prayer, which is a clear instance of what we can categorize as online religion, this site still makes extensive use of embodied experience as an essential component of this prayer. That is, online religion still makes explicit reference to the offline (Campbell 2004, p. 101).

Summing up what I discussed so far, it seems that if we consider online religiosity as the virtual transposition of established cults (or the transposition of ways of being religious), it seems that the whole concept of "online religion" delivers far less than it promises: indeed, the digital religious niche affords unprecedented and (at least in principle) unbiased ground for religious communication, setting up an incredible framework for dialogue with people holding similar and different beliefs, but on the other hand the risk of *amateurization* is always looming over the whole enterprise: to use a Christian metaphor, if it could be said that e-believers are brought back to a climate of fertile confrontation that reverberates the beginnings of the Church, it is also true that the final result is that they end up barricaded each in their own *catacombs*.

This second section, instead, challenged the possibility of online religion by questioning the absence of the body in the virtual conception of the ritual: if the body is involved, it is not fully virtual. Conversely, if the body is *not* involved, then it is virtual alright, but the quality of the religious performance is scarce.

In the next section, though, I will try and answer a different question: "Could we be looking the wrong way? If the Internet is a whole new world, might there be radically different instances of religious behavior that rely on the same mechanisms but concern radically different frameworks?"

11.4 Cyberspiritual but not Religious

So far, it seems that—when dealing with Internet and religion—we are not going to find anything new, except pre-established practices taking advantage (and disadvantage) of enhanced communication systems. To say that explicit online religiosity fails to address *significantly* those human features characterizing religion and spirituality does not imply saying that those features cannot be found address at the intersection of Internet and religiosity.

As a matter of fact, this claim can be exemplified by a few examples of how a tacit supernatural dimension (and its perception) can emerge in Internet-related activities. Furthermore, what should be observed is that Internet seems to afford users with characteristics that are typical of supernatural agents.

11.4.1 Super(natural) Users and Avatars as Semi Gods

Within the perspective of distributed cognition, it is clear that contemporary information technologies empowered human beings in an unprecedented way. It can be argued, though, that this enhancement can be traced back to a quantitative shift rather than a qualitative one. As eco-cognitive engineers, human beings have always exceeded their bodily limits by resorting to external artifacts and manipulations, not only from a physical point of view but from a cognitive one as well. Written language, for instance, allowed to dislocate complex communication in space and time, thus breaking many limits of natural animal communication.

It takes little effort to appreciate how there is a continuum and not a series of ruptures between the first instances of material culture (Mithen 1996) and present day Internet highways connecting billions of individuals worldwide, allowing any kind of data exchange. Conversely, a more challenging reflection could be brought about by observing certain conceptual analogies between the development of cognitive and communicative devices, the evolution of beliefs in supernatural agents and how the constant improvement of cognitive artifacts similarly managed to "supernaturalize" human beings.

As suggested by recent literature about the cognitive origins of religious beliefs (Boyer 2001; Atran 2002; Dennett 2006; Schloss and Murray 2009), some of which were reviewed in the two previous chapters), the notion of supernatural agent developed step by step, thanks to a progressive, intuitive and often unwilling erosion of ontological boundaries (and the related gnoseological ones) regulating our expectations about the presence and the behavior of biological agents.[5] What I mean to stress by recalling this kind of research is the phenomenon's progressive dimension: supernatural agent-concepts were diachronically refined and *updated* consistently with a population cultural development. Clearly, such a slow progression caused the impossibility to appreciate discrete changes from within the process itself (just as parents often fail to appreciate the fact that their children are growing taller unless they notice it thanks to some external mediator, for instance their being able to reach items that were once out of reach), which in turn fostered theoretical hypocrisies like suggesting that research about the origins of religious belief is consistent as far as

[5]This is well exemplified by the numerous connection between Chap. 2 about camouflage and Chap. 10 about the detection of supernatural beings.

primitive cults are concerned, but complex religions such as Judaism or Christianity are a whole different kettle of fish (pointed out for instance by Barrett 2009).

The progressive obliteration of precedent steps is the same that can be observed in the perception of the empowerment achieved thanks to the artifactual distribution of cognitive tasks (calculation, observation, communication, memory storage etc.): many users fail in fact to appreciate the continuum connecting iPhones all the way back to our ancestors' notched animal bones.[6] Therefore, we tend to perceive only the mighty enhancements we benefit of thanks to computers and Internet (think of webcams, video-calls, online shopping, virtual really or the ability to access ever-growing knowledge bases...) by comparing them to an hypothetical *stage zero* of cognitive engineering, forgetting how some of these improvements were actually anticipated by pre-digital ones such as the telephone, an efficient mail system, increased diffusion of literacy and drop in the cost of books and so on.

The bottom-line of this contention is that technology, even if we often fail at appreciating its diachronic and often slow development, indeed made man "supernatural" by letting him transcend some of its natural limits, and this process is reaching an unprecedented magnitude in the digital era. I would like this statement not to be understood in a "weak", quasi positivist way: conversely, I suggest that the current technological development is apexing the process by which human beings are endowed with traits that were once considered as typical of supernatural agents.

Let me review some short examples to make the concept clearer: we can know—and share—information about events that took place on the other side of the globe nearly in real time, and add to them from our own: Greeks and consequently Romans had in their pantheon of deities none less than *Pheme/Fama*, the personification of gossip and rumor, of which Latin author Virgilius eloquently stated "Fama crescit eundo", Rumor grows as it moves on. A beast with many eyes, many ears and many mouths to know and repeat as much as she could: Facebook and other social networking websites and services turn us into a host of minor Fama's (as I tried to show in Chap. 8 and in Bertolotti 2011). We can know and appreciate with our eyes distant sights and people like sorcerers without the need of crystal balls (replaced by monitors), and we can fake our identities and interact with other people with an ability to mislead them that was equaled only by the Olympus Pantheon's passion for disguisement and taking the appearance of humans, animals etc. Such "supernaturalization" does not affect only the social-moral side, but also our ability to cope with the external reality and act on it: via the Internet we can move objects and people from a distance and physically interact with them like genies of ancient tales.[7]

The current epitome of this phenomenon can be individuated in two elements: virtual reality and, in particular, the *avatar*. The concept of avatar, etymologically linked to that of incarnation, represented the essence of virtual social environments

[6]This kind of connection was instead very clear to Stanley Kubrick, as he exemplifies in the epic opening of *2001: A Space Odissey*.

[7]With this respect, it seems interesting to notice how the lexicon of computers and Internet often reflects mythical and supernatural themes: consider for instance words such as *portal, wizard, superuser, trojan horse* and so on.

(Meadows 2008): the avatar is a partial embodiment of the user, and their identities can be told one from the other.[8] Furthermore, the avatar is, or at least used to be, a warrant of anonymity for the user, allowing the Internet to become a reenactment of Victorian morality, where everyone could pursue more or less licit perversions protected by the mask of their avatar. Of course, avatars have reputations as well, but unless data leaks are the case, they do not affect that of their *puppeteers*.

In fact, all is needed is the recourse to the *virtual self*, as Goffman (1961) put it. According to his definition, the idea of a virtual self describes the expectations about the character a person is supposed to have playing a given role. So, one is who she is, and who she is supposed to be when playing a given role in society. In a traditional virtual context one can stress the process of "role distance" as much as she wants: Goffman used that expression to refer to all those cases in which a person distances herself (or who she actually is) from the role she has been assigned to in a given context. This can safely happen in the virtual world as long as no one expects the role of the avatar to coincide with the role of the actual user: conversely, as abundantly illustrated in the previous part, in a social networking website the role of my profile is expected to be the same role I display in real-life interactions.

Yet, the interest of the avatar does not reside uniquely in the moral dimension, but in how the avatar empowers the user to move and act in a different environment, if we consider the avatar not just as the image representing the user on forums and bulletin boards, but the fully-virtualized representation of the user herself in virtual environments, usually consisting of massively multiplayer online role playing games (MMORPG)—like *World of Warcraft*—or complex 3D virtual environments like *Second Life*, which are impossible to categorize exclusively as games or enhanced social environments, since they partake of many different dimensions.

In all of these cases, the Avatar not only allows the user to disembody from her own flesh and blood and re-embody in bytes and pixels, but also to construct habits and expectations about a new peculiar environment. For the present investigation about virtual spirituality, avatars offer a twofold interest. On the one hand, they display many characters that are typical of traditional supernatural agency: if as for certain games (e.g. the aforementioned *World of Warcraft*) it is in the games' intent to recreate a partly mythological scenario including elves, trolls, minotaurs and so on, avatars in *Second Life* are endowed with a number of characteristics that can be defined "supernatural" (they can shift shape and gender, create objects from nothing and yet share them with other avatars, fly etc.), and those traits were not conceived intentionally as to echo mythological beings, but they were just the fittest to allow the best enjoyment of the virtual environment. On the other hand, as far as the spiritual level of avatars is concerned, a kind of *Matrix* paradigm is at stake: if magic is about observing, creating and manipulating distortions in our expectation about ordinary space-time behavior, in a digital environment magic is actually the same thing, but it

[8]The avatar had already been introduced in Sect. 8.6, in Part II. Here, it is interesting to note that the word "avatar" is of Sanskrit derivation, and literally means *to descend down*. It implies the taking of an Earthly form by a Deity. Interesting, then, that it seems to have come to indicate the opposite, in our modern usage.

involves being able to create and exploit distortions in the *code* describing the virtual world. *Cheating* can be an example of virtual magic, when it is not meant in a simply moral but onto-digital sense: users who are able to cheat the system, the code (and not to fool other users) are usually considered by their peers with a mix of awe and fear as it happened for sorcerers.

In conclusion, avatars are a fitting example of how instances of spirituality and religiosity can arise from different grounds than our ordinary life settings: this situation can be interestingly (but maybe pointlessly) "perverted" when forced to connect with *real life* beliefs and practices: in Second Life, for instance, it is possible to find churches and other temples where avatars can gather and worship as in real life. Following my argumentation, though, it is legitimate to wonder whether this is a kind of circular action: what are avatars worshipping, inasmuch as they are already "supernatural" themselves? is this a paroxysm of the failure of wanting to mingle habits of the virtual world with those of everyday "real life," or a failure in the possibility to merely *translate* a behavior from one cognitive niche to the other, as I did show it is the case as far as social cognition is concerned?[9]

11.4.2 Tacit Worship of Internet Highways

If, on the one hand, the enrichment of the religious niche offers to the user an unprecedented quantity of information and powerful research capabilities, on the other hand the extreme autonomy of that same artifactual dimension yields unexpected consequences that should be dealt with. As I already pointed out, the massive diffusion of the so-called *apps*, to browse and navigate the Internet, triggered a shift in the way we look for, and filter, information online: our searches, and indeed our virtual behavior as a whole, are affected by the unapparent algorithms embedded in the app.

This aspect should be considered also in connection with another one, that is *profiling*: it is, summarized to its essence, the series of activities by which the collection of apparently anonymous and irrelevant data can lead to the construction of users' profiles, for instance for security, commercial or safety purposes (Hildebrandt 2008, 2009).

It has been argued that "profiles, which are at the very core of technological possibilities, can be considered as imposed identities, both individually or collectively [and may] generate biased perceptions of individuals" (Gutwirth 2009), but it should also be added that profiling activities impact and bias the very perception we have of the whole Internet environment: the idea would be that being profiled means being observed, and if we are observed then there must be an *observer*. It is exactly the same (partly tacit) inferential pattern I illustrated in the two previous chapters[10] of this part! As contented by common sense and recent research Johnson and Bering (2006), human beings are extremely eager to postulate the existence of superior

[9]In Part II (Chap. 8).

[10]See in particular the model I proposed in Sect. 10.4.

beings especially if these are seen as able to affect their lives in a *moral* way: the Internet as a subject-like observer is indeed often perceived in a moral sense, either as a benevolent entity looking over our safety and making sure we receive any information we need at right time, or as a potentially evil controller, ready to punish us for illegal (or immoral) transactions, copyright infringements and so on.[11]

As some recent studies in paleoanthropology and cognitive science show (Boyer 2001; Atran 2002; Dennett 2006; Schloss and Murray 2009), beliefs in supernatural agents *happened* to us and were not really intentionally chosen, and similarly the most radical atheistic claims cannot eradicate those neural wirings that contribute to originating supernatural beliefs. Likewise, the development of devotional attitudes towards the Internet (which could be seen as echoing the worship of natural elements in ancient cults) could be a mostly tacit activity supported by the way our cognitive systems cope with the (now digital) environment: as sustained by Breton, we face indeed a "cult of the Internet" (Breton 2000).

The distopian novels of the XX century increased our likeliness to entertain such *beliefs* by providing some recognition templates, and they fostered the establishment of the *Panopticon* concept (Foucault 1979). Prima facie, our intuition of the Internet as an omniscient guardian could indeed be traced back to a kind of Panopticon controller (Steinhart 1999), in a more subtle realization of Orwell's *1984* scenario: yet, upon further reflection, we could say another concept could be more useful to identify the kind of omniscience we more or less tacitly appreciate in the Internet, that is the *Panspectron*. It was introduced by De Landa (1991) to indicate the possibility of controlling people's behavior not by actually observing them, but by monitoring and recording the whole electromagnetic spectrum they emit thanks to separate sensors. According to Braman,

> While Jeremy Bentham's concept of the panopticon characterized well information collection in the service of power under modernity, it is De Landa's notion of the panspectron that better captures the situation under postmodernity. The panoptic condition opens with the identification of a subject about which knowledge is desired and then arrays the tools of information collection about it. The panspectral condition, however, is one in which information is collected about everything all the time, with individual subjects becoming visible only when specific questions are asked (Braman 2003, p. 13).

Braman warns us that there is no active omniscience in the Internet: unlike our conceptions of God, the Internet could be seen as an information deity, made of knowledge but devoid of any mind. Even knowledge about single users can only be made actual through the appropriate query. This kind of description should make us aware that tacit devotional feelings towards artifactual niches could be misplaced, and inasmuch as they are misplaced they could be intentionally mislead: for instance, Governments could be interested in fostering citizen's perception of the Internet as an actual guardian knowing, just like the Christian God, what each and every user is doing.

---

[11]Just think of the recent *Datagate* scandal in the USA.

11.5 Conclusion

In sum aim, the aim of this brief chapter was to reflect on the concept of online religion, and provide seminal orientations to guide future studies. The final contention could be that when scholars look for online religiosity they might be looking the wrong way. If religion is a complex phenomenon that originated from the way our cognitive systems coped with environmental constraints, then it is much more interesting to investigate *new* forms of religiosity and spirituality that are being born out of the encounter between our same cognitive endowments and a radically new kind of environment, brought about by the advent of computers and the Internet. Those new forms of cyber-religiosity need not resemble the forms of spirituality we have been used to so far, as long as it is possible to individuate some kind of religious, devotional principle.

The spirit of this new kind of spirituality is well exemplified, albeit provocatively, by the *Cyberpunk's prayer* reported by (O'Leary 2004, p. 45), in which the author expressed her real and sincere devotion and awe for the superior being allowing the prosperity of a new environment which is the digital system.

> Our Sysop, Who art On-Line, High be thy clearance level.
>
> Thy System up, Thy Program executed Off-line as it is on-line.
>
> Give us this logon our database, And allow our rants, As we allow those who flame against us.
>
> And do not access us to garbage, But deliver us from outage.
>
> For thine is the System and the Software and the Password forever.

Notwithstanding its entertainingly bizarre aspect, the contemporary presence of "magical" thinking and technological awareness may come across as just one step short of sheer irrationality. The fact that we let magical explanations emerge concerning certain phenomena, about which we know basically everything *because we completely caused them*, is indeed a cognitive and epistemological riddle. I shall try to illuminate the mystery in the next chapter, by investigating the cognitive architectures that allow us to entertain so firmly beliefs that are actually at best shaky, or even contradictory, about a specific issue. I will do this by fully elaborating on the notion of epistemic and cognitive *embubblement*, a concept that I have already referred to a few times in this book.

References

Anderson C, Wolff M (2010) The Web is dead. Long live the Internet. Wired Magazine Issue: August

Atran S (2002) In gods we trust: the evolutionary landscape of religion. Oxford University Press, Cambridge

Barrett J (2009) Cognitive science, religion and theology. In: Schloss J, Murray MJ (eds) The believing primate. Oxford University Press, Oxford, pp 76–99

Bertolotti T (2011) Facebook has it: The irresistible violence of social cognition in the age of social networking. Int J Technoethics 2(4):71–83

Boyer P (2001) Religion explained. Vintage U.K. Random House, London

Braman S (ed) (2003) Communication researchers and policy-making. The MIT Press, Boston

Breton P (2000) Le culte de l'internet: une menace pour le lien social? La Decouverte, Paris

Campbell H (2004) "This is my church": seeing the internet and club culture as spiritual spaces. In: Dawson LL, Cowan DE (eds) Religion online: finding faith on the internet. Routledge, New York, pp 98–111

Clark A (2005) Word, niche and super-niche: how language makes minds matter more. Theoria 54:255–268

Dawson LL, Cowan DE (eds) (2004) Religion online: finding faith on the internet. Routledge, New York

De Landa M (1991) War in the age of intelligent machines. Swerve Editions, New York

Dennett D (2006) Breaking the spell. Viking, New York

Foucault M (1979) Discipline and punish: the birth of the prison (1975). Vintage books, New York, (Translated by A. Sheridan)

Gibson JJ (1979) The ecological approach to visual perception. Houghton Mifflin, Boston

Goffman E (1961) Encounters. Two studies in the sociology of interaction, Bobbs-Merrill, Indianapolis

Gutwirth S (2009) Beyond identity? IDIS-identity in the information society, pp 123–133

Helland C (2004) Popular religion and the world wide web: a match made in (cyber) heave. In: Dawson LL, Cowan DE (eds) Religion online: finding faith on the internet. Routledge, New York, pp 21–33

Hildebrandt M (2008) Profiling and the rule of law. IDIS-Identity Inf Soc 1:55–70

Hildebrandt M (2009) Profiling and Am I. In: Rannenberg K, Royer D, Deuker A (eds) The future of identity in the information society, Springer, pp 273–310

Johnson D, Bering J (2006) Hand of god, mind of man: punishment and cognition in the evolution of cooperation. Evol Psychol 4:219–233

Johnson D, Krüger O (2004) The good of wrath: supernatural punishment and the evolution of cooperation. Polit Theol 5(2):159–176

Keen A (2007) The cult of amateur. How today's internet is killing our culture and assaulting our economy. Nicholas Brealey Publishing, London

Löveim M (2004) Young people, religious identity, and the internet. In: Dawson LL, Cowan DE (eds) Religion online: finding faith on the internet. Routledge, New York, pp 55–68

Magnani L (2009) Abductive cognition: the epistemological and eco-cognitive dimensions of hypothetical reasoning. Springer, Berlin

Magnani L (2011) Understanding violence. Morality, religion, and violence intertwined: a philosophical stance. Springer, Berlin

Meadows MS (2008) I, Avatar: the culture and consequences of having a second life. New Riders, San Francisco

Mithen S (1996) The prehistory of the mind. A search for the origins of art, religion, and science. Thames and Hudson, London

O'Leary SD (2004) Cyberspace as sacred space: communicating religion on computer networks. In: Dawson LL, Cowan DE (eds) Religion online: finding faith on the internet. Routledge, New York, pp 34–54

Schloss J, Murray MJ (eds) (2009) The believing primate. Oxford University Press, Oxford

Søresen J (2007) Acts that work. Method Theory Study Relig 19:281–300

Steinhart E (1999) Driving to the panopticon: a philosophical exploration of the risks to privacy posed by the highway technology of the future. J Ethics Inf Technol 1(2):155–160

Wilson DS (2002) Darwin's Cathedral. Chicago University Press, Chicago

Wilson, L.: (2003) Interactivity or interpassivity: a question of agency in digital play. Fine Art Forum 17(8)

Chapter 12
Irrationality as an Epistemic Immunization: Cognitive Bubbles

Abstract This chapter focuses precisely on the Epistemic Bubble in order to show how its conceptualization can be further expanded to a broader "Cognitive Bubble:" by this I mean to show how our *thinking of knowing* (or *thinking that we know*) what we *do not actually know* is crucial to many dimensions of human cognition, and not just to the sentential ones (as seemingly suggested by Woods). By introducing the notion of *Religious Bubble*, the same chapter foreshadows the final topic of my dissertation, that is a philosophical approach to religious cognition. More specifically, my research was not aimed at drawing a history of cognitive "progress," which emerges through a juxtaposition of different "stages" of thought (Barnes, Stages of thought: the co-evolution of religious thought and science 2000); if it were, then it would have made sense to begin with religion and the origins of culture, and track the "evolution" leading to science. Conversely, at this point of my research it was possible to framef religion through the conceptual tools developed so far, such as the modeling of external agencies and the construction of cognitive niches. Thus, religion can appear as a model of a class of inferences, traditionally perceived as irrational (or having to do with counterfactual beliefs), but which can be very interestingly studied through the epistemological analysis of traditionally nonscientific domains. Such an outlook can be regarded as analogous to the psycho-anthropo-cogntive effort to frame religion "as a natural phenomenon" (Boyer, Religion explained 2001; Atran, In gods we trust: the evolutionary landscape of religion 2002; Dennett, Breaking the spell 2006): similarly, my aim was to investigate religion as a "philosophical phenomenon" without adhering excessively to what is traditionally understood as *philosophy of religion*. The local interest was in fact to apply an epistemological toolbox in order to study and model the religious inferential regime, how religion could be defined as an ecological-cognitive activity, and how religious pragmatics can regulate behaviors that are typically connected with religion such as sacrifice and forgiveness.

© Springer International Publishing Switzerland 2015

T. Bertolotti, *Patterns of Rationality*, Studies in Applied Philosophy, Epistemology and Rational Ethics 19, DOI 10.1007/978-3-319-17786-1_12

12.1 Introduction

In this chapter I will argue that different fields of human reasoning which frequently instantiate occurrences of *irrationality* are indeed characterized by what could be roughly indicated as epistemic or cognitive immunizations. We saw instances of epistemic immunizations when dealing with *camouflage* (2), *models* (3), with the pragmatic power of gossip (7), and in the analysis of supernatural beliefs (10, 11). If asked to provide an initial definition of epistemic (or cognitive) immunization, one could say that it is *a strong belief that X is the case, which obliterates the absence of sound reasons, within the believer, to believe that X is the case.* This discrepancy could be described as the one between *knowing* and simply *feeling of knowing*: a recent monograph has surveyed this epistemological decoupling as it displays pathological manifestations (Burton 2008), but my intent in this chapter is to extend the investigation of such kinds of decoupling to other, more common aspects of human cognition, and sketch an inferential, epistemic and cognitive model of how it functions.

I will therefore analyze the issue of cognitive immunizations making use of the *Epistemic Bubble* Thesis as proposed by Woods (2005), expanded into a Cognitive Bubble Thesis. I will test the extendability of the Cognitive Bubbles to a shared dimension instead of concentrating on the individual level, then try to apply this conceptualization to fields of human cognition in which immunization processes seem frequent: in the first case I will approach moral reasoning, while the second will set back on the discourse about religion and beliefs in supernatural agents to which this part is mostly dedicated.

12.2 The Epistemic Bubble and the Importance of Doubt

American philosopher Charles Sanders Peirce pointed out how belief formation rests on an emotional factor which cannot be ignored: the irritation of doubt. Such irritation occurs whenever we are in need of knowing something we do not know, or we try to make sense of some new unexplained signs, and so on. The basic concept of *doubt* can be considered as a red line connecting the most primitive kind of pragmatic perplexities with the highest scientific and philosophical attitudes: in a way, this could be seen as a tenet of eco-cognitive epistemology. The cognitive irritation coming from doubt is the state of cognition that prompts us to advance hypotheses and believe them in a wide range of situations: it happens when an agent wonders about who or what produced that crackle she just heard in the bush behind her, but also when she thinks about how to improve the reception of the cellphone she just bought, or when she keeps coming back to guessing what caused the bruises on her neighbor's forehead.[1]

[1] It is commonly debated whether Peirce's definition of doubt comprises only mundane matters or can be extended to higher theoretical fields. In this chapter I will lean towards the second interpretation.

In Peirce's account, thought is an inferential activity, that operates upon signs, whose output—belief—has the main purpose of placating an irritated cognition:

> [The] action of thought is excited by the irritation of doubt, and ceases when belief is attained; so that the production of belief is the sole function of thought. All these words, however, are too strong for my purpose. It is as if I had described the phenomena as they appear under a mental microscope. Doubt and Belief, as the words are commonly employed, relate to religious or other grave discussions. But here I use them to designate the starting of any question, no matter how small or how great, and the resolution of it (Peirce (1931–1958), *How to Make our Ideas Clear*, p. 5.394).

The Peircean perspective was further developed by logicians John Woods and Dov Gabbay, promoters of the agent-based approach to logics: according to this view, human reasoning is mostly about *achieving cognitive targets*, which are marked by cognitive *irritation*. No matter the entity or complexity of the target, human cognition is essentially a more or less advanced problem-solving activity: examples of cognitive targets are knowing whether it is a bear or a boar hiding in the cave, or what powerful entity could have shaped that ridge so that it looks like a human face, or which combination of buttons, if pressed, can get our computer to run faster, but higher theoretical targets can be considered as well, as those who spark the development of scientific models or underpin scientific experimentation.[2] Consequently, what is a belief if not the satisfaction of a cognitive target? The next central issue is that a cognitive agent *like-us* (that is seen as a real agent in an eco-cognitive perspective) is very likely to take any kind of belief for knowledge: as I contended in Chap. 1, cognizing and acting strategically in one's environment require the adoption of pragmatic standards of truth. In a framework of bounded rationality (where time, resources and computational power are limited) even tentative, or questionable beliefs must be assumed as knowledge in order to be the bases for action. Furthermore, Woods argues that, indeed, belief can only *coincide* with a first-person attribution of knowledge.

> (Belief as knowledge ascription). Whenever it is true for Y to say of X that X believes that P, it is also true that X takes himself as knowing that P (Woods 2005, p. 738).

Of course the basic philosophical assumption is that *knowledge* and *belief* do not share the same epistemic status: even a true belief is different from knowledge insofar as knowledge presupposes the possibility of providing the statement with sufficient, relevant reasons. That is, "[o]ne knows that P only if one has at one's disposal a case of requisite strength to make for P" (p. 735).

Such a distinction, which seems perfectly clear from a third-person perspective, is immaterial from the agent's first-person perspective. If we agree with Peirce's intuition about the nature of thought, it follows that *knowledge* is by no means required to placate cognitive irritation. The ideal model of target-attainment would require one and only one epistemic state able to satisfy the cognitive target. Actually, from the real-agent's perspective, a poorer epistemic state can be embraced as long as it fosters the *feeling* of having attained the target by simply discharging (or placating) the original irritation.

[2] cf. Chaps. 3 and 4 in the first part of this book.

Woods, by introducing the concept of Epistemic Bubble, points out how, even though the satisfaction of the cognitive quest is met by a positive final emotional appraisal (i.e. relief from irritation), such a positive appraisal is hardly a symptomatic sign of the proper attainment of the target:

> (Fugitivity of truth). [From a first person perspective] truth is a fugitive property. That is, one can never attain it without thinking that one has done so; but thinking that one has attained it is not attaining it (p. 746).

It is significant to observe how this fallacious inference routinely performed by our cognitive systems is actually an innate kind of *abduction*, which could be modeled as follows:

- Premise 1: *If I know target P* then *My irritation about P is relieved*;
- Premise 2: *My irritation about P is relieved*;
- Conclusion: *I know target P*

For an agent-like-us, *knowing* and *thinking of knowing* can hardly be told one from the other, but they are clearly not the same thing. This leads to the formulation of the Epistemic Bubble Thesis:

> When in an Epistemic Bubble, cognitive agents always resolve the tension between their thinking that they know *P* and their knowing *P* in favour of knowing that *P* (p. 738) [...] A cognitive agent *X* occupies an Epistemic Bubble precisely when he is unable to command the distinction between his thinking that he knows *P* and his knowing *P* (p. 740).

Woods is careful at stressing how this is a *constitutive* constraint of human beings, and not a matter of being more or less intelligent or cognitively unfit. Such disposition can be recognized theoretically and only in a hindsight dimension, but an awareness of the Bubble Thesis can never be operationalized with the scope of avoiding further *ebubblements*. Considering the semi-transparency of the Bubble Thesis, and the inherence of its object to every aspect of human inferential activity, I will from now on refer to it as the *cognitive bubble*. By this, the intention is to acknowledge the "universal" dimension of this kind of ebubblement, coherently with Peirce's universalistic notion of doubt.

12.3 Cognitive Bubbles, the Fixation of Belief and Its Intersubjective Dimension

Woods' account of our cognitive constraints concentrates on individual, isolated agents, but is it possible that we could enrich the Cognitive-Epistemic Bubble Thesis and eventually investigate some of its social implications?

The Cognitive Bubble is clearly a consequence of our cognitive systems dealing with bounded information, lack of time and limited computational capacity; still, another consequence of such limitations is Simon's concept of *docility*, that is "[...] the tendency to depend on suggestions, recommendations, persuasion, and information obtained through social channels as a major basis for choice". Being *docile* allows human beings to cope with their bounded rationality and finite potentials by reducing the importance of first hand experience, on the basis that "(a) social influences will generally give us advice that is for our own good and (b) the information on which this advice is based is far better than the information we could gather independently", thus relying on a more distributed concept of experience (Simon 1993, p. 156). *Docility* is clearly one of the reasons making gossip, as analyzed in Chap. 7 (Part II), a fundamental informational tool. Cognitive niche construction as well, basing on the externalization (hence the sharing) of common knowledge, relies on the fact that human beings are *docile*.[3] This problem leads us immediately to try and understand how information acquired in some way by a cognitive agent becomes a "belief." Peirce considered three main strategies aimed at the fixation of belief: the method of tenacity, that of authority and the method of science. Let us focus on the two first methods: needless to say that the method of authority relies on a social basis, but it can be argued that the method of tenacity as well may benefit from a social ground.

> The instinctive dislike of an undecided state of mind, exaggerated into a vague dread of doubt, makes men cling spasmodically to the views they already take. The man feels that, if he only holds to his belief without wavering, it will be entirely satisfactory. Nor can it be denied that a steady and immovable faith yields great peace of mind. It may, indeed, give rise to inconveniences, as if a man should resolutely continue to believe that fire would not burn him, or that he would be eternally damned if he received his ingesta otherwise than through a stomach-pump. But then the man who adopts this method will not allow that its inconveniences are greater than its advantages (Peirce 1877).

Tenacity could be seen as strictly related to Woods' *ebubblement*. Our constitutive eagerness to take as true something that, in spite of being implausible, is able to calm our cognitive irritation *as something true would*, results in our reluctancy to easily discard such a belief. It can be suggested that not only this is what happens from a cognitive perspective: we also *like and need* this to happen, because no one would be comfortable about spending their life deeply wondering about everything, in a state of perennial indecision. From the very beginning of this book I have argued how our perception (and the underlying inferences) provides us with an array of incomplete appraisals for the sake of fast acting in the surrounding environment.

According to this perspective, the Bubble Thesis can be understood not necessarily as an impairment but rather as a cognitive architecture *benefitting* our cognitive welfare. Some cases of pathological indecision may show the impossibility of proper activation of an Epistemic Bubble: in an evolutionary perspective, Haig speculatively connects the presence of apories within human decision-making systems with the different development (and fitness-related behavior) of genes inherited from one's mother or from one's father, so that as for certain matters, "(some effects) of maternal genes would be opposed by adaptations of paternal genes to increase the suasive power of conscience" (Haig 2005, pp. 21–22). On a different note, van Randenborgh and colleagues bring experimental results suggesting how in particular psychological states (such as dysphoria) processes of rumination—"a phenomenon at the

[3]Docility, indeed, was an issue in the analysis of cognitive niche curation proposed in Chap. 6.

intersection of cognitive and affective processes" (p. 230)—may foster severe indecision and less confidence in one's hardly-reached decisions, triggering states of deeper depression (Randenborgh et al. 2010). Such results seem to avail the necessary nature of the Cognitive Bubble as a protective mechanism in human cognition.

The obstinacy-defense I just sketched out seems corroborated by the fact that, notwithstanding discrete difference in individuals' abilities, human cognitive systems are structured and endowed in a roughly similar way. It is then possible to notice how *ebubblement*, in spite of being a characteristic trait of individuals, can display mechanisms of mutual positive feedback favored by the social background: every agent's Cognitive Bubble is different and exclusive, but many bubbles can show similar traits, and such similarities are reinforced by their very presence.

The possibility of a *common ebubblement* might help shedding light upon a characteristic of many kinds of arguments, such as the religious ones. Philosopher Jennifer Faust examined to what extent they manage to be persuasive (Faust 2008). The problem with arguments about the existence, or the non existence of supernatural beings, but also about differences between cults and confessions, or between political ideologies, is that they often fail to persuade, that is to say, the fail in being recognized as good arguments by the other party.[4]

Faust's account can be proposed using Woods' logical modeling. If we consider parties *X* and *Y*:

- Agent *x* and her peers occupy an Epistemic Bubble *B1* with respect to matter *P*.
  - *Within* the Epistemic Bubble, Agent *x* knows that *P1*, as her *feeling of knowing* is activated by belief *p1*.
  - Agent *x* is unable to command the distinction between her *knowing that P1* and her *thinking of knowing that P1*.
- Agent *y* and her peers occupy an Epistemic Bubble *B2* with respect to matter *P*.
  - *Within* the Epistemic Bubble, Agent *y* knows that *P2*, as her *feeling of knowing* is activated by belief *p2*.
  - Agent *y* is unable to command the distinction between her *knowing that P2* and her *thinking of knowing that P2*.
- According to the initial definition, an agent *knows P* if and only if he can provide a case of requisite strength to make for *P*.
  - From group *X*'s defining Epistemic Bubble *B1*, any case supporting *B2* cannot be accepted as endowed with the *requisite strength* to sustain *P2*.
  - From group *Y*'s defining Epistemic Bubble *B2*, any case supporting *B1* cannot be accepted as endowed with the *requisite strength* to sustain *P1*.

[4]Epistemological and folk-epistemological considerations about an agent's appraisal of her own knowledge cannot be fully separated from the argumentative and rhetorical sphere (Hardy-Vallée and Dubreuil 2010).

Within each group, every agent is locked in her first-person perspective and unaware of her occupying an Epistemic Bubble, but she can readily individuate the very same condition as far as members of an opponent group are concerned, assessing it by a third-person perspective.

According to Faust, this is not a matter of *question-begging* theses: they do not assume what they aim at demonstrating as an hypothesis, but rather they "beg the doxastic question". An argument is accepted as sound with respect to its conclusions only inasmuch one already believes the very conclusions: instead of persuading the opposite party (that is, those who do not already believe in the conclusion), dialectic confrontation between differently configured Cognitive Bubbles produces escalating irritation.

One has little difficulty to imagine such a situation if she considers religious, ethical or political issues. Nevertheless, particular fields of scientific research, often with a high social impact, seem to be however prone to develop mechanisms akin to religion, such as "begging the doxastic question", or displaying incommunicability between opposing opinions: for instance nuclear energy, animal research and experimentation, genetically modified organisms.

12.4 Moral Bubbles and Their Collective Implementation in Ideologies

In the preceding sections I spelled out the Bubble Thesis and some of the cognitive dynamics to which this modeling can be applied. I shall now attempt, more specifically, to extend the application of Bubble Thesis to one of the most characteristic fields of human cognition, that is moral reasoning, and apply the definition of a *Moral Bubble* introduced in Magnani (2011).[5] The moral ebubblement would in fact be homomorphic to the epistemic one: it is the result of impossibility of knowledge as relevant case-making and of a strategic cognitive need to reduce doubt and uncertainty as much as possible. Its very structure reflects the self-immunization postulated in Woods' Bubble Thesis.

> (Immunization) Although a cognitive agent may well be aware of the Bubble Thesis and may accept it as true, the phenomenological structure of cognitive states precludes such awareness as a concomitant feature of our general cognitive awareness (Woods 2005, p. 742).

The fact of being constitutively and easily unaware of our errors is very often bound with the self-conviction that we are not at all aggressive in the argumentations we perform: non-awareness of our error is often accompanied by unawareness of the deceptive/aggressive character of our speeches (and behaviors).

[5] It is worth reminding that my perspective on morality in this chapter is the same as in Chap. 7 in the second part: I do not mean to engage a debate with moral philosophy, but rather *phenomenologically* analyze how cognitive agents behave in a way that can be described as "moral."

It is important to notice how a *Moral Bubble* hypothesis would provide a cognitive account, and not yet another moral one, to the problem of moral violence and its opaqueness, and particularly to a person's self-immunization to her own moral violence (Magnani 2011). Within the Moral Bubble, the moral agent perceives her own moral principles as a given, just as much as a cognitive agents takes her beliefs as positive, genuine truth. We *think we know* the genuineness of our beliefs but actually we believe them just out of tenacity, habits and authority. An attack on our religious, moral or political beliefs triggers an immediate violent response, but not only of a dialectical kind: an overly zealous questioning of beliefs may hit on the hollow foundation on which they stand, thus threatening to *pop* the bubble. Cognitive bubbles are strategically vital to beings-like-us: a threat to our *embubbled* and *embubbling* beliefs is a direct, major threat to our state of cognitive quiet.[6]

In the Moral Bubble hypothesis, the agent is minimally aware of her violence only as far as the *dialectic* level is concerned and she agrees to strike back on the same argument: this can happen in a polite and controlled way, but it is only the tip of the iceberg. The majority of the violent response is *not* a defense of the content of the questioned beliefs, but of the cognitive *tranquillity* that those beliefs allow, within the moral ebubblement. As stated by Peirce "the instinctive dislike of an undecided state of mind, exaggerated into a vague dread of doubt, makes men cling spasmodically to the views they already take" (Peirce 1877). Once a moral belief has been corrupted by doubt, the deployment of a vast amount of cognitive and emotional resources is required: the purpose is to individuate another fitting belief that can restore the previous state of mental welfare.[7] Such violent outburst is not perceived by the agent who performs it, because it is obliterated by the phenomenally unquestioned conviction of the righteousness of her own principles. That is why we are extremely aware of other agents' violence (because they clash with our own bubble) but we are virtually immunized to our own.

Furthermore, such description applies to "relativists" or, in a religious framework, to more or less "militant" atheists: willy-nilly, positions like relativism and tolerance presume as vast an assumption of beliefs as that of any positive credo. And the reason for this is that any moral position, as any theological one, is necessarily "doxastic:"

[6]Concerning the necessary stratification of types of bubbles in the life of beings-like-us, a recent eco-cognitive article aims at showing the interplay between epistemic, cognitive and moral bubbles in the social and political environment (Arfini 2013). Starting with the analysis of the embubblement process in the "New Logic" epistemological framework, the article highlights the analogy between the bubble concept and the similar structure of self-immunization in Derrida's deconstructionist philosophy, which focuses on its political and biopolitical outcomes. Indeed, in the biopolitical environment, Derrida explains the mechanisms and inconsistencies of power plays—especially as far as democracy is concerned—, describing them as self-immunized networks: the research frames this perspective in the cognitive niche theory and sees it as an elaborate system of interconnected cognitive, epistemic and moral bubbles, highlighting the complex entanglement of self-immunized structures in a community system.

[7]Coherently with the view I am arguing for, Woods suggests that a revision of one's Epistemic Bubble is *de facto* impossible: one could only *change* her bubble with another, without actually performing a belief revision: that is possible only for *institutional agents* (i.e. science) which are able to transcend the individual embubbled perspective.

no relevant case can be provided to sustain the belief in God's existence, just as no case of the same nature can be presented to prove God's non-existence. Similarly, any moral of tolerance, or even an *a-moral* position, is ultimately as *moral* as any other positive one, thus prone to the dynamics of ebubblement and self-immunization to one's own violent response to contrasting positions, with the scope of reducing as much as possible the agent's distress caused by epidemic doubt and irritation.

As I said, it can be hypothesized within the Moral Bubble, the moral agent can only perceive her own moral principles as a given, just as much as a cognitive agents takes his beliefs as positive, genuine truth. Similarly, ideologies project a clear coalition level, in which each embubbled individual assures and corroborates the beliefs of his fellows. The whole ideology-projected group becomes blind to their own violence and are able to respond to instances of doubt with the synchronism of one organism and the power of several.

In extremis, should it be impossible to suppress the corrupting belief, the solution rests in the (often physical but sometimes metaphorical) suppression of the corrupted believer: if violence perpetrated outside of the group does not succeed in its scope, it can target one of weakest members of the group itself, labelled as a deviant or a traitor, as contended by Bandura (1999).

As a matter of fact, it is interesting to remember how the Epistemic Bubble is already a cognitive structure aimed at the reduction of irritation caused by doubt: consequently, any willing or unintentional behavior that clashes with one agent's bubble originates a negative emotional appraisal, that can sublimate in the denial of factual evidence[8] contrasting with the bubble. Over and above that, when the ebubblement concerns several individuals united in a single group, the effect of any behavior, perceived as attempting to intrusively pop the bubble, can spark in human agents a particularly violent reaction, that can go beyond a *cognitive* and dialectical violence.

The violent consequences of common ebubblement can be easily spotted in fields such as religion, or politics, and less in science: this is not to say that in science they are totally absent, but the possibility of case-making, of displaying knowledge provided with better epistemic strength and factual experiences to sustain it, reduces the need for a heavy Cognitive Bubble; conversely, whenever cornered, religion and politics *have to* recur to the ultimate weapon of morals, as the strategy of case-making is out of their reach: in fact, requiring believers to provide cases, reasons for their beliefs is usually perceived by the very believers as *morally evil.*

It can be argued that, even if we accept that religious belief display an intrinsically cognitive-perceptional origin and not a moral one (as I will contend in this chapter), their unquestionable moral relevance for beings-like-us allows us to label it as a "Moral Bubble."[9] Religious beliefs most often enclose more or less explicit moral

---

[8]Denial can attend a perceptive and hence, unconscious level. Against, it is not necessarily the case of short-mindedness or stubborn obstinacy.

[9]Johnson and Bering (2006) well demonstrates the moral relevance of supernatural beliefs in the formation of religion as an adaptive process. I will dedicate the final chapter of this book (Chap. 13) to the relationship between religious cognition and violence.

prescriptions, templates (Magnani 2007) or guidelines according to which a believer should behave. Those moral beliefs, shared by the majority of members of the community, can seldom benefit of any relevant epistemic foundation, but on the other hand they seem perfectly sound to those who practice them, and appear to regulate the life of the community: such positive emotional appraisal subtly persuades agents about the truthfulness of their beliefs.

Consider ideology: it is something available "out there", stored in external devices and materials (other people, books, media, etc.) of a given social collective. People readily pick up external ideological "tools" of this kind, then re-represent them internally. The signals consisting in swastikas, red stars, parades, styles of speech and even clothing fashions do not only posses a propagandistic meaning, but they mostly enforce and empower those who *already* believe in it. The historical *topos* about the difference between an army and an armed mob should be clarifying: it is the army *ideology* that allows its superiority. It can be argued that even the military training is only afforded by the presence of a distributed ideology embedded in signs, flags, words, hierarchies and, last but not least, uniforms: the very term *uniform* is significative. The uniform is the *same* for *everyone*, making every subject feel similar, both exteriorly and psychologically equal to the others (Bollas 1993).

That stated, it is possible to further analyze this *shared* dimension characterizing the Moral Bubble: ideologies rely on the distributed dimension of Moral Bubbles and on the mutual reinforcement of moral beliefs. Uniformed individuals (the oxymoron is not intentional but very significative), each in their own *individual* Moral Bubble, act together and combine their roughly similar moral beliefs into a *collective Moral Bubble*, which dramatically empowers its ordinary mechanisms. This collective bubble aims at defusing systematically all potential doubts, adding the action of one individual upon the other to the self-immunization with respect to violence, typical of the Moral Bubble. To sum it up quite briefly, it could be suggested that external (and therefore distributed) socio-cognitive structures no nothing but *mirror* the distribution of cognitive structures such as the ebubblements I have been dealing with so far: could ideologies, and many other collective phenomena, be indicated as distributed eco-cognitive ebubblements?

12.5 Bubbles and Cognitive Firewalls: The Case of Religion

In this final section, I mean to explore whether the Bubble Thesis can be applied to another typical dimension of human cognition which is the proper object of this part: its propensity to hold and maintain beliefs in supernatural agency. My proposal is that such an application is possible insofar as religion seems characterized by a selective switching of certain beliefs as application field varies. Earlier on in this part (Chap. 10), I provided an analysis of how supernatural concepts—once they become part of the cultural patrimony of a group—come to constitute its religion by crystallizing into a set of more or less fixed narratives.[10] These narratives play a role in society as far as they support norms and commitments, which partially explains the persistent success of religion (Boyer 2001; Atran 2002; Wilson 2002; Bulbulia 2009).

For religion to be effective, though, believers are expected to sincerely believe and commit themselves to the positive ontological and moral core of their *credo*. The problem is that a positive commitment to ontological oddities such as supernatural beings, in spite of the social advantages, can be rather dangerous. Atran stressed how this commitment to counterfactual entities should have proved particularly maladaptive for mankind: "[...] to take what is materially false to be true (e.g. people think and laugh and cry and hurt and have sex after they die and their body disintegrate) and to take what is materially true to be false (e.g., people just die and disintegrate and that's that) does not appear to be a reasonable evolutionary strategy" (Atran 2002, p. 5). The key to solving this issue seems to be that humans usually know when to believe (and behave) *as if* counterfactual entities really belonged to their ecologies and when to suspend this belief. As for this matter, Bulbulia claims that religious beliefs are characterized by a mental ®IMAGINE marker, thus illustrated:

> ®IMAGINE [Zugroo is Lord Creator]
>
> practical inference: NOT TRUE [Zugroo is Lord Creator]
>
> workspace inference: CERTAINLY TRUE [Zugroo is Lord Creator] (Bulbulia 2009, p. 63).

The functioning of the ®IMAGINE marker introduced by Bulbulia makes me suggest a further kind of cognitive ebubblement, that is the *Religious Bubble*. If the simple Cognitive Bubble had an illusional dimension, the Religious Bubble has a necessary self-deceptive dimension, which acts to reinforce the ontological commitment onto the belief itself, but simultaneously it defuses the practical inferences about our ecology, enhancing the moral ones.

Further studies should be carried on about this subject, still I mean to suggest that the solution to this conflict could have been coupled with the *confinement* of the supernatural to the dimension of the *holy*, which I will briefly address in Chap. 13. Thus, religion maximized the benefits (for the group) of (individual) commitment to supernatural agents reducing the ecological risks caused by an adoption of magical thinking as a strong Cognitive Bubble. The relegation of what is *holy* to a well-defined dimension of social life, a delineation that is conveyed through education as well, is reflected in the rise of specific *cognitive firewalls*[11] that prevent magical-supernatural concepts and inferences to flood the areas of our mind-brain system dedicated to ecological survival.

---

[10]Remember the *Thunderbird* example, in Sect. 10.4.1.

[11]The notion of "cognitive firewall" belongs to a computational representation of the mind: Cosmides and Tooby define them as "computational methods for managing the threat posed by false, unreliable, obsolete, out-of-context, deceptive, or scope-violating representations. Cognitive firewalls—systems of representational quarantine and error correction—have evolved for this purpose. They are, no doubt, far from perfect. But without them, our form of mentality would not be possible" (Cosmides and Tooby 2000, p. 105).

Objects of belief in supernatural display an inferential richness, i.e. they "invite individual and collective elaboration" (Barrett and Lanman 2008, p. 115), so they can be easily shared within a community. The community cannot warrant for the Religious Bubble as much as it does for the Moral Bubble: Moral Bubbles rest entirely on a group dimension, while the cognitive firewalls preventing the overflow of religious beliefs seem to evolve differently on an individual base. As a matter of fact, people display different strength as for cognitive firewalls, and even present-day religious believers may belong to a magic dimension—that is, deprived of any cognitive firewall. It must be pointed out that, even if certain individuals remain in an ancestral magic dimension as for themselves (a consequence of "weaker" cognitive firewalls), they work as a further costly commitment for the whole religious group (Atran 2002). In other words, their cognitively unrestricted religiosity manages to reinforce and enforce the beliefs of the rest of the groups, yet without majorly affecting the way of coping with ecological material reality displayed by more cognitively restrained believers. Normal believers might not agree with fundamentalists, or the so-called "extremists," nevertheless their eagerness corroborates the beliefs of the whole group.[12]

12.6 Conclusion

In this chapter I have analyzed a series of rational domains in which cognitive and epistemic immunizations are fundamental. I challenged the original definition of Epistemic Bubble, in an attempt to shape a broader concept that I defined *Cognitive Bubble*, to highlight its fundamental presence in most kinds of human cognitive processes. I described how, consistently with other claims about constraints displayed by human rationality, the Cognitive Bubble is not a phenomenon concerning only isolated individuals but can instead be augmented by mechanisms of positive feedback so to form a kind of collective ebubblement, which clearly appears in the rhetorical dimension characterizing social interactions. Such a theoretical apparatus could be applied to other dimensions of human reasoning in which immunizations and selective switching of acquired knowledge seem to play a crucial role: moral reasoning and religion. As for the former, the Moral Bubble hypotheses provides cognitive and emotional arguments for the (hard to perceive yet irrefutable) doxastic nature of moral stances, and for the opacity of those violent behaviors that moral beliefs trigger so easily; whereas, as for the latter, the presence of a Religious Bubble can be hypothesized to provide an account of the undying presence of counterfactual—and yet relatively harmless—beliefs concerning supernatural agents.

---

[12] This mechanism could be linked to the sacrificial mindset I shall analyze in the next chapter.

So far I introduced and outlined a possible explanation for the human tendency to entertain conflicting, or apparently irrational, beliefs about particular matters. I will dedicate the following and final chapter to analyzing how, in a mental framework that can be loosely defined as "religious," this tendency to overcome doubt and avoiding epistemic burdens at the same time is enacted in two apparently opposed behaviors: sacrifice and forgiveness.

References

Arfini S (2013) Bolle epistemiche, scienza e credenza. In: Magnani L (ed) Introduzione alla new logic. Logica, filosofia, cognizione. Il Melangolo, Genova

Atran S (2002) In gods we trust: the evolutionary landscape of religion. Oxford University Press, Cambridge

Bandura A (1999) Moral disengagement in the perpetration of inhumanities. Pers Soc Psychol Rev 3:193–209

Barnes MH (2000) Stages of thought: the co-evolution of religious thought and science. Oxford University Press, Oxford

Barrett J, Lanman JA (2008) The science of religious beliefs. Religion 38:109–124

Bollas C (1993) Being a character. Psychoanalysis and self experience 1992. Routledge, London

Boyer P (2001) Religion explained. Vintage UK. Random House, London

Bulbulia J (2009) Religiosity as mental time-travel. In: Schloss J, Murray MJ (eds) The believing primate. Oxford University Press, Oxford, pp 44–75

Burton RA (2008) On being certain. St. Martin's Press, New York

Cosmides L, Tooby J (2000) Consider the source: the evolution of adaptations for decoupling and metarepresentation. In: Sperber D (ed) Metarepresentations: a multidisciplinary perspective. Oxford University Press, Oxford, pp 53–115

Dennett D (2006) Breaking the spell. Viking, New York

Faust J (2008) Can religious arguments persuade? Int J Philos Relig 63:71–86

Haig D (2005) Intrapersonal conflict. In: Jones M, Fabiam AC (eds) Conflict. Cambridge University Press, Cambridge, pp 8–22

Hardy-Vallée B, Dubreuil B (2010) Folk epistemology as normative social cognition. Rev Philos Psychol 1:483–498

Johnson D, Bering J (2006) Hand of god, mind of man: punishment and cognition in the evolution of cooperation. Evol Psychol 4:219–233

Magnani L (2007) Morality in a technological world. Knowledge as duty. Cambridge University Press, Cambridge

Magnani L (2011) Understanding violence. Morality, religion, and violence intertwined: a philosophical stance. Springer, Berlin

Peirce CS (1877) The fixation of belief. Pop Sci Mon 12:1–15

Peirce CS (1931–1958) Collected papers of Charles Sanders Peirce. In: Hartshorne C, Weiss P (eds), vols 1–6. Harvard University Press, Cambridge (vols. 7–8, Burks, A. W., ed)

Simon H (1993) Altruism and economics. Am Econ Rev 83(2):156–161

van Randenborgh A, de Jong-Meyer R, Hüffmeier J (2010) Rumination fosters indecision in dysphoria. J Clin Psychol 66(3):229–248

Wilson DS (2002) Darwin's Cathedral. Chicago University Press, Chicago

Woods J (2005) Epistemic bubbles. In: Artemov S, Barringer H, Garcez A, Lamb L, Woods J (eds) We will show them: essays in honour of Dov Gabbay (volume II). College Publications, London, pp 731–774

Chapter 13
The Cognitive Impact of Religious "Rationality": On Forgiveness and the Sacrificial Mind

Abstract The aim of this chapter is to apply many of the tools and insights developed alongside this book on two major issues in which religion plays a crucial role: namely sacrifice (in particular self-sacrifice) and forgiveness. Sacrifice will be analyzed from a chiefly epistemological and pragmatic perspective, whereas I will adopt with forgiveness a frame relating to evolutionary studies and social cognition. The interest connecting those two fields is that they both, in different ways, partake of the topics of *violence* and *irrationality*: it is therefore extremely interesting to read them within the eco-cognitive perspective that has characterized this book since the beginning. As I did for the genesis of supernatural beliefs, I will attempt at problematizing the perceived irrationality of a given trait, or behavior, in order to understand how it can be traced back to another *pattern* of rationality: when, as in the case of the sacrifice of intellect, an "irrational" point will be located, it will be my aim to circumscribe it as fittingly as possible. Therefore, my intention is to analyze two central aspects of religious behavior relying on the notion of violence (for sacrifice) and its retention (for forgiveness), thus using violence as a kind of eco-cognitive prism. Following the philosophical orientation indicated by Magnani's *Understanding Violence* (Magnani 2011), I will argue in favor of the *epistemic* and *pragmatic/heuristic* roles of violence (and of its retention). In the first section, in fact, I will contend that the religious sacrificial mindset fascinatingly weaves together a theoretical non-understanding (in the form of an "epistemic" violence tracing back to the *sacrificium intellectus*) with a pragmatic understanding, in order to achieve the further possibility for an eco-cognitive acting. In the second and final one, conversely, I will challenge *forgiveness* as a heuristic violence committed against a norm, part of a set of rules enabling the functioning of a the cognitive niche, for the sake of the functioning of the cognitive niche itself. In sum, my final contention will be that *violence* and *irrationality* are, paradoxically, the necessary and strictly linked counterweights that are able to make religion (as I have analyzed it in this part) livable and even profitable from the perspective of an eco-cognitive rationality.

© Springer International Publishing Switzerland 2015

T. Bertolotti, *Patterns of Rationality*, Studies in Applied Philosophy, Epistemology and Rational Ethics 19, DOI 10.1007/978-3-319-17786-1_13

13.1 Introduction

The aim of this final chapter is to apply many of the tools and insights developed alongside this book on two major issues in which religion plays a crucial role: namely sacrifice (in particular self-sacrifice) and forgiveness. Sacrifice will be analyzed from a chiefly epistemological and pragmatic perspective, whereas I will adopt with forgiveness a frame relating to evolutionary studies and social cognition. The interest connecting those two fields is that they both, in different ways, partake of the topics of *violence* and *irrationality*: it is therefore extremely interesting to read them within the eco-cognitive perspective that has characterized this book since the beginning. As I did for the genesis of supernatural beliefs, I will attempt at problematizing the perceived irrationality of a given trait, or behavior, in order to understand how it can be traced back to another *pattern* of rationality: when, as in the case of the sacrifice of intellect, an "irrational" point will be located, it will be my aim to circumscribe it as fittingly as possible.

Therefore, my intention is to analyze two central aspects of religious behavior relying on the notion of violence (for sacrifice) and its retention (for forgiveness), thus using violence as a kind of eco-cognitive prism. I will argue in favor of the *epistemic* and *pragmatic/heuristic* roles of violence (and of its retention). In the first section, in fact, I will contend that the religious sacrificial mindset fascinatingly weaves together a theoretical non-understanding (in the form of an "epistemic" violence tracing back to the *sacrificium intellectus*) with a pragmatic understanding, in order to achieve the further possibility for an eco-cognitive acting. In the second and final one, conversely, I will challenge *forgiveness* as a heuristic violence committed against a norm, part of a set of rules enabling the functioning of a the cognitive niche, for the sake of the functioning of the cognitive niche itself.

In sum, my final contention will be that *violence* and *irrationality* are, paradoxically, the necessary and strictly linked counterweights that are able to make religion (as I have analyzed it in this part) livable and even profitable from the perspective of an eco-cognitive rationality.

13.2 Sacrifice and Epistemic Violence

To deal with sacrifice is much alike to taking a walk on a philosophical tightrope. This seems to be the case for a number of reasons, which incidentally are the same reasons that made sacrifice such a fundamental object for philosophical reflection—be it religiously, ethically or intellectually oriented. Keenan's introduction to the topic of sacrifice brilliantly sums up one of the chiefly problematic aspects of sacrifice:

> In the genealogy of Western sacrifice, one can trace an increasing interiorization, spiritualization, and dialecticization of sacrifice. Throughout this genealogy, sacrifice has predominately been understood as a necessary passage through suffering and/or death (of either oneself or someone else) on the way to a supreme moment of transcendent truth. Sacrifice effects the revelation of truth that overcomes the negative aspect of the sacrifice. In a word, sacrifice

> pays. One gets a return on ones investment. But this economical understanding of sacrifice only makes "sense" if it is pushed to its "logical" extreme. *Ironically, it is as if the economical understanding of sacrifice inevitably unworks itself. The work of sacrifice unworks itself. To be what it "is" sacrifice must sacrifice itself* (Keenan 2005, p. 1, added emphasis).

A bit like Heisenberg's electron, sacrifice "unworks" itself especially when put under scrutiny. At least from a philosophical perspective, sacrifice should not pay, it should be made *per se*: yet in the very first pages of Keenan (2005) we are shown that already in the Gospels, one of the main sources of the Christian conception, sacrifice *both* pays *and* should not be made because it pays. If sacrifice is fully brought back within the domain of the rational logos, then we lose the transcendental dimension that makes it such a fascinating topic (both existentially and philosophically), while if we abandon completely the way of the logos, then we must also relinquish any hope of *understanding* sacrifice by means of conceptualization, as Bubbio (2004) explains.

It could be added that sacrifice is indeed a land of confusion: furthermore, any investigation of sacrifice—and therefore of self-sacrifice—must come to terms with hurting personal or doctrinal sensibilities, and therefore can easily run the risk of being either pointlessly controversial, or plainly *doxastic*. If it is not possible, therefore, to be impartial in this field of studies, I shall try to be at least honest, and start by declaring the philosophical weaponry I will use, aside the eco-cognitive perspective.

The main *dramatis personae* at play in my exploration of sacrifice and self-sacrifice are Derrida, received in De Vrie's reading, and Girard. Understanding sacrifice (and self-sacrifice) as a liminal phenomenon (that is involving boundaries, and the possibility of crossing them), I resorted to Derrida's speculations because, especially in Derrida (1999), the French philosopher provides what could be called a "vectorial" conception of sacrifice, that is something that (at least intentionally), originates in a *here* and proceeds to a *there*. I decided to rely on De Vrie's interpretation of Derrida (De Vries 2002) inasmuch as, in the tradition concerning philosophy of religion, his stance—while not exempt from legitimate criticism—clearly deals with the violence presumed by Derrida's *vectorial* conception of sacrifice without reducing this violence to something else. Girard's contribution concerns the necessary existential level of sacrifice, the dynamical/narrative life-oriented approach: my interest will be in spelling out how the scapegoating process can become intertwined with the kenotic dimension of self-sacrifice, together with the relevance of mimetic desire as far as it concerns the ultimate sacrifice of intellect. It could be strange to juxtapose Girard to a deconstructionist account of sacrifice, but I suggest that the two visions might support each other providing a fuller understanding of the violent dimension of self-sacrifice.

Much of the evolutionary and cognitive literature on religion I have already referred to in this part considers sacrifices, and self sacrifices, within the "costly commitment" theory. Atran offers an exhaustive sum of the paleo-anthropological discourse:

> Because human representations of agency and intention include representations of false belief and deception, human society is forever under threat of moral defection. Simple consent among individuals seldom, if ever, successfully sustains cooperation among large numbers of people over long periods of time. Emotionally hard-to-fake and materially costly displays

> of devotion to supernatural agents signal sincere willingness to cooperate with the community of believers. […] Commitment is useless unless it is successfully communicated. It follows that displays of credible commitment are as significant as the behavioral commitments they are meant to signify. Displays also are more frequent than actual commitments, because they are usually less costly. Nevertheless, for cooperation to work, displays of commitment must be thoroughly convincing, and to be convincing people must be willing to make the ultimate sacrifice, however rare (Atran 2002, p. 114).

In spite of the soundness of this explanation, it is a functional one. It explains what the function of sacrifice came to be, and not what is going on in the believer's rationality when she is performing a sacrifice. With this latter respect, it is a kind of *explaining away*.[1]

The eco-cognitive approach to philosophy, which does not prescind from the fact that cognitive agents are *real* agents immersed in a there and then, may be particularly sensitive to the issue of sacrifice inasmuch it would seem, at least prima facie, that the agent—mostly in self-sacrifice—acts against the fulfillment of her own goals. Consequently, the eco-cognitive and epistemological approach, albeit unusual, may indeed offer and unbiased view of self-sacrifice, reducing the irrationality to the cost of admitting the violence of the action,[2] and abstaining from a mainstream comfortable conception exemplified for instance by a recent article about the nature of self-sacrifice (Dalferth 2010). The individuated tendency is a rebuttal of both a deconstructionist and a mimetic appraisal of religion and its ambivalent conception of sacrifice, that deprives religion of its essential violent core: not only violence and killing are *collateral* elements of sacrificial practices, but Dalferth's description of self-sacrifice for an altruistic scope reduces the unpleasant, painful consequences for the self (they can involve bloodless limitations of different kinds, physical or moral injuries or even death) to a merely contingent effect of one person committing her own life for the benefit of others.

I suggest that the conception of self-sacrifice just mentioned is faulty because of two principal aspects. The first one seems to be a matter of sensitivity as far as the topic of violence is concerned: elements such as violence and killing cannot be so easily removed from the understanding of sacrifice without providing a stronger alternative justification of why they are so—historically and anthropologically—widespread, and the result might consists in unawarely dealing with something which is not sacrifice anymore. The second faulty aspect could be more of a moral contention: the philosophers' main unease seems to reside in the impossibility of conciliating an act of violence (even if self-oriented) and an act of love, yet again this concern seems

---

[1]Albeit Atran nicely sums up many tenets of evolutionary and cognitive studies of religion, it is impossible to accuse him of *explaining away* anything. Atran's book fully acknowledges the complexity of the religious phenomena and states that, albeit it can be analyzed fruitfully by evolutionary and cognitive sciences, it is not possible to provide a simple evolutionary explanation of religion, for instance as an adaptation.

[2]A similar result was achieved in the previous Part II when analyzing gossip and other forms of social cognition: the "idle talk" bias was reduced, at the cost of highlighting the embedded potential of strategic violence.

to be a blind refusal of accepting within philosophy the challengingly paradoxical nature of sacrifice and self-sacrifice.

I shall hereby develop the argument in the following way: first, I will suggest a definition of violence which does not *make* self-sacrifice necessarily violent, but rather aims at being consistent with the common sense conception of sacrifice *as* actually violent (if self-sacrifice was not violent, then it would not be perceived as something rare and everyone would be committing non-trivial self-sacrifices wherever we laid our gaze). Framing this initial claim within the vectorial conception of sacrifice offered by Derrida (and exemplified by De Vries), I will individuate in the violence against intellect (*sacrificium intellectus*) the core of the violent dimension of self-sacrifice, insofar as the author of the sacrifice does not limitedly commits *part* of her understanding in the sacrificial practice, but all of it, since she is both the agent and the patient of sacrifice. At this point, I will have gathered enough material to spell out two fundamental violent aspects of self-sacrifice. The first concerns the *exemplar* self-sacrifice in Western tradition, that is Jesus Christ's. The self-sacrifice committed by God's own λόγος is the epitome of sacrifice as *sacrificium intellectus*, therefore the highest gradient of intellectual violence. At the same time, this is crucial as it further corroborates the interpretation of Jesus' sacrifice as *the last* sacrifice: no mimetic attempt to reenact. His sacrifice can hold the comparison under the fundamental aspect of intellectual violence. The second "mirror" to reflect about the violence of self-sacrifice will concern the sacrifice enacted by superheroes in popular culture, namely Batman, and the extent to which the sacrifice of intellect is at play when kenotic self-sacrifice and scapegoating processes become hard to tell one from the other (i.e. when the *hero*'s commitment seems to reverberate internally the external blame).

13.2.1 Sacrificial Violence: Imitation and Limitation

In a famous interview, the french scholar René Girard maintained that "mimetic desire, even when bad, is intrinsically good, in the sense that far from being merely imitative in a small sense, it's the opening out of oneself" (Adams and Girard 1993, p. 24). The concept of mimetic desire, and the consequent violence, are a fitting background for an investigation of self-sacrifice and its possible perversions.

Girard illuminates this matter by reflecting upon the very etymology of the word *sacrifice*: it originates from the latin word *sacrificium*, which is in turn composed by the union of the verb *facere* and the attribute *sacer, sacrum*, that is, *to make something/someone holy*. But once again, the very word *sacer* has a violent connotation insofar as it recalls the act of ritual *slaughtering* (Girard 1977).

> Religious interpretations of the sacrificial crisis have always been inclined to categorize its accompanying phenomena either as "good" or "bad." The farther one progresses, the more pronounced becomes the tendency to make separate and distinct entities of the two aspects of the sacred. In Latin, for example, *sacer* has retained its original double connotation, but the need to isolate the beneficent aspects of the word has resulted in an offshoot, *sanctus* (p. 279).

Religion is commonly perceived as relating to practices involving sacrifice: as I will show, sacrifice is not exclusive to the religious framework, even though some of its main characteristics do share much with a religious conception: I will make use of the Girardian theory to operate a sort of reverse engineering of self-sacrifice. That is to say, holding the sacrificial dimension as a core, I will explore the religious dimension of sacrifice to approach the most powerful religious self-sacrifice, and from this I will move to non-religious self-sacrifice.

Religious violence scares us more than many other kinds of violence. The very word "religion" makes us think about altars, prayers, priests and temples, but also about sacrifices, crusades, inquisition, *jihad* and *mujaheddins*, penance, sects and other elements—all conveying a strong connotation of violence. Here I am not committing a hasty and insulting generalization: it might be claimed, on the contrary, that many forms of religion and spirituality, for instance Buddhism,[3] do not cause suffering *per se*. Nevertheless, as contended in Magnani (2011, Chap. 6, Sect. 5), the peaceful nature of oriental religions might be a wishful myth, as opposed to the violence experienced by Westerners regarding their own religious history, or ultimately a dangerous misunderstanding generated by the juxtaposition of the oriental framework with the Western-Christian one. Žižek (2009) observes, commenting on Buddhism, that a blindly benevolent and peaceful attitude towards the world could be phenomenologically impossible to distinguish from merciless universal indifference. Taylor's analysis of phenomena of social isolation in Japan sheds further light with this respect (Taylor 2006): a world of peaceful indifference is indeed a world without mimetic rivalry (and without the consequent violence), but also a social landscape devoid of compassion and reaching out towards others.

In my perspective, the eco-cognitive link between violence and hurting others could be considered in the wider sense of *limitation* of potentialities not only with respect to life and physical integrity, but also to the freedom of acting "differently," even if the religious imposition foresees no patent harmful consequence: we could say that limitation is also an obstacle to further *imitation* on the victim's part, a way of thwarting the replication of mimetic desire. From this perspective, self-imposed vegetarianism is a mitigated form of violence, for instance, as well as forbidding oneself from killing a cow when it would provide more than one satisfying and protein-rich meal: vows and the celebration of chastity are embraced by the same category. The Jewish observance of the *Sabbath* can be considered as the "negative sacrifice" par excellence.

Seligman et al. (2008) contend that saying "please" and "thank you" is a simple form of ritual, but a further reflection can be drawn: those simple utterances are a form of sacrificial limitation as well (even if not in a primarily religious sense), inasmuch as a subject is induced to proffer them even if she might think differently, or would rather not interrupt her activity e.g. carry on eating, playing etc. We have

[3] I do not feel it proper to adopt the canonical objection of pop-theology on the matter, which typically consists in denying that Buddhism and other Asian religions are actually religions, but rather philosophies: Buddha, *karma* etc. are clearly supernatural concepts, which allow us to label as religions those doctrines that deal with them.

to continuously undergo these limitations to our freedom in order to make any social setting livable, just as it happens within a family: "Everyone loves the others. That is not the point. Instead, the problem is to get everyone to act *as if* they love one another" (p. 25). The more or less violent (if only symbolical) limitations to our freedom are not about fostering sincere behaviors—as humans we are already quite skilled at that—but to maintain their phenomenological appearance, a much harder yet more vital task: this also contributes to laying some pragmatic obstacle to the overflowing of mimetic violence. In this particular perspective, it is easy to understand how the whole establishment of morality is an entanglement of ritualistic behaviors involving different kinds of limitations which make the (social) world livable, predictable and, alas, very often perceived as violent.

13.2.2 The Sacred as a Violent Epistemic Fence

One of the ideas supporting this analysis is the suggestion that religion, as a complex cultural phenomenon, developed mainly as a cognitive fence around *magic*, to prevent it from overflowing into other domains than the peculiar domain of magic itself (Bulbulia 2009): it is the process I described in the first two chapters of this part. By "magic" I refer to the wide range of beliefs concerning agents, operators, correlations, relationships and so on, which violate to different extents our expectations about the world. The cognitive origins of these beliefs have been intensely studied over the past few decades (Boyer 2001; Atran 2002; Dennett 2006): notwithstanding their differences, many of these approaches agreed in showing how supernatural and magical concepts resulted from inferential process that were fine-tuned for everyday ecological problem-solving necessities (chiefly the detection of biological agency).[4] A reality-check, though, called for these theories to be implemented with an explanation of how this kind of belief got eventually *limited* inasmuch as we manage to engage different modes of behavior, sometimes behaving *as if* the magic was real and other times as if it was not.[5]

Religious violence is overwhelming because, being an expression of the magic, it apparently occupies a zone beyond the reach of our possibility of understanding. To quote the rich imagery of De Vries (2002, p. 288), the otherness of magic entails a necessarily violent dimension insofar as the ontology of the natural world becomes a "hauntology"[6] where, in the tradition of Derrida's writings, the object of belief is so intrinsically *other* that it becomes the Other of the Other, lest it loses the Alterity that defines it. Then, a sacrifice is the necessarily violent act by which

[4] cf. Chap. 10.

[5] This topic, elaborating on some insights by Bulbulia (2009), was developed in Chap. 12.

[6] The word *hauntology* is a blend of the verb *to haunt* and the suffix *-ology*, originally coined by Derrida: it aims at labeling the region of what *neither is nor is not*, the typical condition associated with the specter. It constitutes a kind of suspended, fluctuating and unintelligible region that *haunts* our usual ontologies (Derrida 1994).

something (in one's *ecology*) is made holy, that is partaking of yet another original violence. As we will see, sacrifice is oriented towards magic, but every time one directs himself towards magic, towards divinity, he also directs himself towards the violence intrinsically embedded within it. This can be understood holding in mind the eco-cognitive origins of the magic itself, as the product of particular endowments aimed at our biological survival. Religious doctrines, rituals, symbols, artifacts could therefore play the pivotal role of setting a cognitive boundary between what is *Other*, *divine*, *magic*, and *what is not*. More precisely, the *sacred* is the element allowing this juxtaposition and contraposition between something understandable—such as culture—and something *conceived as unconceivable*.

The holy can be said to partake of both the cultural, intelligible dimension and the magical domain. Holiness may concern a wide range of distributed objects, and can be more or less reified: a physical item can be sacred, but also a place, a person, an image, a word, a phrase, a ritual. Holiness, in fact, seems to embody not so much an ontological interruption but rather a cognitive, gnoseological one: in our natural ecologies, sacredness is a self-transcending property in front of our intellect, inasmuch as it connects something which is there with something which is *Other*.

Furthermore, holiness almost always involves a subsequent *feeling of privations*: what has been instituted as holy usually loses its ecological role, or keeps it but in a metaphorical sense. This is clear, for instance, as far as animals in religious understanding are concerned.

> The animal nature of these creatures is striking and significant. At one and the same time, it points away from itself and is mingled with humanity and even with divinity. In fact, the animals in the procession either do not behave like animals or are not real animals at all. This characteristic – animality suspended and reinterpreted – is typical rather than peculiar when animals or images of animals appear in religious settings. Thus the religious significance of these animals does not lie primarily in their inherent animal nature but in that to which it gives added meaning. There is a synergetic effect between the actual animal and the being with which the animal is combined or connected – be it a human or a god (Gilhus 2006, p. 94).

That is because the institution of the sacred as *the* meaningful hiatus requires an act that replicates in a mimetic fashion a distinctive characteristic of the level it connects with, namely the absolute Other: this effective (performative) act is sacrifice. If divinity is the absolute other, which cannot be understood, sacrifice must be an act that transcends its own understanding, replicating the cognitive and physical violence entailed by the *gnoseo-onto-hauntological* (to use De Vries' terminology) region of magic. Sacrificial violence, aimed at isolating by means of sharp cuts the region of the *sacred*, can display different degrees of intensity. It can be a form of mutilation (as in the case of circumcision, as far as humans are concerned) or partial bleeding (when blood is requested for the ritual), yet most sacrifice involves killing the "offering." This represents the most impressive and meaningful disruption of the biological order, and the performative efficacy of sacrifice is thus maximized.

Seemingly ignoring the whole Girardian theses (and thus committing a kind of "Girard*icide*," to the extent that the French scholar is not even mentioned *en passant*) Dalferth—referring to a sacrificed animal—contends that "the killing of the sacrificial animal plays only a subordinate role in the procurement of blood for effecting the symbolic incorporation" (Dalferth 2010, p. 79). The contemporary wishful account of sacrifice begs another question from scholars and, in an epistemological outlook, is questionable from the point of view of simplicity: if death is not necessary to the scope of sacrifice, one wonders why bloody sacrifices have always been performed in the history of mankind. The empirical evidence about contemporary bloodless sacrifices does not help this perspective, all the more because one could wonder why sacrifices involving death were more frequent at stages of mankind characterized by a far lesser availability of food resources. Of course, what distinguishes a sacrifice from a mere killing is the presence of a ritualized dimension, which relies on the violent underside. As I will suggest, this violent dimension is not necessarily bloody *per se*, but can involve many kinds of violence still symbolically connected to physical pain and injuries.

A sacrifice, within an indistinguishable ritualistic dimension, is an act which deals with boundaries: it sets them, and yet it may transcend them as well, thus acting as a vector between different planes, both ontological and ethical: from the *here* to the *Other*, from the *worst* to the *best*. Boundaries do exist, even if within different ontological regions, and it requires some energy to cross them. The performative effect of the ritual rests in its capability to make the world "livable," first of all from a social point of view, but also with respect to our possible understanding of the world.

> We are constituted on our boundaries, that is to say, constituted on a plane we do not totally control, one that is always also open to the other, to the stranger, to what is different and unknown and beyond the controlling power of the center. This is what makes boundaries dangerous. Rather than trying to eliminate boundaries or to make them into unbreachable walls [...] ritual continually renegotiates boundaries, living with their instability and labile nature, [...] allowing us to live in [the world] by creating temporary order through the construction of a performative, subjunctive world. Each ritual rebuilds the world "as if" it were so, as one of many possible worlds (Seligman et al. 2008, p. 11).

The necessary violent dimension of sacrifice is the price to pay, as suggested so far, to be able to deal with boundaries and, according to Seligman et al., dealing with boundaries requires a confrontation far lesser than crashing into them: this confrontation is allowed by the violence of sacrifice and its ritualized dimension.

In the previous Chap. 11, I have introduced the notion of *religious niche*, developed upon the concepts of *ecological* and *cognitive niche* (introduced in Part II, Chap. 5): Seligman's description of ritual (and hence sacrifice as a peculiar subset of rituals) evokes indeed a superimposition of liminal meaning upon one's actual ecology. In this sense, sacrificial activity does indeed create, by separating it, a new ecological region a cognitive agent may deal with—for good, or for bad. A key element, so far, is the ambiguous strategic value of this eco-cognitive region individuated by ritual and sacrifice as a mean to encapsulate overwhelming magical intuitions: as already suggested when dealing with the curation of cognitive niches, the balance to be struck is between an overflowing of magic, and a suicidal raise of sacrificial, border-construction activity.[7]

---

[7]For cognitive niche curation, and its relationship with religion, see Part II, Chap. 6.

One could condense this necessarily strategically ambiguous conception of sacrifice, and consequently of religion, borrowing from De Vries (2002, p. 287) his interesting neologism "perverformativity," a fusion of the words performance and perversion. The "pervertibility" of any sacrificial religious performance rests in its original constitution: turning to divinity is conceived as *good* in the performer's mind, yet this turn can sometimes requite a turn to violence. Again, we could be challenged on this assumption by proof of non violent or bloodless sacrifices but violence does not need to be addressed towards the victim of the sacrifice. Furthermore, it seems short-sighted to see only one victim in the sacrifice, for instance the creature being slaughtered.

It is important to bear in mind that there is yet another kind of violence—more pervasive and even easier to obliterate—present in every sacrificial scenario: that is, every sacrifice is a *sacrificium intellectus* insofar as it is a violence towards comprehension,[8] in a mimetic fashion that reproduces the violent cognitive puzzlement induced by magic *per se*. Derrida offers another powerful insight as far as this aspect of sacrifice is concerned, that is his *à Dieu/adieu* figure (Derrida 1999). Any act of Faith, any kind of sacrifice, be it violent, bloodless or merely intellectual, implies saying *adieu*—farewell—to the very object of the act, because it is being dedicated, consecrated in this turn towards God, in the moment of the *à Dieu*—to, towards God. The violence of total privation (i.e. death or irreversible departure) of the object can be mitigated by a partial privation of some characteristics of the object, which becomes something else from what it was before. Yet, on a similar ground, the *adieu* moment concerns the executor as well, as long as she is saying farewell to the object and to her possibility of understanding: not only her cognitive habits are disrupted by the object becoming at once the same and something else, but also the very end of the process transcends the possibility of her intellect.

Sacrifice, as I contended before, originates in our natural ecology but tends toward the Other. This cognitive *adieu* is absolute, total: "[...] to say adieu, if only for an instant, to the ethical order of universal laws of human rights by responding to a singular responsibility toward an ab-solute other—for example, the other par excellence, God—implies sacrificing the virtual totality of all innumerable others" (De Vries 2002, p. 159). The *à Dieu/adieu* figure then is not a bi-polar one: furthermore, not only the farewell can be interpreted in different, yet coherent, ways. The second pole is polysemic as well, since the French preposition "à" can be a locative and dative proposition, but can denote possession as well. French maintained the *dativus possesivus* from latin: *c'est à moi* means "it belongs to me," hence *à Dieu* signals what belongs to God. Sacrifice makes an object to partake of *what belongs to God*. But if it is up to God, it is *not* up to me anymore. The object does not belong to me, to my ontology, to the gnoseological regions I am empowered to navigate: it is now beyond the eco-cognitive borders set by sacrificial activity, or even better, it is now part of the sacrificial border needed to encapsulate the Other.

Derrida's reflection is crucial to help us understand Keenan's contention that sacrifice "unworks" itself, especially when we attempt to understand it. Its vectorial

[8] Tertullian's formulation, *"certum est, quia impossible est"* is exemplar.

character, its continuos shift among different, heterogenous planes challenges any conceptualization that is not willing to shift and bend together with its object. Its vectorial nature, a transiency fueled by the further sacrifice of intellect, begs for a dynamical (yet not mystical) understanding.

13.2.3 Self-Sacrifice and the Scapegoating of Intellect: Christ, Batman, and Girard

From the perspective I just delineated, it should follow that not only self-sacrifice can be perceived as violent just as any other kind of sacrifice (because it could not be otherwise), but from an epistemological and eco-cognitive perspective, it is even a more ferocious kind, inasmuch the subject—since she coincides with the object of sacrifice—forsakes on the sacrificial altar not only her understanding of the peculiar sacrifice, but also her understanding as a whole, her being a sentient, living subject. Even if there can be more violent kinds of exogenous sacrifices, from the perspective of epistemic violence, self-sacrifice is the very sublimation of violence.

We can consider the case of Maximilian Kolbe, an Auschwitz prisoner who voluntarily took the place of another prisoner to whom he was totally unrelated and died in his place: it has been contended that committing oneself to death—out of an act of love—"is neither the end nor a means of what one does, but is rather taken as an unavoidable collateral damage, so to speak, in abiding under all circumstances by the love of one's neighbor" (Dalferth 2010, p. 84). It seems quite hard, from a philosophical point of view, to agree with this contention. Kolbe's extreme love for humanity claimed the forsaking of his own humanity (through the forsaking of his life) as a means to achieve a "local best." This leads me into noticing yet another typical feature of self-sacrifice: as a performative gesture, self-sacrifice aims at a consequence (be it in the long term or in the short term).[9] With this respect, we can

[9]From an ethical point of view, it could be extremely dangerous to ever label death as a "collateral damage", especially because—out of unselfish love—one might achieve a greater good by killing somebody else, and not oneself: if one's death is merely a collateral damage, it becomes surprisingly simple to condemn a single person to die if this would save many others. Agreeing with this perception might just solve centuries of moral debates, dating at least back to Thomas Aquinas and the *Second Effect* theory! Furthermore, the collaterality of "death out of love" sinisterly echoes the typical story of violent outcomes of *psychopathological* "fatal love" (Buchli 2006): out of an apparently unselfish "love bubble" the killer consciously and responsibly chooses to be *possibly (yet automatically)* violent, and—even after actually committing the crime—still thinks of himself as a loving person, notwithstanding all the violence he perpetrated. When the killer provides an explanation of his "irresistible impulse," which he could allegedly not control (because they are out of the reach of his loving conscience), he consciously affirms he killed his woman *because* he loved her. On yet another level, the same paradoxical quintessence of unselfish violent love can be found in the archetype of the revolutionary leader, as resumed by Ernesto Che Guevara himself: "Let me say, with the risk of appearing ridiculous, that the true revolutionary is guided by strong feelings of love. It is impossible to think of an authentic revolutionary without this quality. This is perhaps one of the greatest dramas of a leader; he must combine an impassioned spirit with a cold

suggest that self-sacrifice is about a declaration of humanity. One's total sacrifice of her own humanity is at the same time a cry for hope for humanity as a whole. In this sense, it could be possible to define kenotic self-sacrifice through a neologism, stressing how it maximizes dimension of *hopenness*: that is to say, self-sacrifice totalizes how a sacrifice embodies an *openness* to the worst, but this is fully sublimated in favor of *hope* for humanity, inasmuch the subject abandons any possibility of future action as an individual self.

Here it seems possible to individuate the core of the conjunctive dimension of sacrificial rites, as suggested by Seligman et al. (2008): kenotic self-sacrifice, out of love, individuates the most powerful shared "as-if." The injection of hope in the system is obtained by a testimony (a "costly commitment", in the evolutionary and cognitive lexicon) creating a *shared conjunctive belief*, depicting the world "as if" it was worth that one commits her life for the situation to improve: this makes sense also when understood as a dramatic niche-curation activity, an attempt to reframe the negative trend of the niche itself. Self-sacrifice is not suicide out of despair, which would be a consequence of feeling doomed. Conversely, it is a performative way of signaling that things might improve. Hope is what is needed most during crises, and a crisis dimension can be a constant in many self-sacrifice scenarios (otherwise there would be no need for self-sacrifice, as a matter of fact): the scapegoat phenomenon can be frequently individuated in crises as well, as shown by Girard (1986).

During a crisis, variants and deviants are hastily identified and, moreover, deviants are often turned into *scapegoats*, considered to be responsible for the crisis itself. The repetition of this phenomenon was proved to persist over the centuries as a structuring principle in mythology and popular culture (Girard 1986): a sign of how this kind of thought is deeply nested in human mind. Indeed, Girard individuated several stereotypes defining a situation of scapegoating: for the present analysis the *collective murder* is extremely interesting. It is a situation in which the whole community, transfigured into an angry mob, physically suppresses the scapegoat: death punishment can be regarded as the ultimate form of negative assortment. If the life of an individual is taken, *a fortiori* he is expelled from the community with no hope of further reintegration. Even if some acts of bullying and mobbing can actually end up with a murder, usually the collective homicide is mitigated into an isolation and expulsion of the deviant from the community.

The self-sacrifice of Jesus Christ is, according to Girard, the ultimate sacrifice. This is why the narratives of the Gospels stress how unfounded were the accusations that lead to Jesus' condemnation, which ultimately consisted in the execution of an innocent person: Jesus was killed for *no reason* at all. The sacrifice of Jesus represents the *exemplar* of self-sacrifice: Jesus accepted to become *The Scapegoat* for the sinful condition of humanity after the Fall. In Western imagery, we now associate the image of the Lamb with meekness and gentleness (William Blakes' poem "The Lamb"

(Footnote 9 continued)
mind and make painful decisions without flinching one muscle. Our vanguard revolutionaries […] cannot descend, with small doses of daily affection, to the places where ordinary men put their love into practice" (quoted in Anderson 1997, pp. 636–637).

exemplifies brilliantly this conception), but when John the Baptist—upon seeing Jesus approaching—said "Behold, the Lamb of God, who takes away the sin of the world!" (John, 1:29) he was not thinking of meekness and tenderness. Rather, what he referred to was the tradition of the scapegoat as defined in the Jewish tradition (Leviticus, 16): a goat was ritually invested with the sins of the community of Israel and then sent into the desert, usually to certain death.

It is tempting to connect the dots laid out by the Girardian account and thus face a slight theological challenge: within a religious dimension, that is accepting that Jesus is the son of God and God himself, His self-sacrifice empowers the already violent dimension of an atonement rite. Also from the perspective of epistemic violence, John the evangelist presents us with a challenging issue: the famous opening of the Gospel reads that "In the beginning was the Word, and the Word was with God, and the Word was God" (1,1). In the Greek original, the term for *Word* is the philosophically pregnant λόγος. The verse 29 states that "The Word became flesh and made his dwelling among us," and we all know the epilogue of such dwelling. What if the λόγος in Its denotation of reason, logic, discourse, understanding gives up Its own existence for the sake of humankind? Then, not only the Sacrifice of Christ is the primer for all self-sacrifices, but also represents the peak of sacrifice as *sacrificium intellectus*. I had suggested, with Derrida and de Vries, that every sacrifice is a sacrifice of understanding which stirs up the violence of irrationality inasmuch it performatively *bridges* what we know with the Other: in self-sacrifice, the individual's understanding is put *in toto* on the altar and killed, but in the sacrifice of Christ, which is clearly the highest expression of unselfish love, the irrational (undifferentiat*ed* and therefore undifferentiat*ing*) violence of men leads to the suppression of cosmogonical understanding, rationality, λόγος as a whole.

Girard's account is optimistic, in a way. His mastery in identifying the intrinsically violent core of many cultural practices climaxed in his brilliant reading of Jesus' self-sacrifice as the sublimation of the scapegoat mechanism. On the other hand, he might have been (wishfully) too eager in assuming the once-and-for-all*ness* of this event. He acknowledges that the sacrifice of Jesus did not put an end to mimetic desire bursting into victimizing violence, but he claims that God's own scapegoating, a sort of short circuit, revealed the absurdity of the mechanism as a whole and made it easier to individuate and correct. This might be true, but it is of little consolation compared to the awareness of so many violent events characterizing Western society which took place after the self-revealing sacrifice of Jesus, such as witch-hunt, Inquisition, pogroms, not to mention the atrocities perpetrated during the 20th century, which could be hardly said not to partake of scapegoating dynamics. Awareness and understanding help once we can look back, but they have not seemed useful to prevent such collective outbursts of violence so typical of human nature. If we reject this awareness—about the inescapability of violence even in the course of religious history and once it has been openly individuated and explained—we are left with other possibilities, such as Benjamin's messianic hopes (which turn again towards an Other, and delineate another pervertible "standing for the best") (Benjamin 1978) or Milbank's rebuke against the earthly religious institutions, seen as responsible of the social perversion Milbank (1990).

If I have dealt so far mostly with the religious conception of sacrifice, it does not mean that all self-sacrifices must be seen within a religious framework, as witnessed by the aforementioned case of Maximilian Kolbe. Neither it means that all sacrifices partake of a religious nature: quite the contrary, as suggested by Girardian theory, it is religion that is characterized by a sacrificial dimension. The common denominator of sacrificial dynamics is the will to resolve a distressful situation, a crisis, a conjuncture whose outcome is not known but unlikely to be merry: in other words, the apparent irrationality of sacrifice is a way of addressing a cognitive puzzlement, which can be caused by the cognitively-generated magic, but not exclusively by that. The sacrificial dimension, which is to become religious, is the *standing* for the best, and not the best itself: it is a vector connecting different planes, fueled by—and opening to—violence. As maintained by Girard, "mimetic desire, even when bad, is intrinsically good, in the sense that far from being merely imitative in a small sense, it's the opening out of oneself. […] Extreme openness. It is everything. It can be murderous, it is rivalrous; but it is also the basis of *heroism, and devotion to others, and everything*." (Adams and Girard 1993, p. 24, added emphasis). The relationship between self-sacrifice, mimetic rivalry and heroism will be addressed in the closing of my analysis of self-sacrifice.

The fundamental difference between the ordinary treatment of a deviant and the punishment of a scapegoat during a crisis consists in the fact that the scapegoat is held responsible for the crisis itself. Furthermore, scapegoats—overwhelmed by gossip and mobbing—eventually believe to be guilty, participating in their very same punishment, as shown by the tragedy of Oedipus.[10] It is a clear situation of magic thought: not only the deviant is considered to be the cause of the entire crisis, but his condemners actually perceive his suppression as the end of the crisis. Curiously, we could say that a fertile ground for the emergence of a scapegoat is just as good for the emergence of a hero: sometimes—if not often—those two characters match.

The topic of painful self-sacrifice, even if not necessarily deadly, has often emerged from popular culture: a significant example, this time not from religion, can be found in the Batman movie trilogy directed by Christopher Nolan. It is hard not to understand the three movies as extremely Christological: then again, it might be suggested that—even though Christ's (self)sacrifice is indeed the first and most complete unmasking of the sacrificial system—the two narratives are similar insofar as they deal with the same core. Batman's trilogy depicts perfectly the unfolding (and resolution) of a scapegoating mechanism: *Batman Begins* (2005) deals with the public adoption of the hero, *The Dark Knight* (2008) with his public disgrace and (self)scapegoating while the last episode, *The Dark Knight Rises* (2012), tells about his rehabilitation through the unveiling of the sacrificial mechanism. This one-two-three movement can be considered as a model underdetermining many super-hero narratives.

The second episode of the trilogy, *The Dark Knight*, provides us with some extremely interesting insights. The protagonist—Batman—embodies a perfect

[10]This issue has frequently emerged in Part II when analyzing social cognition. It could be particularly interesting to match the current discourse with the reflection offered in Sect. 7.4.3.

sublimation of a scapegoating mechanism: in a crescendo of mimetic rivalry[11] (driven by envy and desire), the people of Gotham start blaming Batman as the cause of the violence he is trying to fight and repress. At the end of the movie, the avenger sacrifices himself for the city's crisis, accepting the blame in order to protect the reputation of the district attorney who had managed to spark some hope in the citizens (the very politician whom, after personal disgrace following the action of the Joker, turned into the super-villain Two-Face). The situation is thus resumed by Batman's only true ally, police officer Gordon:

> Because [Batman is] the hero Gotham deserves...but not the one it needs right now. So we'll hunt him, *because he can take it*. Because he's not our hero...he's a silent guardian...a watchful protector...a dark knight.

Of course, Bruce Wayne is not giving up his life as Jesus did, but he is giving up is reputation as a hero: the citizen would like to kill him (collective murder) or at least to capture and sentence him. Batman does not act this way for a later reward, but because of the love he feels for this Other, which is nothing less than the eco-cognitive niche revolving around the city of Gotham: his fellow citizens that deserve a brighter future. In a clearly self-sacrificial dimension, Batman accepts to be hunted (and potentially killed) by those who should be his allies because *he can take it*, to the point of valuing the Other's hope over *his* truth. Is this not a most clear example of the "standing for the best" mentioned above? Truth is understanding, and in this situation it is an understanding about one's righteousness which becomes the object of the *sacrificium intellectus*. Furthermore, this violent sacrifice is performed by the same intelligence which sacrifices itself, and this sacrifice is the only possible mean to attend the goal (in this case, letting people "believe in Harvey Dent"), and not a collateral effect. The sacrificial "openness to the worst" (truly *hopenness*, in this case) is well represented in Gordon's final speech, that merges the semantics of the hero (which is a character everybody should be able to look up to) with those of mystery and obscurity.

For the sake of completeness, it should be mentioned that the closing chapter of the trilogy fittingly echoes the sacrificial ending of *The Dark Knight*: before the *Rise*, we are shown an Oedipal Bruce Wayne, crippled both in body and mind. The self-sacrifice was complete, and *worked* (as a matter of fact, many sacrificial dynamics indeed *appear* to be effective): Batman's taking the blame for the death of Harvey Dent allowed the police to fight back crime with far greater efficacy. The emergence of a new villain will be necessary to spark the re-birth of the hero, who must recover from a partly self-inflicted *damnatio memoriae* (what better kind of collective murder can be used against a sacrificial victim that cannot be apprehended?). Paradoxically,

[11] A whole study should be dedicated to the Girardian aspects of Batman, both in comics and movies. The mimetic struggle, for good and for evil, is amazingly in plain sight: Batman's resources and allies partake of his very nature (*Bat*girl, *Bat*cave, *Bat*mobile, *Bat*cave, *Bat*wing, *Bat*cycle, to name just a few), while for most of his foes (and some friends) the *topos* of the *double* is mightily exploited in a series of couples and dichotomies: Batman/Bruce Wayne, Batman/Robin, Batman/Joker and so on. The undifferentiated force of the crowd (the *citizens* of Gotham) is yet another constant in the narration.

it will be the villain himself to unveil Batman's self-sacrifice: the truth will plunge the scandalized citizens in further discomfort. Batman's *kenotic* withdrawal for the good of the city was indeed an act of self-sacrifice, but it disturbingly reverberates the acceptance of the city's blame: it is hard to draw the line, in this case (but only in this case?) between a willing self-sacrifice and an acceptance (undergone "standing for the best") of a scapegoating mechanism. Spiderman's famous sentence comes to the mind, with this respect: "With great powers come great responsibilities", and great responsibilities immediately echo Levinas' unforgettable words:

> Responsibility is what is incumbent on me exclusively, and what, humanly, I cannot refuse. I am I in the sole measure that I am responsible, a non-interchangeable I. I can substitute myself for everyone, but no one can substitute himself for me. Such is my inalienable identity of subject. It is in this precise sense that Dostoyevsky said: "*We are all responsible for all men before all, and I more than all the others*" (Levinas 1985, pp. 96, 99, 101).

Batman can be considered as a model of superhero, and superheroes can be considered the model of the exceptional actions we sometimes undertake, and thus help us reflect about it. To express it in Girardian terms, the sacrificial superhero is an interesting blend of two fundamental figures: on the one hand, he partakes of Oedipus, believing his guilt and accepting his responsibilities, yet on the other hand he fights for justice taking a stand for the best, and this fighting back is a way of claiming, pragmatically, his innocence until the final stand, like the Biblical character of Job (Girard 1987).

Indeed, one of the aims of this analysis of sacrifice was to provide a Girardian perspective of self-sacrifice able to make room for its possible perversion and negatives sides, so that it would be sensible with the eco-cogntive perspective I have adopted in this book. When dealing with religion, violence is an issue that is often explained away, nevertheless if one wants to spell out the inferential dimension underdetermining religious behavior, it should not be enough to offer a functional view of violent behaviors such as sacrifice: the presence of violence should be acknowledged, respected and explained—instead of being *explained away*: this position has already been put into practice by the eco-cogntive perspective by Magnani's *Understanding Violence* and the debate it sparked.

In this sense, the nature of sacrifice and self-sacrifice, is *prima facie* a paradox to understanding (Bubbio 2004), but it must be acknowledged in all its indispensability and horrid nature: it is not a paradox if a paradox is something else, *alien* from our moral and existential lives, and then prone to be sacrificed. As I suggested in the past pages, it is indeed possible to offer a non-functionalist explanation of religious violence (or at list of part of it) so that it is coherent with the study of different styles of rationality. It is along the same lines that I will tackle the final issue of this book, always relating to religious behavior, by offering an inferential and cognitive account of *forgiveness*.

13.3 Forgiveness: Between Religion and Cognitive Science

Forgiveness, even though not exclusive to the religious mindset, has often been framed as a religious topic, especially as far as the Judaism and Christianity were concerned. The literature on the topic is nearly endless, but it seems more proper to set off from an eminently religious aspect, in order to connect with the discourse I have been setting up in this book so far. The Gospels do indeed present us with forgiveness in all of its problematic nature:

> Then Peter came to Jesus and asked, "Lord, how many times shall I forgive my brother or sister who sins against me? Up to seven times?"
>
> Jesus answered, "I tell you, not seven times, but seventy-seven times. Therefore, the kingdom of heaven is like a king who wanted to settle accounts with his servants. As he began the settlement, a man who owed him ten thousand bags of gold was brought to him. Since he was not able to pay, the master ordered that he and his wife and his children and all that he had be sold to repay the debt. At this the servant fell on his knees before him. 'Be patient with me,' he begged, 'and I will pay back everything.' The servant's master took pity on him, canceled the debt and let him go. But when that servant went out, he found one of his fellow servants who owed him a hundred silver coins. He grabbed him and began to choke him. 'Pay back what you owe me!' he demanded. His fellow servant fell to his knees and begged him, 'Be patient with me, and I will pay it back.' But he refused. Instead, he went off and had the man thrown into prison until he could pay the debt. When the other servants saw what had happened, they were outraged and went and told their master everything that had happened. Then the master called the servant in. 'You wicked servant,' he said, 'I canceled all that debt of yours because you begged me to. Shouldn't you have had mercy on your fellow servant just as I had on you?' In anger his master handed him over to the jailers to be tortured, until he should pay back all he owed. This is how my heavenly Father will treat each of you unless you forgive your brother or sister from your heart" (Mt. 18:21–35).

Why should this wonderful message of forgiveness be seen as problematic? Because, relying on divine deterrence (Johnson and Bering 2006), Jesus preaches that forgiveness should be granted always (this is exemplified by "seventy-seven times"), and the following parable is convincing as it shows that one should forgive since she is always in the situation of needing forgiveness herself. From a naïf moral point of view, this is a highly motivating and inspiring preaching, but if we consider it from the inferential perspective, and frame it within the cognitive niche construction theory, it becomes slightly problematic. As I developed some issues relating to cognitive niches in Part II, I referred to how cognitive niches embody axiological systems (initially proto-moral and then fully moral) stating more or less explicit rules and regulations to which the group supporting the niche (and those supported *by* the nice) should abide: the information shared in the cognitive niche also prescribes how deviants from such rules should be dealt with. As with all niche construction activities, the aim is to preserve the knowledge distributions as long as they manage to instruct an adequate level of welfare: modifications in the set of rules and regulations could easily cause undesirable changes in the group's performance. To make a computational metaphor, the cognitive niche can be compared to a software, or an operating system: removing, or tampering with certain lines of coding will usually impact (often in a negative way) the overall performance of the system.

How does all this refer to the religious insistence on forgiveness? As a matter of fact, a systematic recommendation to forgive should quickly deadlock whatever moral paradigm: from an abstract point of view, without contemplating the actual cognitive acts (and their pragmatic effects), to every moral rule R would correspond a $non(R)$ rule—which stands for the forgiven violation—without this situation affecting the overall health of the moral system. Where should we look for the solution to this dilemma? A quick answer would be to postulate a bubble-like model (akin to the religious bubble I sketched out earlier, in Chap. 12) to select on which occasions forgiveness should and should not be granted: this, nevertheless, would beg the question about how this selection would work. Would it operate on culture? or the local group? on the agent's past? In my opinion, it is much more interesting to tackle the issue from another side, and see it from a structural perspective: I will try to spell out what characteristics of moral inferences allow the non-self-defeating features of forgiveness. That is to say, I will dedicate the end of this chapter to understanding how an injunction of religious rationality, whose strategical benefit for the group is questionable (at least from a theoretical perspective), manages to preserve the welfare of the cognitive niche. In order to do so, I will start by referring to one of the harshest critics of forgiveness.

In 1887 Friedrick Nietzsche composed and published one of his most polemical yet insightful books, *On the Genealogy of morality*. The second essay, entitled "'Guilt', 'bad conscience' and related matters" provides an interesting discussion about the forgiveness (moral and judicial) of wrongdoers, which is seen as something "beyond the law". Nietzsche says:

> As the power and self-confidence of a community grows, its penal law becomes more lenient; if the former is weakened or endangered, harsher forms of the latter will re-emerge. The "creditor" always becomes more humane as his wealth increases; finally, the amount of his wealth determines how much injury he can sustain without suffering from it. It is not impossible to imagine society so conscious of its power that it could allow itself the noblest luxury available to it, – that of letting its malefactors go unpunished. "What do I care about my parasites", it could say, "let them live and flourish: I am strong enough for all that! " [...] Justice, which began by saying "Everything can be paid off, everything must be paid off", ends by turning a blind eye and letting off those unable to pay, – it ends, like every good thing on earth, by sublimating itself. The self-sublimation of justice: we know what a nice name it gives itself – mercy; it remains, of course, the prerogative of the most powerful man, better still, his way of being beyond the law (Nietzsche 2007, pp. 47–48).

Nietzsche correctly acknowledges how forgiveness is essentially a context-dependent matter: it rests on a number of factors but the *conditio sine qua non* is that the forgiver must be able to sustain, in terms of psychological, social and physical resources, the act of forgiving. As for this, there is not much difference between forgiveness and tolerance (which conveys more of a passive idea): both forgiveness and tolerance have a cost. The feeling of tolerance and forgiveness presupposes an emotional cost because the very ideas of tolerance and forgiveness entail the fact that one is not *automatically* supposed to tolerate and forgive a certain thing or event (Smith 1997): this is true of the evangelical prescription as well, since forgiveness is presented as something that must be granted *a number of times*, each time by giving one's assent, since forgiveness is not a perennial state of mind.

Nietzsche, though, contends that once the members of a community can afford it, they will necessarily and uncritically perform all of the forgiveness they can: this is, needless to say, the exact nemesis of *morality* and justice, that is to say of the niche-enforcement activities that manage the orthodoxy of the niche and hence warrant for its welfare. Not only, through forgiveness morality becomes its own self victimization—*self-sublimation* in Nietzschean words: a morality that forgives *automatically* is a morality that engages in a sophisticated oeuvre of self-destruction. Let us consider this violent yet interesting claim: it is sensible because an act of forgiving, not as mere tolerance but as an active practice (something like embracing and lifting someone back up to her previous status within the moral group), always involves—from a theoretical point of view—a second infringement of the norm inasmuch as the forgiver *rehearses* in her mind the action of the wrongdoer.

Let me spend a few words on this complex, yet rewarding, process. The forgiver must consider, as if they were both before her, the (infringed) norm and the wrongdoer. She can decide to support the reasons of the latter (that are after all the reasons for forgiveness) against the norm: this is obtained by the *rehearsal* of the violation and hence by the second violation of the norm. By forgiving, thus, the forgiver as well places herself *outside* of morality to join the wrongdoer. This physical exit of both actors from the original state is well exemplified by the parable of the Prodigal Son in the *New Testament* (Luke 15:11–32): the parable is strictly about forgiveness and it is important to notice how, at the end of the episode, when the prodigal son comes back to his original group, the forgiver/father does not wait for him but runs *out* to meet him. Jesus Christ! on forgiveness

> But while he was *still a long way off*, his father saw him and was filled with compassion for him; *he ran to his son*, threw his arms around him and kissed him (v. 20).

Thus he makes explicit the act of placing himself out of the original morality as well! This clearly reverberates through the words of the "good" son, who perceives his father's forgiveness as a violence against him, the one who had always behaved according to the morality of the family.

> The older brother became angry and refused to go in. So his father went out and pleaded with him. But he answered his father, "Look! All these years I've been slaving for you and never disobeyed your orders. Yet you never gave me even a young goat so I could celebrate with my friends. But when this son of yours who has squandered your property with prostitutes comes home, you kill the fattened calf for him!" (vv. 28-30).

As Tara Smith puts it, "[…] forgiveness is the judgment that a person's immoral action should not be treated as proof of a grave moral defect or an irredeemably bad character. The person offering forgiveness in effect says: 'I will not write you off on the basis of this incident'" (Smith 1997, p. 37). In other words, the forgiver's action is indeed a second attack on the moral norm: the forgiver says, "I care about the norm but I care more about you, about our relationship, your relationship with the group, etc.".

This situation can be understood as a double stake. The person who is receiving apologies from another person has to choose between two opposing alternatives: on the one hand, forgiving immediately affects one's sense of *position in the field*, which means that forgiving immediately projects an imaginary coalition conflicting with the one the forgiver belongs to. This implies—as I have just stressed—a sort of temporal disengagement from her morality. On the other hand, failing to offer forgiveness literally means cutting off a social bond established with the person to be forgiven. As we will see in the following, *moral viscosity* is a major factor in defusing the various double stakes one might face.

In this perspective, Nietzsche's account seems exact: *systematical* forgiveness is indeed the ultimate self-mortification of morality. One must take a step backwards, though, and understand if such a systematic nature of forgiveness is really the case: in the philosopher's *Genealogy*, morals originate from a *rebellion* of the slaves to their ancient, mighty masters. This conception of the origins of morality is often advocated by its denigrators as a mere instrument of thwarting and repression, yet, as Žižek points out,

> The ultimate irony is that this "critique of ethical violence" is sometimes even linked to the Nietzschean motif of moral norms as imposed by the weak on the strong, thwarting their life assertiveness: "moral sensitivity', bad conscience, guilt-feeling, as internalized resistance to the heroic assertion of Life. For Nietzsche, such "moral sensitivity" culminates in the contemporary Last Man who fears the excessive intensity of life as something that may disturb his search for "happiness" without stress, and who, for this very reason, rejects "cruel" imposed moral norms as a threat to his fragile balance (Žižek 2004, p. 2).[12]

As we read in the *Genealogy*, morality is the rule of the weak, who—out of resentment—label as *evil* the prerogatives of their former masters. Weaklings are not *able* to contrast those who do them wrong, so their resentful withstanding of any abuse is "given good names such as 'patience', also known as the virtue; not-being-able-to-take-revenge is called not-wanting-to-take-revenge, it might even be forgiveness" (Nietzsche 2007, p. 28). In other words, what is enacted by the "weak" is simply a revolution within the structure of the cognitive niche.

Nevertheless, the *automatic* character of morality seems to be overstressed in the Nietzschean account. If it were indeed as depicted, forgiveness as a moral norm would indeed have annihilated morality itself long since: yet the moral groups embodying Jewish and especially Christian thoughts on the goodness of forgiveness have been thriving for more than three thousand years, now. This is the *paradox* of forgiveness: on one hand forgiveness reiterates the attack upon the moral norm and sets the forgiver *out* of the group as the wrongdoer, on the other hand both the forgiver and the wrongdoer are found back in the group, and morality survives in spite of the double attack and ultimately forgiveness receives moral praise.

---

[12]On another account, Nietzschean reconstruction is strikingly similar to the result of recent findings in cognitive paleoanthropology: according to such research, the rise of egalitarian moralities was permitted by the development of remote killing artifacts and techniques, that would make it harder for alpha males (better hunters) to bully feebler individuals and administer the game to their own liking (Bingham 2000; Boehm 2002).

It is important to take into the account how forgiveness is hardly ever a moral injunction like the other norms. If a set of moral norms were accompanied, *ceteris paribus*, by a final norm imposing forgiveness of any wrong deed, then this would really mean a short-circuit of the whole moral institution! One must not confound *forgiveness* with its, alas, widespread degeneration supported by *do-goodism*: an indiscriminate exercise and recommendation of general tolerance that is as self-pleasing as it is self-deceiving. This is often the accusation directed at Christian/Catholic eagerness to forgive, and probably the target of Nietzsche's polemic, but it is a degenerated interpretation of the recommendation to forgive.

Forgiveness is a *plus* rule that situates itself on another level with respect to the rest of moral norms, if only for the fact that it *may* refer to any other moral norm. It is wrong to steal and yet a thief can be forgiven, it is wrong to lie and yet a liar can be forgiven, it is wrong to commit adultery and yet an adulterer can be forgiven, and so on. Forgiveness is a *possibility*, however encouraged it must remain in the realm of what *can* be: forgiving requires a complex emotional progression that could never assume the characteristic of compulsion. Besides, our cultural backgrounds are full of characters that are deemed morally righteous even if they opposed severity and harsh justice to a will to forgive. Not to forgive, as a matter of fact, is usually morally deprecated only as far as it betrays incoherence: it is *wrong* for me not to forgive one who has indulged in the same wrong deed that I indulged in, especially if I have been forgiven for that wrong myself (as preached in the parable of the unworthy servant). Apart from this case, one can ironically assess that whereas violence—that is morality through its related punishments and generated conflicts—is always a duty in the sense that it is unescapable even for the perpetrator (and its sublimation may lead to further violence and perversion, as extensively shown in Magnani 2011), forgiveness…is no such duty! Forgiveness is *not* a duty and it is not usually perceived as such.

The paradox of forgiveness, worth further investigation, consists in the fact that in all of the cases I just mentioned, in spite of the double attack on the rule (from the wrongdoer and the forgiver), the rule survives and is still *morally* believed by the forgiver (and probably by the wrongdoer): stealing, lying, and being unfaithful is *still* wrong. A situation of *do goodist* indiscriminate tolerance, it is important to state this once again, cannot even be analyzed under this schema because the wrongdoer knows from the beginning that her deeds will not be recriminated, let alone punished. Forgiveness, on the other hand, implies that the forgiver will achieve a deeper level of understanding, with respect to the moral infringement that has been perpetrated: it should have generated resentment and a will to perform a just punishment of the violator but, just because of forgiveness, the forgiver will not recur to punishment and will attempt to placate resentment.

The solution to this paradox is foreshadowed by a term introduced by Lahiti and Weinstein (2005): *moral viscosity*. Connected to the concept of *moral character*, that is the sum of (usually positive) moral traits one person builds as a result of her ethical education, moral viscosity represents the tendency of moral beliefs to *stick* to their holder in spite of the changing contexts. Morality is not a solipsistic matter but always rests on a shared and distributed dimension, as contended in Part II. The issues

I have just debated regarding the violence of forgiveness (forgiver and wrongdoer both exiting the community) have an extreme relevance for the whole moral group, where every individual is intertwined with the others in a holistic fashion. From this point of view, the offense to the norm perpetrated by the forgiver is felt by all other individuals who partake of that moral. Yet, the *viscosity* of morality prevents the whole group-morality from being excessively jeopardized by any attack on the shared norms.

It is tempting to draw a sharp juxtaposition between moral viscosity and forgiveness: if we start from the individual moral agent, we can argue that viscosity is essentially about self-forgiveness. We occasionally behave in a way that does not comply with our own moral requirements, yet we manage to get over it. Even if we consciously break a norm or *disengage* it,[13] that viscous norm remains bound to our patterns of moral inferences and we can maintain our commitment to our group's morality thus preserving its effect in coalition enforcement. Clearly, this is not the case of a schizophrenic or split personality, neither I am referring to a permanent or reiterated state like *bad faith*, nor with plain hypocrisy. Viscosity is what allows us to deal with and overcome single moral inconsistencies: "I cheated on my partner once but I still think that cheating is wrong", "My coworker dropped his ten-dollar bill and I took it but I still think that stealing is wrong". It is important to bear in mind how this is an extremely cultural issue and not a global psychological one. In far-eastern culture, for instance, practices such as *seppuku* (the Japanese ritual form of suicide) inform us of a different sensibility as regards personal failures and self-forgiveness.[14] Other cultures would relate to a less viscous nature of morality with a symbolic elimination of the wrongdoer: this is the case of the already amply discussed *scapegoat* ritual, originally described in the *Bible* (Leviticus, 16), as a ritual in which the guilt of wrongdoers would be *passed on* to a third creature, namely a goat, that was then to be expelled from the group and chased into the wilderness.

If we get back to our analysis of the interpersonal dimension of forgiveness we see how the dynamics of viscosity can help us understand and solve the initial paradox. If viscosity represents single beliefs sticking to an agent, in a distributed framework of moral beliefs viscosity is what keeps single agents stuck to morality, and hence to the cognitive niche projected and supported by the moral in question. Of course the degree to which viscosity can cover for an abrupt modification in the observance of a norm depends on the "power and self-confidence of a community" (Nietzsche would agree). If the offense is too dire (there are plenty of unforgiven wrongs) or the number of perpetrators is too elevated there can be only two solutions: either the wrongdoers are definitively expelled from the group or the whole moral institute crumbles with the dispersion of the same group. It is important to stress how the viscosity of morality is not the theoretical counterpart to simple hypocrisy, it does not allow morality to withstand an unlimited number of violations without displaying actual impoverishment: as a matter of fact, the level of viscosity is an extremely contextual factor so that an offense for which viscosity can act as guarantee on one

[13]The disengagement/reengagement of morality is described in Magnani (2011, Chap. 5).

[14]Further details on this topic are illustrated in Fuse (1980).

occasion, can prove to be fatal (for the wrongdoer) in another situation. Furthermore, there are also cases in which the violation is so significant that forgiveness cannot be granted without the forgiver permanently joining the wrongdoer(s) *outside* of the original group.

When the *viscosity*[15] of morality allows and supports forgiveness, the whole process can be thus summarized:

1. As the wrongdoer infringes the moral rule, she immediately places herself outside of the moral group;
2. Because of the theoretical dynamics of forgiveness which I exemplified before, the forgiver reenacts the violation of the norm and for an instant (though this cannot be transposed into actual, physical time) takes a stand with the wrongdoer against the norm, exiting the moral coalition as well;
3. Morality, due to its characteristic of *viscosity*, acts as a blubbery substance and reconstitutes its structure before the rest of the community feels the *wounds as their own*;
4. Viscous moral remnants upon the forgiven wrongdoer merge with those of the forgiver so that they can both be re-absorbed into the main viscous unity, that is the original moral/cognitive niche.

The aim of this conclusive chapter was to analyze two important issues, usually connected with the religious mindset, relying on the theoretical framework that I developed and laid out so far in the book. Religious sacrifice (and all the more self-sacrifice), and religiously-informed forgiveness can be considered as two major cores of irrationality. Whereas the former is more readily identified with irrationality in its violent dimension, I nevertheless suggested that forgiveness too may be eccentric to a clear logic as far as it concerns the respect for a norm, and the contemporary respect for the norm-breaker.

By framing the investigation in an eco-cognitive (and hence naturalistic/inferential) perspective, I was able to make the case that these two behaviors still responded to the same *patters of rationality* individuated through the book: the tendency to make sense and model one's surroundings in order to operate on them, and the necessity to develop cognitive niches in order to achieve a cognitive proficiency, is more manifest in scientific rationality, but once we learn to appreciate it where it is the clearest, it can be seen as driving social cognition and magical/religious thought too. As showed by this final chapter, the multilayered violence of the sacrificial mindset can be seen as a way of coping with the cognitive weight imposed to one's rationality by the commitment to religious ontologies that, as I explained in the previous chapters, are an unavoidable product of how our cognitive endowments are geared to cope with the external world. Similarly, the "irrationality" of forgiveness (which has angered and scandalized past and presents Pharisees, judges, philosophers and common people

[15]This rough parameter, which is more qualitative than quantitative, can be traced back to different influences. The religious one is fundamental: it is easy to argue that Western society's eagerness to forgive is heavily dependent on the importance of the theme of forgiveness in early Christian writings (McCullough and Worthington 1999). The philosophical contribution, later absorbed into an *idéologie spontanée*, plays a crucial role as well.

making them resonate with Nietzsche's arguments) is a constant and fitting reminder that, even if our current epistemic proficiency is heavily relying on cognitive niche construction and the axiologies they imply, *cognitive niches were made for men, not men for cognitive niches.*

References

Adams R, Girard R (1993) Violence, difference, sacrifice: a conversation with René Girard. Relig Lit 25:9–33
Anderson JL (1997) Che Guevara: a revolutionary life. Grove, New York
Atran S (2002) In gods we trust: the evolutionary landscape of religion. Oxford University Press, Cambridge
Benjamin W (1978) The critique of violence. In: Dements P (ed) Reflections: essays, aphorisms, autobiographical writings (trans: Jephcott). Hartcourt Brace, New York. First appeared in the Archiv für Sozialwissenschaft und Sozialpolitik, 1921, 3
Bingham PM (2000) Human evolution and human history: a complete theory. Evol Anthropol 9:248–257
Boehm C (2002) Variance reduction and the evolution of social control. Paper presented at annual workshop on the co-evolution of behaviors and institutions, Santa Fe
Boyer P (2001) Religion explained. Vintage U.K. Random House, London
Bubbio PD (2004) La ragione e il suo altrove. Città nuova, Rome
Buchli E (2006) Il mito dell'amore fatale. Baldini Castoldi Dalai, Milan
Bulbulia J (2009) Religiosity as mental time-travel. In: Schloss J, Murray MJ (eds) The believing primate. Oxford University Press, Oxford, pp 44–75
Dalferth I (2010) Self-sacrifice. Int J Philos Relig 68:77–94
De Vries H (2002) Religion and violence: philosophical perspectives from Kant to Derrida. Johns Hopkins University Press, Baltimore
Dennett D (2006) Breaking the spell. Viking, New York
Derrida J (1994) The spectre of Marx (trans: Kamuf P). Routledge, New York
Derrida J (1999) Adieu à Emmanuel Levinas [1995] (trans: Brault PA, Naas M). Stanford University Press, Stanford
Fuse T (1980) Suicide and culture in Japan: a study of Seppuku as an institutionalized form of suicide. Soc Psychiatry Psychiatr Epidemiol 15(2):57–63
Gilhus IS (2006) Animals, gods and humans: changing attitudes to animals in Greek Roman and early christian thought. Routledge, New York
Girard R (1977) Violence and the sacred [1972]. Johns Hopkins University Press, Baltimore
Girard R (1986) The Scapegoat [1982]. Johns Hopkins University Press, Baltimore
Girard R (1987) Job: the victim of his people [1985]. Stanford University Press, Stanford
Johnson D, Bering J (2006) Hand of god, mind of man: punishment and cognition in the evolution of cooperation. Evol Psychol 4:219–233
Keenan DK (2005) The question of sacrifice. Indiana University Press, Bloomington
Lahiti DC, Weinstein BS (2005) The better angels of our nature: group stability and the evolution of moral tension. Evol Hum Behav 2:47–63
Levinas E (1985) Ethics and infinity [1982] (trans: Cohen RA). Duquesne University Press, Pittsburgh. Conversations with P. Nemo
Magnani L (2011) Understanding violence. Morality, religion, and violence intertwined: a philosophical stance. Springer, Berlin
McCullough ME, Worthington ELJ (1999) Religion and the forgiving personality. J Pers 67(6):1141–1164
Milbank J (1990) Theology and social theory beyond secular reason. Blackwell, Maiden

Nietzsche F (2007) In: Ansell-Pearson K (ed) On the genealogy of morality [1887] (trans: Diethe C). Cambridge University Press, New York

Seligman A, Weller R, Puet M, Bennett S (2008) Ritual and its consequences: an essay on the limits of sincerity. Cambridge University Press, Cambridge

Smith T (1997) Tolerance and forgiveness: virtues or vices? J Appl Philos 14(1):31–41

Taylor M (2006) Strategies of dissociation: a mimetic dimension to social problems in Japan. Anthropoetics 12(1)

Žižek S (2004) A plea for ethical violence. Bible Crit Theory 1:02-1-02-15

Žižek S (2009) Violence [2008]. Profile Books, London

Chapter 14
Patterns of Rationality: Concluding Remarks

The search for repeating patterns of rationality was nested in a comprehensive framework, compatible with, and inspired by, eco-cognitive epistemology. My goal was to juxtapose and compare three broad and distinct areas typical of human rationality—namely science, social cognition (understood as cognitive niche maintenance) and religion—in order to analyze the inferential regimes by which they are characterized, and to show how it is possible to witness in different fields of human cognition a similar will to develop models of the "environment," be it physical, practical, social and so on, and also the constant attempt to "fill the blanks."

The first part consisted in the epistemological core of book, in which I outlined and set to operation the tenets of the eco-cognitive approach to epistemology. The main task was to focus the analytical attention onto the *reality* of the epistemic agents, and their being immersed in an environment on which their goals and strategies depend. This allowed me to tackle from a novel perspective a fundamental issue of contemporary epistemology, namely the nature and function of models and representations. The idea was to rely on various insights from cognitive science and stress the relationship between the idea of scientific model and that of mental representation: the novelty consisted in connecting the idea of scientific representation with the naïf representational intelligence ranging from the animal proto-mind all the way to human rationality. The idea, inherited from philosophy of biology, of a pragmatic and strategic *raison d'être* of the modeling faculty could provide the backbone of a new conception of modeling that is consistent with a dynamic understanding of the history of scientific endeavor. This strategic understanding of modeling is a cross-domain specificity of human rationality, and does not concern science alone, even if in science it becomes the object of a precise reflection and precise cares (for instance in the experimental framework I described at the end of the first part). Conversely, the necessity of making models of our surroundings and the way they operate and react is indeed a common denominator of many aspects of human rationality, even those usually considered to be "less rational:" this was the topic I explored in the second and the third parts of the book.

The second part focused on two main issues: the construction and maintenance of cognitive niches, and the importance of social cognition as preliminary and ancillary

© Springer International Publishing Switzerland 2015

T. Bertolotti, *Patterns of Rationality*, Studies in Applied Philosophy, Epistemology and Rational Ethics 19, DOI 10.1007/978-3-319-17786-1_14

work for the establishment and maintenance of cognitive niches. These two aspects are strictly interconnected, also with the strategic dimension of rationality and the production of models of one's surroundings, also from a social perspective: if, one the one hand, robust cognitive niches endow their users with powerful supports for their cognitive activities (through more or less permanent epistemic externalizations), on the other hand it is required that potential members engage and maintain a framework of social monitoring (hence of mutual modeling) in order to warrant for the maintenance of an optimal ground for the elevation of the cognitive niche. It was interesting to see how the analysis of apparently different shades of rational behavior could actually be framed within a single theoretical approach, without the need of a change in conceptual tools or specific language. Therefore, I was able to nest in the second part a deep analysis of gossip not so much as a phenomenon needing a moral or evolutionary explanation, but rather an epistemological one, which I was able to provide as part of the global "sense-making" activities advocated by the eco-cognitive view on epistemology. Activities such as gossip, necessary as the "ground level" of cognitive niche construction, do indeed exhibit similarities in their inferential patterns if compared with other, non social-based, forms of cognition. My outlook on cognitive niches was able to include human beings' ability (or lack thereof) to interact with new members of the cognitive niche, following the disruption of the traditional difference between *biota* (niche-constructors) and *abiota* (niche-constructed): I briefly illustrated how the advances in technology leading to the development of artificial cognition insert in the cognitive niche a new artificial epistemic actor able to affect the cognitive niche in a "rational" way, which is not always consistent with the good of its creators.

I eventually analyzed the *nemesis*, so to say, of rationality in the final part, as I dealt with what I called *the eco-cognitive epistemology of irrational beliefs*. After having subdued the moderate, occasional, irrationality of social cognition, I applied the conceptual toolbox I had developed along my research to a somewhat stronger dimension of irrationality, the one traditionally associated with religion and magic. Once again, unsurprisingly at this point, the rational divergence of such beliefs can be accounted for in an eco-cognitive perspective: in opposition to a *veristic* epistemologic conception, I was able to analyze and spell out the knowledge-forming capabilities, and the pragmatic utility, of those inferences grounding magic and religion, disregarding the fact that such beliefs may not be correspondently "true."

The kinds of beliefs produced by magic or religious inferences do indeed correspond to a sense-making struggle very much alike to the one we see at play in other frames, for instance when survival is at stake, but also—as I contended in the first part—they can be tracked back (or forward) to the same *patterns of rationality* displayed in scientific endeavor. Human rationality is not a set of sealed-off components, each deployed in (and only in) its own field. Rather, we rely on its many components in a wide array of circumstances, also implicating different truth-regimes. If we consider our skills in distributed and hypothetical cognition, we can see that they are employed with no qualitative disruption both in science and religious thinking, to make a case concerning two opposites, but only a quantitative difference in care, attention, corroboration and goals. Consider, as an example, the famous "Lion Man" artifact (in Sect. 10.5): it is a fitting example to show how the distribution of cognitive tasks suddenly opens new epistemic pathways, allowing us to make inferences that would not be possible if the artifact was not available. The artifact was indeed a necessary condition for our ancestors to develop the mental notion of such "Lion Man". *But soft!* I had used, at the very beginning of this book, Holland's definition of a model as something *which allows us to infer something about the thing modeled.* Will this let us argue that there is a very strong bound connecting two poles as remote on from the other as the Hohlenstein Stadel *Lowenmensch* and, say, Bohr's model of the atom? Backed by the research I carried out and I presented in this book, I would answer affirmatively. Without forgetting the obvious differences, it is important to notice how they both are the output of a sense-making rationality, which operates through the same channels, such as the development of models that have a poietic impact on one's reality (the model allows us to "know" the atom, just as the figure allowed our ancestors to "know" the Lion man, by creating them to a certain extent), which is enacted through the distribution of cognition in one's cognitive niche, and the latter—in turn—can be achieved though a shared and controlled effort in social cognition.

In conclusion, most of this book could probably be placed under the current label of *social, or applied, epistemology*. Still, the use of this disciplinary term requires a clarification, since two main different approaches go under the same definition. One the one hand, Steven Fuller's foundational outlook is more sociologic and is therefore structured as as *sociology of science* (Fuller 1991): a number of illustrious authors, such as Helen Longino, can be ranked among the lines of this outlook, which could be resumed as (a) strictly connecting the word *epistemology* with *philosophy of science* and (b) focusing on the social mechanisms underpinning the development of scientific endeavor. The *ethnographic/anthropologic* approach to science (N.J. Nersessian, B. Latour, K. Knorr Cetina et al.) can be seen as related to this connotation of *social epistemology*, together with a certain perspective on *folk-science* stressing the social dimension.[1] The other connotation of social epistemology is more linked to the Cartesian tradition seeing epistemology as the theory of assessing what is rationality, differentiating knowledge from mere belief, and the appraisal of knowledge-related (epistemic) practices: Alvin I. Goldman could be taken as the representative of this second perspective, in his contention that "social epistemology is a branch of traditional epistemology that studies epistemic properties of individuals that arise from their relations to others, as well as epistemic properties of groups or social systems" (Goldman 2010, p. 1). The research collected in this book partakes of both approaches: the first part (especially Chaps. 3 and 4) engages indeed the "biological ancestors" of scientific modeling (framing the analysis in strategic interaction, both at scientific and pre-scientific level), and explores the fact that a

[1] F.C. Keil observes, for instance, that "[the] success of science arises from the ways in which scientists learn to leverage understandings in *other minds* and to *outsource* explanatory work through sophisticated methods of *deference* and simplification of complex systems," (Keil 2010, p. 826, added italics) and suggests that similar procedures are well exemplified by laypeople's knowledge-gathering methods (folk-science).

correct understanding of experimentation should not be separate front the understanding of the social ground it is embedded into: such an outlook is more akin to the first meaning of "social epistemology". Conversely, many issues in the second and third chapter (dealing with cognitive niche construction and maintenance, social cognition and religion) are affronted with a philosophical perspective closer to Goldman's social epistemology, save for one aspect. The latter characterizes his social epistemology as *veritistic*: "Veritistic epistemology (whether individual or social) is concerned with the production of knowledge, where knowledge is here understood in the 'weak' sense of true *belief*. More precisely, it is concerned with both knowledge and its contraries: *error* (false belief) and *ignorance* (the absence of true belief). The main question for veritistic epistemology is: Which practices have a comparatively favorable impact on knowledge as contrasted with error and ignorance? Individual veritistic epistemology asks this question for nonsocial practices; social veritistic epistemology asks it for social practices." (Goldman 1999, p. 5). Coherently with the ecological/cognitive perspective I adopted in my research, the veritistic attention fell outside of my project which was rather focused on epistemic *pragmatics* as inherited from the Peircean tradition: basing a significant part of my research on cognitive and evolutionary studies, and on agent-based logical frameworks, I characterized my research in social epistemology as chiefly *descriptive* rather than normative, focusing my interest on aspects connected with the social emergence of *successful and persistent beliefs*, notwithstanding their relationship with a strong notion of Truth. Thus, the ecological/cognitive framework allowed me to undertake an "individualized" socio-epistemological analysis, that would therefore consider social epistemic dynamics both as influenced by, and influencing, the cognitive and epistemic performances of single cognitive agents. That is to say, their recurring patterns of rationality.

References

Fuller S (1991) Social Epistemology. Indiana University Press, Bloomington

Goldman AI (1999) Knowledge in a social world. Oxford University Press, Oxford

Goldman AI (2010) Introduction. In: Haddock A, Millar A, Pritchard D (eds) Social epistemology. Oxford University Press, Oxford, pp 1–28

Keil F (2010) The feasibility of folk science. Cogn Sci 34:826–862

Index

© Springer International Publishing Switzerland 2015

T. Bertolotti, *Patterns of Rationality*, Studies in Applied Philosophy, Epistemology and Rational Ethics 19, DOI 10.1007/978-3-319-17786-1

MIX
Papier aus verantwortungsvollen Quellen
Paper from responsible sources
FSC® C105338

If you have any concerns about our products,
you can contact us on
ProductSafety@springernature.com

In case Publisher is established outside the EU,
the EU authorized representative is:
Springer Nature Customer Service Center GmbH
Europaplatz 3, 69115 Heidelberg, Germany

Printed by Libri Plureos GmbH
in Hamburg, Germany